Building Pathology and Rehabilitation

Volume 9

Series editors

Vasco Peixoto de Freitas
Aníbal Costa
João M.P.Q. Delgado

More information about this series at http://www.springer.com/series/10019

Aníbal Costa · António Arêde
Humberto Varum
Editors

Strengthening and Retrofitting of Existing Structures

Springer

Editors
Aníbal Costa
RISCO, Department of Civil Engineering
University of Aveiro
Aveiro
Portugal

António Arêde
CONSTRUCT-LESE, Civil Engineering Department, Faculty of Engineering
University of Porto
Porto
Portugal

Humberto Varum
CONSTRUCT-LESE, Civil Engineering Department, Faculty of Engineering
University of Porto
Porto
Portugal

ISSN 2194-9832 ISSN 2194-9840 (electronic)
Building Pathology and Rehabilitation
ISBN 978-981-10-5857-8 ISBN 978-981-10-5858-5 (eBook)
https://doi.org/10.1007/978-981-10-5858-5

Library of Congress Control Number: 2017946958

Printed on acid-free paper

This Springer imprint is published by Springer Nature
The registered company is Springer Nature Singapore Pte Ltd.
The registered company address is: 152 Beach Road, #21-01/04 Gateway East, Singapore 189721, Singapore

Preface

Existing construction structures have always been a topic of concern for owners, stakeholders, technicians and researchers, though for different interests and motivations. Apart from other reasons such as aesthetics, socio-economic and cultural values, the role of load-bearing structures in any existing construction probably is, or should be, the major issue when dealing with conservation, preservation and/or rehabilitation of all types of built heritage.

Quite often, conservation and preservation of existing constructions may not be likely to require strengthening or retrofitting of their structures, thus keeping the intervention at the maintenance level. However, when structural repair or rehabilitation is imposed, either due to any kind of external occurrence causing damage or required by modifications of construction configuration or usage, structural retrofitting and/or strengthening might be unavoidable.

Clearly, this issue is extremely wide, in the sense that structural interventions in existing constructions are very much dependent on a large number of different conditions and factors. Of course, the motivation for such interventions has to be first referred since it is likely to influence all the subsequent process and options: for instance, structural strengthening due to load increase resulting from construction usage modifications can lead to interventions very different than those required to make it prepared to withstand relevant seismic demands. Also, different types of constructions, e.g. buildings or bridges, will require different approaches for structural strengthening or retrofitting, despite similar technologies might be adoptable and, obviously, the main constructive materials will be determinant for the type of intervention. Furthermore, the structural safety format framework for existing structures is still an open and non-consensual issue, requiring particular attention for its crucial importance in current design practice; in fact, it is not uncommon that designing strengthening solutions for existing structures to comply with code standards developed for the design of new structures might become practically and/or economically unfeasible. Last but not least, often different types of strengthening technique solutions can be proposed for a given case under analysis, for which the designer can be faced with difficulties on performing their evaluation in order to make a rationally sustained option.

Considering the above mentioned, the reader immediately realizes the challenge of organizing a book addressing the wide-spectrum topic of strengthening and retrofitting of existing structures. Amongst a few options for the book organization, the editors finally considered that, possibly, the most logical and clear one would involve a first level of book division according to the type of structures, namely buildings and bridges, followed by a second level of chapter sequence related with the structural material.

According to this option, the book first provides a general overview of the motivations, concepts and approaches for structural strengthening and retrofit, which constitutes the introductory chapter. Subsequently, due to their particular importance and peculiarities, historical buildings and cultural heritage monuments are focused in what concerns conservation issues and structural interventions, as addressed in the second chapter. The book then includes six chapters, which go into detail on strengthening and retrofitting options, solutions and techniques, according to the type of construction material, namely stone and brick masonry, adobe, timber and reinforced concrete, the latter more thoroughly addressed depending also on the strengthening material, consistently with its widespread use and importance in current construction from early–mid twentieth century. As for bridges, the book includes three chapters, focusing the strengthening of reinforced concrete, masonry and steel bridges. As a common issue to buildings and bridges, ground and foundation systems are also addressed, concerning their reinforcement and rehabilitation, in a specific chapter. Two final chapters are included, one presenting and discussing the safety assessment of existing structures, particularly under seismic action given its importance and specific issues, while the other addresses tools to prioritize possible strengthening techniques.

It is worth noting that, within the particular case of seismic rehabilitation, the solutions based on seismic isolation concepts are presently well established resorting to appropriate devices developed for practical applications. Although this could suggest having included a chapter fully dedicated to seismic isolation, the editors considered that such an option would have gone beyond the book scope, since it constitutes a quite specific topic, for a particular type of structural loading, which would have required an extensive written piece of text for meaningfulness. Therefore, seismic isolation is considered in the book as a possible and viable option in seismic rehabilitation works, with a few practical application references included in some chapters.

In several chapters, beyond the description of strengthening and retrofitting techniques, examples of application in real case studies are also included, thus providing a more practical view of the proposed solutions.

All in all, it is the editors' conviction that the book provides a broad overview of the solutions' spectrum for strengthening and retrofitting of existing structures, from simple and well-established procedures to more recent and cutting-edge solutions, giving the reader important information and inspiration for the adoption and implementation of adequate interventions on existing structures, without disregarding the compatibility concerns with the original materials, structural components and systems.

Aveiro, Portugal — Aníbal Costa
Porto, Portugal — António Arêde
Porto, Portugal — Humberto Varum

Contents

Structural Strengthening and Retrofit; Motivations, Concepts and Approaches

Giorgio Macchi, Gian Michele Calvi and Timothy John Sullivan

1 Introduction

The need for strengthening and repair of buildings and civil engineering works may arise when they have been damaged in such a way that they are no longer fit for their normal use. In such cases the structure cannot afford, with an accepted reliability, a further sequence of the same action or of other accidental actions and consequently, the risk of loss of lives and the risk of further structural and contents damage, would be unacceptable. A strengthening intervention that is able to restore an acceptable level of safety against such actions is called *retrofitting*.

The need for strengthening or retrofitting may also arise for buildings or structures which were not damaged previously. This occurs, for instance, when an appropriate assessment shows that the building would not resist expected accidental actions with an acceptable reliability. In this case the building does not need repair, but requires retrofit.

In most cases, the level of risk that will be accepted for some form of future damage is mainly limited by the necessity to reduce the risk of loss of life. However, the acceptability of damage is lower when the building has a great importance because of its function (strategic buildings); therefore, sophisticated

G. Macchi (✉)
University of Pavia, via Villa Mirabello 3, 20125 Milan, Italy
e-mail: giorgio12011930@hotmail.it

G.M. Calvi
Scuola Universitaria Superiore IUSS Pavia, Palazzo del Broletto - Piazza della Vittoria n. 15, 27100 Pavia, Italy
e-mail: gm.calvi@iusspavia.it

T.J. Sullivan
Department of Civil and Natural Resources Engineering, University of Canterbury, Private Bag 4800, Christchurch 8140, New Zealand
e-mail: timothy.sullivan@canterbury.ac.nz

A. Costa et al. (eds.), *Strengthening and Retrofitting of Existing Structures*,
Building Pathology and Rehabilitation 9, https://doi.org/10.1007/978-981-10-5858-5_1

tools for strengthening and retrofit are most commonly adopted for strategic buildings.

Other types of valuable buildings, for which it is common to accept a higher risk of collapse or even limited damage, are historical (e.g. Heritage Buildings) with high aesthetic value. In such cases, it will additionally be necessary to identify non-invasive retrofit solutions.

The choices in the field of strengthening and retrofitting imply high costs, particularly when seismic risk has to be taken into account; therefore, optimisation strategies are suggested.

Every strengthening intervention requires a previous set of investigations and analyses in order to get a reliable Diagnosis of the structure, i.e. to get its numerical model and therefore its behaviour at the Ultimate Limit State and at the Serviceability Limit State.

The actions to be taken into account are defined by national standards, such as the Eurocodes EN1991 [1] and EN1998 ([2] for seismic actions) and by additional National Annexes. The tools for the Diagnosis are not within the scope of this book. This book instead aims to deal with the methods of retrofitting structures of different types of construction (masonry, timber, reinforced and pre-stressed concrete, steel, steel and concrete composites), and different types of buildings and civil engineering works (old buildings, modern buildings, bridges, foundations), see EN1990 [3].

2 Motivations

2.1 Conservation: A New Need

A recent and widely accepted set of structural Codes, the EUROCODES, devotes an entire volume (EC 8) to seismic design, and its Part 1998-3 to "Assessment and Retrofitting of Buildings", with 3 Annexes: A for reinforced concrete (RC) buildings, B for steel buildings, and C for masonry buildings. This shows how retrofitting existing structures has taken an important role in structural engineering, that reflects the progress made in the industry over the last decades and centuries.

The use of iron ties or metal clamps to connect stones goes back to the ancient Egyptians, and to the monuments of the Roman Forum and their following anastylosis works. Forged iron beams were used in India (1240 A.D.) in the Konarak temple and, later on, steel profiles were used in the famous Balanos' restorations of the Athen's Acropolis (in the years 1898–1909). However, we had to reach the year 1831 for the statements of Hugo [4] in the novel "Notre Dame de Paris", advocating for the first time the conservation of the old gothic churches. He supported Heritage Conservation in France, together with Prosper Mérimée.

Ruskin [5] suggested monuments conservation in 1849 (in "The Seven Lamps of Architecture—The Lamp of Memory"), conceiving it as the opposite of restoration

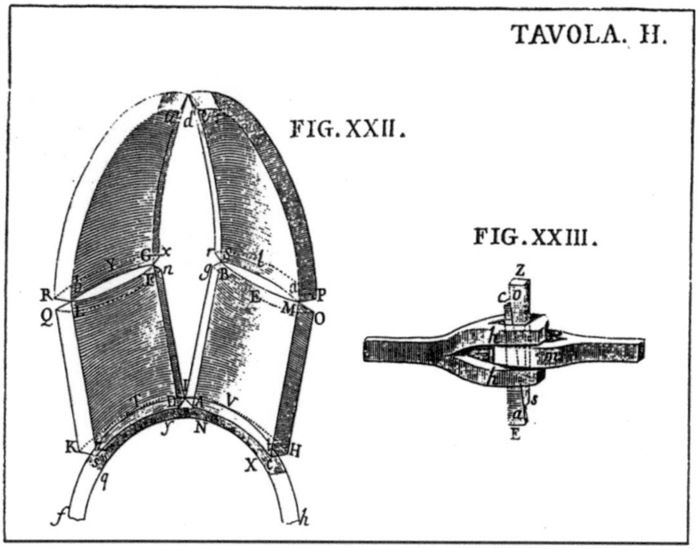

Fig. 1 New chains of the Dome. St. Peter's Basilica in Rome [6]

(the practice of restoration of his time). He wrote that masonry should be kept compact using "iron".

Five new iron chains were in fact applied by Vanvitelli [6] to the Great Dome of Saint Peter's Basilica in Rome, in 1743, far before the above mentioned theoretical statements (Fig. 1). Such extraordinary intervention was conceived to save the dome, on which several wide cracks appeared in the year 1740 and a suspicion of total collapse was aroused by most mathematicians. Nevertheless, it represents a case of strengthening a Heritage Building according to some of the most recent criteria for conservation: the new iron chains are not invasive, respective of the original technique, and are reversible.

Similarly, an exemplary intervention is the huge 3 m thick buttress built by Raffaello Stern [7] for the stability of the Colosseum in Rome in the year 1807: the solution is preserving the damage suffered by the building due to earthquakes, which are still very evident, and is clearly differentiated from the monument, as it is built in brickwork instead of stone (Fig. 2). One year before, in 1806, Rondelet [8] also strengthened the central pillars of the Panthéon in Paris by means of a thick layer of new stone masonry in order to double the insufficient cross section (Fig. 3).

Fig. 2 Brickwork buttress of the Stern restoration (1807) of the colosseum in Rome [7]

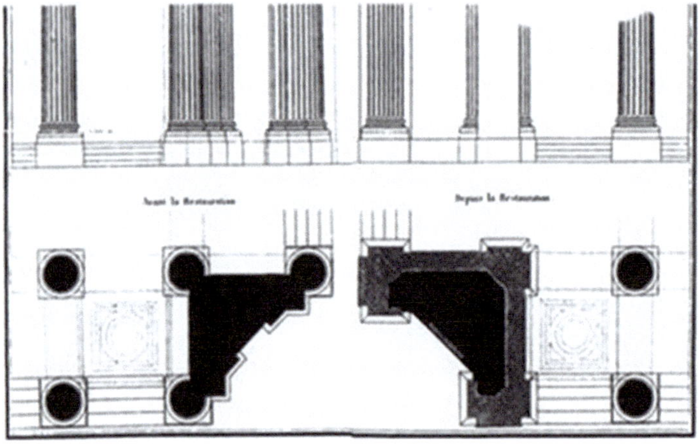

Fig. 3 Panthéon of Paris; Addition of new masonry [8]

2.2 Conservation Codes

Vitruvius did not speak of Conservation in his treatise "De Architectura". In old times, the buildings of important towns, including churches and art masterpieces, were often destroyed during wars and the ruins became a thick layer in the ground as the years went by. During the French Revolution (1789) the Bastille was razed to the ground, and the single stones were sold as souvenirs.

The codification of criteria to be followed for the Conservation of Heritage Buildings took place only between the 19th and 20th century, when proponents of European culture felt the responsibility to leave the precious architecture as heritage to future generations. Such a process, initiated by poets and writers as Hugo [4], Mérimée [9] and Ruskin [5], led to a philosophy of conservation through the works of Camillo Boito, Alois Riegl, then Gustavo Giovannoni and the Charter of Athens in 1931, and finally with Raymond Lemaire, Piero Gazzola, Roberto Pane and others, who codified it at the end (in 1964) in the Charter of Venice, the widely accepted Code of ICOMOS, and therefore of UNESCO.

Afterwords, in 1994, the Charter of Nara extended the concept of Authenticity to fragile monuments which can be preserved only by periodical reconstruction, as required mainly by Eastern Countries, as Japan (e.g. for the Imperial Villa of Katsura in Kyoto).

Another important improvement was added in 2003, the ISCARSAH Guidelines [10], which provide more detailed rules for strengthening interventions.

According to Hugo [4], (the war on the demolishers *in French*), there were three main reasons for the ruin of the ancient masterpieces:

- "time, putting wrinkles on their face;
- the revolutions, with blind and furious acts of vandalism;
- the fashion, which did worse than the revolutions, by cutting, dismantling, killing the building both in the form and in its symbol, in its logic and in its beauty".

The above mentioned Charters, together with many national guidelines, show that people became more conscious of the unity of human values with regard to ancient monuments as a common heritage and that is our duty "to hand them on in the full richness of their authenticity". The motivation of Heritage Conservation may be considered a matter of the "Ethics of responsibility" (see Jonas [11]), in particular of the "Principle of precaution", which requires Prevention, i.e. measures to adopt today in view of foreseeable calamities.

Today, the main causes of damage to Heritage Constructions are:

- wars;
- fashion;
- time (including accidental and destructive actions such as: earthquakes, fire, wind, tsunami, floods, settlements, landslides, progressive collapse, impact and vibrations, explosions, material deficiencies, human error).

The effects of earthquakes are, however, a predominant cause of damage having the consequence of loss of lives. Therefore, some parts of this text are mainly aimed to seismic protection, but are useful for every kind of damage.

The scientific community has drawn up a list of Values to be considered in the choice of Heritage Buildings to be protected:

- cultural history milestones;
- rarity;

- research potential;
- representativeness;
- aesthetic qualities;
- creative or technical achievements.

The Charters then define the main rules to be followed in strengthening interventions:

- implement Conservation works instead of Restoration works;
- respect the original material, and every new material used for integration should always be recognizable;
- every reconstruction work should be ruled out a priori (only anastylosis may be permitted);
- compatibility and durability of new materials should be assured;
- invasivity of the intervention should be avoided;
- provide recourse to sciences and techniques which can contribute to the safeguard of the structure, but the applications should be proved by experience;
- reversibility, or the possibility to treat again, should be searched for (examples of reversible solutions include works done on the Tower of Pisa, Fig. 4, and the Pavia cathedral, Fig. 5).

The choice of the strengthening intervention is treated in Chapter "Strengthening of Stone and Brick Masonry Buildings" (Concepts).

Fig. 4 Leaning Tower of Pisa; Pretensioned circumferential tie realized with a removable stainless steel band

Fig. 5 The cathedral of Pavia (1488—today)

Factors guiding the definition of the physical properties of strengthening interventions may be found in Chapter "Seismic Retrofit of Adobe Constructions" (Approaches).

2.3 *Seismic Strengthening*

Wars have been seen to cause the highest losses of human lives, (millions, in World War I and in World War II), but also earthquakes, for many Countries, cost dearly in terms of human lives and loss of buildings. The earthquake of Tangshan (China), in 1976, caused the loss of 240,000 lives; the town of Bam, in Iran, was razed to the ground in 2003. It should be remembered that the victims are not killed by the earthquake, but by the collapse of inadequate buildings. The same applies to Heritage Constructions: their inadequate strength is the cause of their loss. The above considerations explain why earthquakes are calling more and more the attention of society and of the governments of seismic-prone Countries to require extensive interventions for an adequate reduction of seismic risk.

As buildings (Heritage Buildings included) are usually designed for vertical actions (permanent loads and other gravity loads), their resistance against the prevailing horizontal components of the earthquake action may be insufficient in many cases. Furthermore, earthquake actions are dynamic by definition, a feature which makes most traditional building materials insufficient. History and the progress of geodynamics studies is providing a level of knowledge of seismic phenomena which tends to reduce their "accidental" character. However, a prudent attitude is still needed, and strengthening of buildings and other constructions is necessary or, when strengthening is not sufficient, actions should be reduced by means of dynamic measures such as base isolation or energy dissipation.

3 Concepts for Structural Strengthening and Retrofit

3.1 Consideration of Loading Scenarios

In order to establish a suitable retrofit strategy, one must first establish the loading scenarios that will need to be addressed by the eventual retrofit measures. Figure 6a illustrates a possible loading scenario, expressed as a 3-dimensional plot of force, displacement and time (or return period, considering risk evaluations conducted over a specific design life). This general expression of loading is capable of representing all typical loading scenarios encountered in engineering; a specific gravity load would be plotted as a horizontal surface of uniform force (with displacements depending on the stiffness of the structure), the displacement imposed by a shift in foundations change could be represented as a vertical plane (with the force depending on the stiffness of the structure), whereas the specific scenario plotted in Fig. 6a is somewhat representative of seismic demands, recognizing that earthquakes will impose both displacement and force demands on a building as a function of the building period.

The time axis in Fig. 6a is particularly relevant for risk considerations; clearly, the longer the design life then the more likely it becomes that the structure is subject to large magnitude loading. Building codes typically address this point by specifying the design life of a building and establishing loads that ensure certain

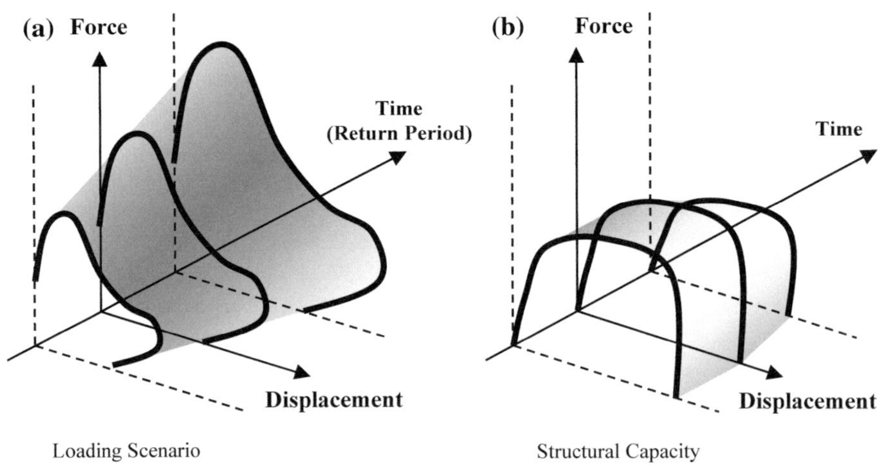

Fig. 6 Conceptual illustration of **a** loading scenario and **b** structural capacity, expressed in terms of force-displacement-time axes

reliability requirements are satisfied. This then simplifies the engineering problem down to one of considering a certain combination of force and displacement demands at (say) a serviceability or ultimate limit state design level. However, as the importance of sustainable engineering becomes ever more evident to society, it is apparent that the time-axis should not be forgotten; for instance, life-cycle concepts can be conceptually introduced through consideration of the time-axis, underlining the need for any strengthening or retrofit intervention strategy to assess not only the up-front costs but also maintenance costs and the final impact on performance over time. This is particularly relevant to the identification of optimum retrofit solutions, as explained further in Sects. 3.3–3.7.

3.2 *Evaluation of Structural Capacity*

The assessment and retrofit of a structure will inevitably require that the capacity of a structure, in either a pre- or post-retrofit condition, be compared with relevant loading scenarios (described in the previous section). To this extent, it is apparent that a general definition of capacity should again consider the three axes of force, displacement and time, as shown in Fig. 6b. Capacity curves, that will be structure-specific, may be relatively simple planar surfaces (consider a structure that responds elastically and then fractures in a brittle manner) or non-linear, such as the surface depicted in Fig. 6b. The particular force-displacement response shown Fig. 6b is intended to represent a lateral force-displacement (pushover) capacity curve for a building or bridge structure. One could consider different damage states along the curve that could be realised by different seismic loading scenarios, and that would require repair works that vary in both complexity and cost. In addition, one could consider the capacity of both structural and non-structural elements (partition walls, ceilings and the like). These concepts will be called upon further in later sections when identifying different retrofit approaches.

The time axis is again considered relevant to the conceptual definition of structural capacity here, since it allows for corrosion effects, the degradation of material properties with time, and fatigue effects, amongst other things, to be accounted for. Furthermore, by making a time-based assessment of demands versus capacity, it is apparent that one could consider how a planned intervention or maintenance operations could impact on the overall structural performance. These concepts are at the heart of new emerging considerations for optimum retrofit strategies, as explained in the next section.

3.3 *Optimum Use of the Available Resources*

The general criterion that should guide any decision related to structural strengthening, should be, in principle, the optimization of the resources to be invested compared with the benefits to be obtained, considering the protection of human life as well as economic and social aspects [12].

This conceptual statement may find an operational translation in the calculation of an average *expected annual loss* (EAL), expressed as a percent fraction of the total cost of reconstruction.

While this parameter is relatively straightforward to be evaluated for economic losses in ordinary buildings, its numerical quantification may become more difficult, and in some cases even impossible, when dealing with indirect losses related to social aspects and protection of culture and architectural heritage and even more so when the protection of human life is considered.

Though the conceptual validity of the general approach still remains, it is understood that in such cases a second, less quantitative, criterion shall be considered, that will lead to some sort of implicitly or explicitly defined *social agreement*, not necessarily based on scientific evaluations.

As an example, consider the problem of defining an accepted value for the probability of annual occurrence of a human casualty as a consequence of a specific hazard in a specific geographic area. For instance, take the case of tsunamis in Japan: is it acceptable to fix such a probability at 10^{-4}, i.e. to have averagely one victim every 1000 years? How has such a choice been affected by the event in 2011? Does this choice justify the construction of a 400 km long wall 14 m high along the coast? How would a different numerical choice affect the height of the wall?

Uneasy questions, which will not be further discussed here and instead, from this point, the focus will be on more standard problems for which it is assumed that some economy-based logic can be applied.

In this framework, the evaluation of an EAL requires four conceptual steps, as briefly described in the following sections.

3.4 Hazard Analysis for Uncertain Actions Such as Earthquakes

The aim of hazard analysis for a site is to estimate the rate of exceedance of any situation that may challenge a construction and its content and that shall be characterized by some intensity measure.

For example, in the case of earthquake hazard, the "situation" will be a ground motion at the site of the construction, characterized by a given annual probability of exceedance or an average return period. The intensity measure traditionally used to represent the seismic hazard has been the peak ground acceleration (PGA), possibly associated with a spectral shape to immediately estimate a structure acceleration. Once made clear that displacements are more relevant than accelerations, it appeared rational to shift to a displacement response spectrum as a key intensity measure, though this is not yet fully transferred into the practice.

Sticking with the example of seismic hazard, the problem to be addressed is how to define the seismic intensity parameter based on seismic sources and their assessed potential to induce given magnitude earthquakes with specific recurrence intervals.

The standard approach is a site specific *probabilistic seismic hazard analysis* (PSHA), based on the separate consideration of all possible earthquakes that could occur within a given region, their frequency of occurrence and the levels of ground motion intensity they could produce at a given site.

Often a PSHA is only carried out to calculate the PGA at a given return period and this is used to anchor a spectral shape provided in a code. It is obvious that in this way the resulting spectrum is not of uniform hazard, as the shape of a response spectrum in the code is fixed regardless of the return period or location, rather than changing its shape with these two parameters, influenced by the magnitude of the earthquakes which contribute most to the hazard [13].

Regardless of the procedure used to calculate PGA and spectral shape for a given return period at a given site, structures will eventually be subjected to a specific damaging earthquake event and thus, considering the geographical variation of the ground shaking and the different response of each building to its specific action, a spatial variation of performance and damage has to be expected. However, with the magnitude and location of the next earthquake being unknown, it may seem rational to design all structures to ground motions which should have the same probability of exceedance.

From a logical point of view, it appears that there are some potential flaws in this way of reasoning briefly described above. Consider, for example, a case in which the hazard evaluation is based on seismogenic zones. Since low magnitude earthquakes, say, e.g., M5, which have perhaps a low damage potential, are much more frequent than high magnitude ones, say M7, the uniform hazard spectrum may be dominated by small earthquakes, particularly for uniform hazard spectra produced for a short return period (or for a long return period in low seismicity areas). In some seismic areas (e.g. in Italy), close events are dominating the hazard and a large magnitude earthquake is normally probabilistically relevant only as a far distance event. However, it is unquestionable that in case of a M7 event it is likely that some structure will be close to the epicenter. Even if the case of a strong, close event should in general influence appropriately the results of a PSHA, particularly when extremely large return periods are considered, the results obtained do not consider the actual values of the intensity measure that would result at the epicenter. The values close to a fault are somehow chopped off and not considered for design or assessment purposes. A structure is not likely to be subjected to a large number of strong events during its life, thus responding with different performances to each one of them. The consequence is that, when a large magnitude earthquake occurs, some structures close to the epicentre are likely to be hit by actions much larger than those used for design and inappropriately expected with a given return period.

Similar considerations apply to other hazards, such as tropical cyclones, industrial accidents, terrorist attacks, and may be treated in a similar way. Slightly different cases are those related to gravity, always present, where hazard may be induced by structural aging (e.g. in case of ancient or old construction) or by unexpected loading (e.g. in case of unexpected increase of the trucks weight and frequency on a bridge).

3.5 Structural Analysis

The objective of structural analysis is to obtain some engineering demand and capacity parameters as a function of hazard parameters.

Relevant parameters to assess damage (the next step) can be assumed to be the value of the inter-storey drift, either to estimate the non-structural damage (in a direct way) and the structural damage (estimating the elements curvature, or strain, demand and the possible attainment of brittle failure modes). Other parameters of interest may be, for example:

- the floor or deck vertical deflections;
- the foundation relative displacements or rotations;
- the vibration of some structural element;
- the floor accelerations (that may be relevant for some class of non-structural elements, see Ramirez and Miranda [14]; noting that quite inadequate models are currently recommended in most codes to estimate floor accelerations, see [15, 16]).

In seismic assessment, traditional methods have been based on a comparison between base shear capacity and base shear demand, deducted from the acceleration spectrum assuming a period of vibration and a force reduction (or "behaviour") factor. Both parameters can be calculated applying more or less sophisticated approaches, i.e., using empirical equations to determine the period of vibration and evaluating the applicable force reduction factor on the basis of building typology or, at the other extreme, deriving both factors from a pushover analysis, potentially including an assessment of possible brittle failures of elements that should terminate the analysis. In this last case a force based assessment procedure can provide accurate results, but are essentially limited to one single performance, i.e. the probability of reaching a collapse limit state.

The possibility of performing non-linear time history analyses to assess different performances implies the use of sets of accelerograms for each one of the return period earthquakes of interest, essentially obtaining for each return period a pass/fail result. This pass/fail result may actually be quite tricky since often, and more so for long return periods, some seismograms may fail the structure and some not. What an engineer is supposed to do in those cases is not at all obvious.

Considerations to apply Direct displacement-based concepts to practical building assessment were published in the nineties (e.g.: [17, 18]) while a more systematic and comprehensive presentation is much more recent [19]. The inclusion of probabilistic concepts in Direct displacement-based assessment is the subject of present studies [20].

It is interesting to note that a displacement-based seismic assessment may be conceptually driven by the structural response. In other words while the more traditional question to answer is "*what will be the response of the structure to a given input ground motion*" in this framework the proper question is "*what will be the ground motion that will induce a given performance*". While this may appear an academic distinction, it is a very practical and effective change. Actually, the characterization of hazard is normally represented with somewhat smooth functions, while abrupt changes may characterize any kind of performance function, for example in the form of points where a significant change in stiffness is expected or, more dramatically, where some sort of local or global failure is predicted. The implication is that imposing a given return period to design strengthening may imply that an inconvenient point in the performance function will result (this applies to a smaller extent in a probabilistic environment, where a range of performance levels may correspond to different records, all consistent with the same return period). Essentially, while the definition of a given return period or a given yearly probability of exceedance is conventional, and consequently irrelevant, the association of a performance to physical events is possible and desirable.

This concept applies, to a much larger extent, to the choices related to a strengthening intervention since the economic resources required to reach a given performance are normally changing with finite steps. As discussed later, the selection of the strengthened structure's performance cannot be rationally defined without considering the discontinuity in some cost-benefit function.

In modern codes, an association between level of knowledge about the structure to be assessed and the protection factors to be applied to obtain a certain level of protection is often explicitly defined. From a conceptual point of view this is clear and rational: if one is more confident on the predicted response he or she can apply smaller protection factors. In general the confidence level is expressed as a function of the type of analysis (often neglecting the fact that a more refined analysis does not necessarily imply a higher confidence in the results) and the available data on geometry, detailing and material properties.

The confidence about the predicted response is not necessarily a function of the number of physical tests on structural materials and soil.

Again, the selection of quality and number of investigations to be performed should be driven and justified by the preliminary assessed response and possible strengthening choices rather than generically imposed in a document. For example, when considering the possibility of adding a shear wall system to a frame building the focus should probably be on displacement compatibility and diaphragm action capacity of the floor, while in case of a possible insertion of a base isolation system the level of shear capacity to be checked for the structure is limited by the shear transmitted through the isolation devices.

The use of some sort of back analysis can be a powerful tool to design strengthening interventions, not necessarily limited to seismic problems [21].

3.6 *Damage Analysis*

The objective of damage analysis is to associate response parameters (e.g., inter storey drift or floor acceleration, obtained from structural analyses as discussed above) and expected damage. This is commonly done through the use of non-linear fragility functions that indicate the probability of reaching a certain damage state as a function of the engineering demand parameter. Detailed discussions on this issue can be find in Mitrani-Reiser [22] and Ramirez and Miranda [14].

In a general sense, one can separate non-structural and structural damage, to consider the first one sensitive to inter-storey drift, for a certain share, and to floor accelerations, for the complementary share, and to relate structural damage to the inter-storey drift demand alone, including the assessment of brittle failures. Subsequently, assuming that non-linear correlations between drift and drift-related non-structural damage and, similarly, between floor acceleration and acceleration-related non-structural damage are known, one can take the structural analysis results and compute the likely damage expected for the whole building (considering both structural and non-structural elements). Furthermore, note that these non-linear fragility functions can be modified to suit the element at hand without losing generality and probabilistic evaluations of reaching certain damage states can be computed.

3.7 *Loss Analysis*

In the framework of performance-based assessment, the objective of loss analysis is to calculate the probable repair cost for each level of damage state defined in the previous step. In principle, a loss estimate should include consequences such as deaths, repair costs and downtime consequences (the well known *3D approach*, see Fajfar and Krawinkler [23]). However, as anticipated, an accepted value for the death toll can only be derived from societal agreement (to avoid collapse may be crucial in this respect and this performance may be considered as a special case), while the problem of evaluating losses associated with downtime can be more easily related to monetary parameters. In its extreme simplification, this could be done assuming that the cost of repair will be proportional to damage, for example associating a (larger) fraction of the value of the building to non-structural content and a (smaller) fraction to structures. A separate assessment will consider the potential losses associated with the impossibility of using the facility, the interruption of production, the loss of clients, etc.

Again, this can be elaborated and probabilistic aspects can be included without losing generality.

Returning now to the concepts introduced in Fig. 6, it is apparent that one could compute the annual probability of a certain magnitude of load (hazard analysis), use structural analysis to establish the likely response of the structure subject to that load (structural analysis), relate the structural response to damage states for both structural and non-structural elements (damage analysis) and finally compute the likely loss for the load. Integrating the results of such loss estimates over all relevant load levels, one can obtain the expected annual loss, which is considered a particularly useful performance measure exactly because it relates the impact of loading to a monetary figure (that will be easily understood by lay persons) not just for a single event but with account for the time axis shown in Fig. 6.

4 Strengthening and Retrofit Approaches

4.1 Retrofit Scenarios

The best retrofit approach will depend on the problem at hand. As new materials are created and technological solutions develop, engineers are being offered an ever increasing number of options for the retrofit of an existing structure. This section will briefly review and discuss the different retrofit strategies that are typically considered, making reference to the concepts introduced in the previous section and to the two hypothetical case study structures indicated in Figs. 7 and 8.

The existing multi-span bridge indicated in Fig. 7a represents a scenario in which, after an inspection, it is recognized that the foundation of a central pier along the bridge is deteriorating and that repairs are required. Note that this scenario is somewhat similar to the *Ponte della Becca* (Peck Bridge) in Pavia. As seen in Fig. 7b, in this case it is worth comparing loads with demands in force versus time axes. The building indicated in Fig. 8a instead represents an existing building that has been assessed for seismic loads and deemed be at risk of forming a soft-storey mechanism (in which deformations concentrate in a single storey, at ground level in

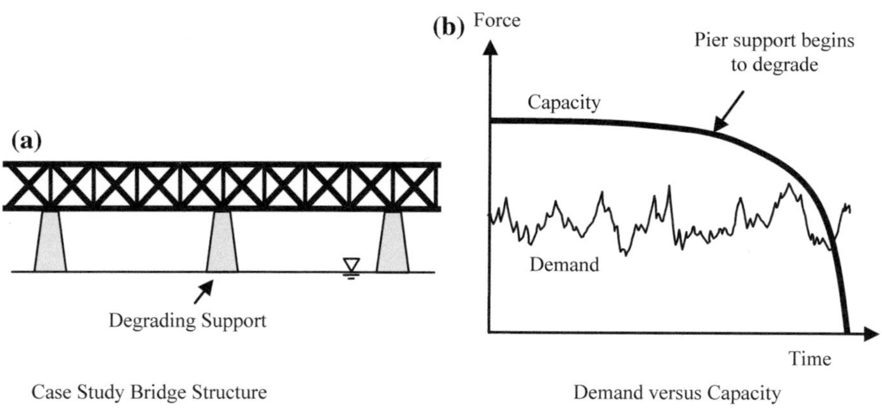

Fig. 7 **a** A case study bridge structure with degrading support, and **b** comparison of the demands and capacity, expressed in terms of force-time axes

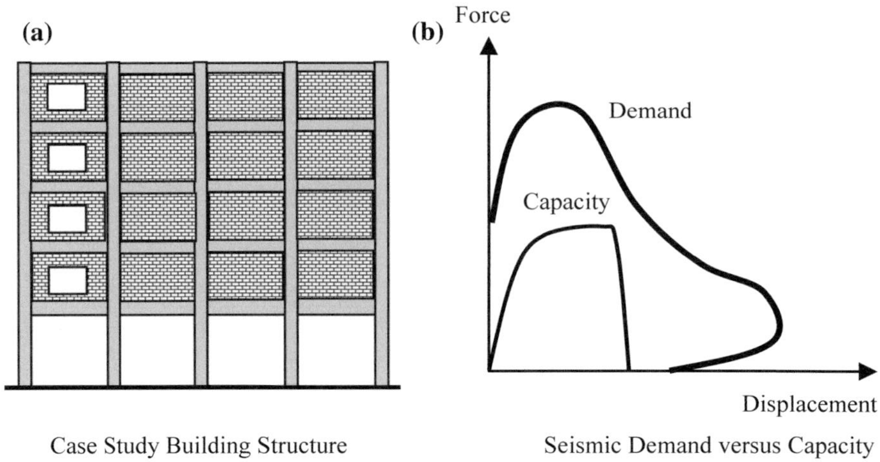

Fig. 8 **a** A case study building prone to the formation of a soft-storey during earthquake shaking, and **b** comparison of the demand and capacity, expressed in terms of force-displacement axes

this case) and poses an unacceptably high risk to the occupants of the building. This scenario is instead representative of a multitude of residential buildings in Italy and other seismically active regions around the world. As seen in Fig. 8b, in this case it is advantageous to consider the earthquake loading (associated with a certain return period) and capacity in force-displacement axes.

The following sections will now explore different retrofit strategies for these and similar systems. Detailed considerations for the design of retrofit solutions are not within the scope of this chapter and are instead the subject of the latter chapters of this text.

4.2 Adding New Structural Elements

A basic strengthening approach relies on the insertion of additional elements reacting to horizontal or vertical actions. For bridge systems, this could involve adding supplementary structural elements to the deck, new piles connected into the existing foundations or new piers altogether. As illustrated in Fig. 9, for the case study bridge system mentioned earlier, the insertion of new piers could be an effective solution (and was in fact the solution adopted for the Peck bridge in Italy) since it ensures a good level of resistance is provided (as indicated in Fig. 9b) and reduces the load on the deck (not shown in Fig. 9b for simplicity). Furthermore, it can be executed with respect for the retrofit charter (described in Sect. 2.2).

For building structures, the addition of new structural elements may consider the use of steel braced frames (which can be connected into either steel or concrete

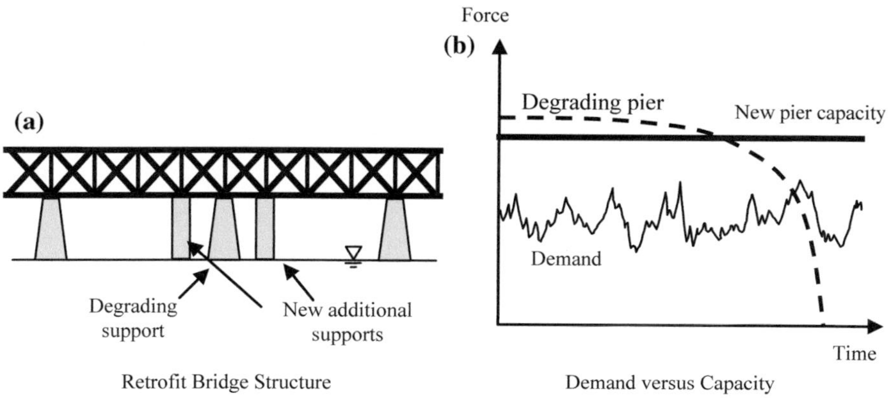

Fig. 9 **a** Retrofit of the case study bridge via the introduction of new pier supports, and **b** comparison of the demand and capacity, illustrating that the retrofitted building has sufficient capacity

buildings relatively easily) or concrete walls (possibly obtained by strengthening masonry panels), that could be inserted in the interior of a building or outside it.

If the primary reaction system of the original building was already made by walls, then the purpose of additional walls could be to increase strength and stiffness and to regularize the torsional response, thus reducing the expected damage even at relatively low displacement demand. If the original structural system was based on frames, the introduction of much stiffer elements may completely change the response, arriving at the limiting case in which the original frame provides only a negligible contribution to the response, and the only fundamental requirement will be that its displacement capacity will be larger than the displacement demand associated with the response of the new wall system. For example, considering the case study building presented earlier, the introduction of new concrete walls and foundations within the building could add considerable strength and stiffness to the system, as illustrated in Fig. 10.

In all cases in which new elements are added to a structure it is imperative that all parts of the new load-paths formed are checked; for instance, in the case that walls are added to a building, the capacity of the foundations corresponding to the new walls and the capacity of the horizontal diaphragm to transmit the action, globally and locally, need to be checked.

The average costs of these kinds of intervention can be estimated preliminarily assuming some total length of the wall or framed systems and assigning an average unitary cost. To this cost it may be necessary to add the sum required to strengthen the floor diaphragms, if needed, and the cost of removal and replacement of non-structural finishing material, that may be extensively required when the walls will be added internally or when an extensive floor strengthening will be required.

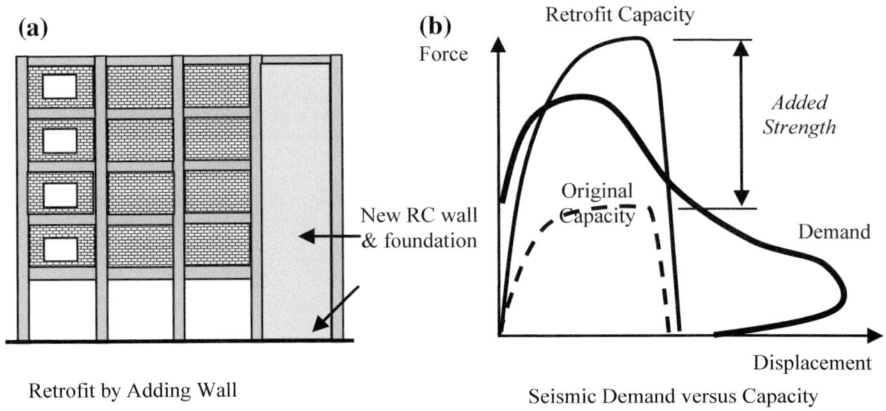

Fig. 10 **a** Retrofit of the case study building via the addition of a new RC wall and foundation, and **b** comparison of the demand and retrofit capacity, illustrating how the added strength provides the existing building with sufficient capacity

4.3 Strengthening Existing Elements

Another basic strategy to improve the response of a building counts on the application of capacity design principles to reduce the likelihood of brittle failures. In this context it is thus possibly required to increase the strength of some element in a selected way, to favour ductile damage modes. For example, it is typical to increase the shear strength of columns and beams to obtain flexural failure modes, to increase the strength of external joints and to increase the flexural strength of columns to shift the formation of plastic hinges to the beams. This last example does not aim to avoid a brittle collapse, but to prevent the formation of a soft storey. In some case (possibly academic), weakening ductile modes has been considered instead of strengthening brittle ones.

These kinds of intervention tend to modify in a significant way the last part of a pushover capacity curve (and possibly the associated deflected shape), increasing the displacement capacity of the structure. However, they may have a negligible effect on the first part of the curve (the modification of stiffness up to yielding is not significantly affected) and on the yield strength (since, in general, shear and flexural failure modes have similar strengths).

Typical interventions are based on external jacketing or wrapping of an element or part of it, using carbon or glass fibres, steel plates or thin layers of reinforced concrete. For the bridge case study shown in Fig. 7, one could consider underpinning the foundation of the existing pile. However, such an operation may not be particularly practical in the case mentioned, due to the presence of the river and deteriorating condition of the existing pier. For the building structure shown in Fig. 8, one might aim to strengthen the columns at the ground storey via jacketing or fibre wrapping to avoid brittle shear failures. Alternatively, the aim might be to strengthen the columns sufficiently so as to reduce the likelihood of the soft-storey mechanism forming. However, this type of intervention is not likely to be very effective in such cases since it may simply lead to formation of a soft-storey at the upper levels, without improving the demand to capacity ratio.

Local strengthening of acceleration-sensitive non-structural elements can also be an effective retrofit measure. For instance, bracing of suspended ceiling systems is known to reduce their vulnerability significantly, similarly for sprinkler systems (refer to FEMA E-74 [24] for a number of practical means of reducing the vulnerability of non-structural elements).

It is obvious that the cost of strengthening an element will vary significantly, as a function of its geometry and of the applied technique. However, analysing a large number of cases, reasonable average costs, that could be used for rough first estimates, have been estimated.

If the fraction of the total number of elements to be strengthened has been guessed or calculated, it is possible to estimate the total cost of the structural intervention.

4.4 Locally Increasing the Deformation Capacity

If it is assumed that all the possible brittle failure modes have been eliminated by a proper application of capacity design principles, i.e. by an appropriately selected local element strengthening, the displacement capacity of the structure can be limited by insufficient curvature (and consequently rotation) capacity in critical section of columns and beams.

An insufficient rotation capacity of columns might be detected in case of a soft storey formation, or exclusively at the column base. Note that a soft storey mechanism is not always unacceptable, it depends on the associated storey rotation capacity (including second order effects) and the associated global displacement capacity relative to the seismic demand.

An intervention aimed at increasing deformation capacity in RC structures is normally based on confining measures, to avoid bar bucking and to increase the compression deformation capacity of concrete. Fibre wrapping and steel encasing are thus again the typical choices. The effects on strength and stiffness will be even more negligible than in the previous case and the effects will be still limited to the last part of the pushover curve. This can be seen for the case study building structure in Fig. 11, where the use of fibre wrapping to increase confinement to columns leads to a significant increase in the system displacement capacity, such that the demand is less than the retrofit capacity.

The interventions in this case can be limited to the critical zones of the elements and as such, the cost per structural member is therefore lower, but of the same order of magnitude compared to that discussed in the previous section. In the discussion of relative merits of different strengthening choices no distinction will be made about these two kinds of strengthening for what concerns costs.

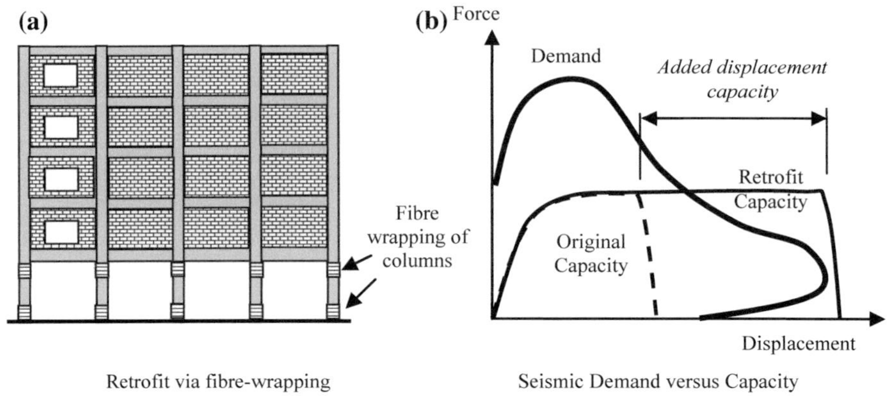

Fig. 11 **a** Retrofit of the case study building using fibre-wrapping to provide confinement to the columns, and **b** comparison of the demand and retrofit capacity, illustrating how the added displacement capacity provides the building with sufficient capacity

Another useful retrofit intervention in existing buildings may actually focus on improving the deformation capacity of non-structural elements. It is well recognized that both masonry infills and lightweight partitions will typically be able to sustain only low drift demands (as low as 0.2–0.3%) before incurring damaged. Consequently, retrofit measures could consider modifying the connection details of the non-structural elements (or substituting them completely as part of a refurbishment process) so as to increase the load required to cause building damage.

4.5 Introducing Isolation Systems

The insertion of an isolation system, at the base or at some height of the building, can often be a last- recourse intervention to improve the seismic performance of the building. The essence is that in this way the maximum shear that will pass through the system is governed by the system capacity (see, as a general reference, Christopoulos and Filiatrault [25]). As an example, imagine using friction pendulum devices [26]: if the dynamic friction coefficient is in the range of 5%, it is likely that the maximum base shear force on the building will be in the range of 10% of gravity.

For earthquake events, or portions of events, that will not induce this level of acceleration, the structure is responding like a fixed base structure, while for any value of acceleration exceeding this value the difference in base shear, and consequently in structural drift demand, will be marginal.

The consequence is that for very frequent events the effects of the isolation system will be negligible, but the damage minor, while for a large variation of demand, provided that the displacement capacity of the isolation system is not reached, the damage will still be minor.

The key issue with existing structures is the possible technique to insert the isolation system and, consequently, the problems related to the relative movement between the isolated part and original part of the building, i.e. how to create the necessary gap, and the interaction between installations connected to the two sides of the gap.

There are cases where the presence of some distance between in–ground structure and external retaining walls allows a relatively simple cut of the columns and the insertion of the isolators. In other cases it has been possible to create an independent foundation, possibly on piles, and to uplift the whole building to insert the isolation devices between the now existing double foundation system.

There are a fairly wide range of isolation devices available in the market, including high damping rubber bearings, lead-rubber bearings and friction pendulum devices. Beigi et al. [27] also recently proposed an innovative brace device that maintains the isolating effect offered by soft-storey mechanisms but increases column deformation capacity and reduces p-delta demands.

The cost required for the isolation of a structure may vary significantly as a function of the actual situation and, obviously, the ingenuity of the solution. Consider for example that cutting a column (to insert an isolation device) at the base, at the middle or at the top, will completely change the bending moment diagram generated by the same shear force, and will consequently induce the need for strengthening the column itself or other parts of the structure. Consider as well that using a single or double sliding surface device and positioning the concavity upward or downward may change the elements on which the relative movement will induce a bending moment due to the eccentricity of the vertical reaction.

The design of this kind of strengthening should in general include the procedures for substituting a device, to re–centre the building after a strong event and possibly to test the response of the upgraded system.

4.6 Reduction of Demands

It is obvious that a retrofit intervention may operate reducing the demand rather than increasing the capacity.

Typically this may be obtained increasing the dissipation capacity of the building, i.e. increasing the equivalent viscous damping level to be used to correct the acceleration and displacement spectra.

Normally this kind of provision is not adopted as a single measure, but is coupled with other interventions, possibly strengthening some member to avoid brittle failure, or inserting new steel braced frames, or in connection with the creation of an isolation system, to reduce the displacement demand. In this last case the problem of locally adsorbing a large fraction of the total shear may require a local intervention on the foundation or on the upper structure, normally a RC wall.

In the case of a damped steel braced frame the devices can be inserted in the diagonal bracings, not only to add damping, but also to keep the maximum shear force under control. The cost associated with the introduction of dampers is extremely variable.

Figure 12 illustrates how viscous dampers might be used to reduce the demands on the case study building. It can be seen that by adding viscous dampers (in the configuration shown) the lateral force-displacement capacity of the building is not affected but the retrofit solution is anyway successful since it reduces the demands below the capacity.

Another example of an intervention oriented towards the reduction of demand is to restrict usage of the building or bridge, so as to limit the mass and loads imposed. For the bridge example referred to in Fig. 7, traffic restrictions could be introduced so that only light vehicles are allowed to cross the bridge (as occurred for the Peck bridge in Pavia, Italy). Alternatively, change in use of a building structure, or substitution of an existing floor with lightweight concrete may imply lower gravity loads and lower seismic demands. On a similar note, another more innovative approach to reducing demands is to consider the introduction of a tuned mass. The general concept is simple: if the building can be regarded essentially as a single degree of freedom system with most of its mass associated to the first mode of vibration, adding a tuned mass that vibrates with a similar period of vibration, but in the opposite phase, will induce a favourable reduction of shear force at all instants. This of course applies when the system responds essentially elastically. A complete description of the approach for seismic applications [28] is obviously more complex, but the optimal ratio between the first period of vibration of the building and the period of vibration of the tuned mass can be calculated as a function of the tuned mass divided by the participating mass of the first mode of vibration of the building

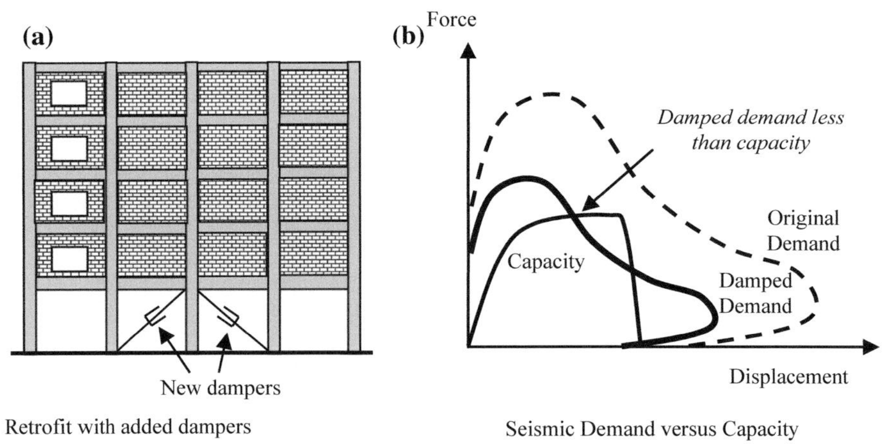

Fig. 12 **a** Retrofit of the case study building by adding viscous dampers, and **b** comparison of the demand and retrofit capacity, illustrating how the damped demands imply that the retrofit building has sufficient capacity

and of the damping ratio of the building. Again, it is difficult to give generally reliable figures of cost, while the reduction of shear forces and displacement demand may be in the order of 50% in very favourable cases.

Additional reading material relevant to the subject of Structural Strengthening and Retrofit; Motivations, Concepts and Approaches, can be found in references [29–42].

References

1. CEN EC1 (English): "Eurocode 1: Actions on structures—Part 1-1: general actions—Densities, self-weight, imposed loads for buildings" EN 1991-1-1, Comite Europeen de Normalization, Brussels, Belgium; 2002.
2. CEN EC8: Eurocode 8—Design provisions for earthquake resistant structures, EN-1998-1:2004: E, Comite Europeen de Normalization, Brussels, Belgium; 2004.
3. CEN EC0 (English): "Eurocode—Basis of structural design" EN 1990:2002, Comite Europeen de Normalization, Brussels, Belgium; 2002.
4. Hugo V. Notre Dame de Paris. Paris: Hetzel; 1831.
5. Ruskin J. The seven lamps of architecture. London: Ward, Lock; 1911.
6. Vanvitelli L. Lettera a Poleni. Rome, 7 September. MS. Cod. Marc. It. IV, 1743; 680 (=5559), 3 p., 1 pl.
7. Price NS, Talley Jr, MK, Vaccaro AM, editors. Restoration and anti-restoration, introduction to part V, in historical and philosophical issues in the conservation of cultural heritage. Getty Publications; 1996. ISBN 0-89236-398-3.
8. Rondelet J. Discours pour l'ouverture du cours de construction. Paris: Imp. de Fain; 1806.
9. Mérimée P. Notes d'un voyage dans le Midi de la France, Paris (reprinted 1989), Editions Adam Biro, Paris.
10. ICOMOS: ISCARSAH Guidelines. 2014. https://iscarsah.files.wordpress.com/2014/11/iscarsah-guidelines.pdf.
11. Jonas H. The imperative of responsibility, in search of an ethics for the technological age. The University of Chicago Press, Chicago 60637; 1984. ISBN 0-226-40597-4.
12. Calvi GM. Choices and criteria for seismic strengthening. J Earthq Eng. 2013;17(6):769–802.
13. Iervolino I, Chioccarelli E, Convertito V. Engineering design earthquakes from multimodal hazard disaggregation. Soil Dyn Earthq Eng. 2011;31:1212–31.
14. Ramirez CM, Miranda E. Building specific loss estimation methods and tools for simplified performance-based earthquake engineering. Technical Report No. 171, John A. Blume Earthquake Engineering Center, http://blume.stanford.edu, Stanford University; 2009.
15. Sullivan TJ, Calvi PM, Nascimbene R. Towards improved floor spectra estimates for seismic design. Earthq Struct. 2013;4(1).
16. Calvi PM. Relative displacement floor spectra for seismic design of non structural elements. J Earthq Eng. 2014;(18)7.
17. Priestley MJN. Displacement-based seismic assessment of reinforced concrete buildings. J Earthq Eng. 1997;1(1):157–92.
18. Calvi GM. A displacement-based approach for vulnerability evaluation of classes of buildings. J Earthq Eng. 1999;3(3):411–38.
19. Priestley MJN, Calvi GM, Kowalsky MJ. Displacement-based seismic design of structures. Pavia: IUSS Press; 2007.
20. Welch DP, Sullivan TJ, Calvi GM. Developing direct displacement-based procedures for simplified loss assessment in performance-based earthquake engineering. J Earthq Eng. 2014;18(2):290–322.

21. Nascimbene R, Fagà E, Cigada A, Vanali M, Moratti M, Pinho R, Calvi GM. Construction of a scaffolding for the major spire of the Milan cathedral: modeling, analysis, verification and dynamic identification (in Italian, English summary). Progettazione Sismica. 2012;4(1):15–34.
22. Mitrani-Reiser J. An ounce of prevention: probabilistic loss estimation for performance-based earthquake engineering. Ph.D. Dissertation. Pasadena, CA: CalTech; 2007.
23. Fajfar P, Krawinkler H. Performance-based seismic design concepts and implementation. In: Fajfar P, Krawinkler H, editors. Proceedings of the international workshop, Bled, Slovenia, 28 June–1 July. PEER Report 2004/05, The Pacific Earthquake Engineering Research Center; 2004. ISBN 0-9762060-0-5.
24. FEMA E-74: Reducing the risks of nonstructural earthquake damage—A practical guide. FEMA E-74 Document, Federal Emergency Management Agency, Washington, DC; 2011.
25. Christopoulos C, Filiatrault A. Principles of passive supplemental damping and seismic isolation. Pavia: IUSS Press; 2006. p. 480.
26. Calvi GM, Pietra D, Moratti M. Criteri per la progettazione di dispositivi di isolamento apendolo scorrevole. Progettazione Sismica, EUCENTRE, Pavia, Italy. 2010;03:7–30.
27. Beigi H, Christopoulos C, Sullivan T, Calvi G. Gapped-inclined braces for seismic retrofit of soft-story buildings. J Struct Eng. 2014;. doi:10.1061/(ASCE)ST.1943-541X.0001006,04014080.
28. Sadek F, Mohraz B, Taylor AW, Chung RM. A method of estimating the parameters of tuned mass dampers for seismic applications. Earthq Eng Struct Dyn. 1997;26:617–35. doi:10.1002/(SICI)1096-9845(199706)26:6<617:AID-EQE664>3.0.CO;2-Z.
29. Comerio MC. Resilience, recovery and community renewal. In: Keynote paper, 15th World Conference of Earthquake Engineering, Lisbon, Portugal; 2012.
30. Elnashai AS, Salama AI. Selective repair and retrofitting techniques for RC structures in seismic regions. Research Report ESEE/92-2, Engineering Seismology and Earthquake Engineering Section, Imperial College, London, UK; 1992.
31. Fardis M. Seismic design, assessment and retrofitting of concrete buildings: based on EN-Eurocode 8. Springer Science & Business Media, 25/07/2009—Technology & Engineering; 2009.
32. FEMA P-58-1: Seismic Performance Assessment of Buildings: Volume 1—Methodology. FEMA P-58-1, Prepared by the Applied Technology Council for the Federal Emergency Management Agency, Washington, DC; 2012.
33. FEMA P-58-3: Seismic Performance Assessment of Buildings, Volume 3—Supporting Electronic Materials and Background Documentation: 3.1 Performance Assessment Calculation Tool (PACT), Version 2.9.65, Federal Emergency Management Agency, Washington, DC; 2012.
34. Fib: Seismic assessment and retrofit of reinforced concrete buildings. Fédération Internationale du béton, Bulletin. 2003;24.
35. Fib: Seismic Bridge Design and Retrofit—Structural Solutions. Fédération internationale du béton, Bulletin 2007;39.
36. Jirsa O. Divergent issues in rehabilitation of existing buildings. Earthq Spectra (EERI). 1994;10(1):95–112.
37. Pinho R, Elnashai AS. Repair and retrofitting of RC walls using selective techniques. J Earthq Eng. 1998;2(4):525–68.
38. Price NS, Talley MK, Melucco VA. Historical and philosophical issues in the conservation of cultural heritage. Los Angeles: Getty Conservation Institute; 1996. ISBN 0892363983.
39. Priestley MJN, Seible F, Calvi GM. Seismic design and retrofit of bridges. New York: Wiley; 1996.
40. Rodriguez M, Park R. Repair and strengthening of reinforced concrete buildings for seismic resistance. Earthq Spectra (EERI). 1991;7(3):439–59.
41. Roeder CW, Banerjee S, Jung DR, Smith SK. Role of building foundations in seismic retrofit. Earthq Spectra (EERI). 1996;12(4):925–42.
42. Thermou GE, Elnashai AS. Seismic retrofit schemes for RC structures and local-global consequences. Earthq Eng Struct Dyn. 2005.

Cultural Heritage Monuments and Historical Buildings: Conservation Works and Structural Retrofitting

Romeu Vicente, Sergio Lagomarsino, Tiago Miguel Ferreira, Serena Cattari and J.A.R. Mendes da Silva

1 Introduction

Historical constructions are an important part of the cultural heritage, because of their architectural value and evidence of building techniques. Their conservation over the centuries is a responsibility of our society, in order to pass on to future generations.

It is worth noting that the structural safety of historical constructions to permanent long-term actions, in many cases, has been proved over time. The diagnosis of the present conditions of the building can be made by a complete interdisciplinary knowledge based on historical notes, technological survey, non-destructive testing procedures and the interpretation of crack and decay patterns.

R. Vicente (✉)
RISCO, Department of Civil Engineering, University of Aveiro,
3810-193 Aveiro, Portugal
e-mail: romvic@ua.pt

S. Lagomarsino · S. Cattari
Department of Civil, Chemical and Environmental Engineering,
University of Genoa, 16145 Genoa, Italy
e-mail: sergio.lagomarsino@unige.it

S. Cattari
e-mail: cattari@diseg.unige.it

T.M. Ferreira
ISISE, Department of Civil Engineering, University of Minho,
4800-058 Guimarães, Portugal
e-mail: tmferreira@civil.uminho.pt

J.A.R. Mendes da Silva
Department of Civil Engineering, Faculty of Science and Technology,
ADAI/LAETA, 3030-790 Coimbra, Portugal
e-mail: raimundo@dec.uc.pt

A. Costa et al. (eds.), *Strengthening and Retrofitting of Existing Structures*,
Building Pathology and Rehabilitation 9, https://doi.org/10.1007/978-981-10-5858-5_2

Slow and inevitable aging processes might affect the current structural stability due to different possible origins: (1) material deterioration; (2) anthropic modifications, particularly in the urban environment; and (3) climate and environmental changes. Monitoring is a necessary stage in all three cases, both through advanced instrumentation techniques and qualified visual inspections, in order to detect when interventions are needed. Usually, deterioration processes can be slowdown and serious damage can be prevented by a periodic maintenance.

On the contrary, the preservation from natural hazards requires a preventive assessment aimed at the specific vulnerability and risk to different events (floods, earthquake, fire, biological, etc.), which cannot be based only on a qualitative approach and the observation of the building behaviour from the past. In particular, earthquake represents the main cause of damage to masonry structures and, due to the high return period of severe events in a prone earthquake area, a direct proof of safety is usually not available for the specific case. Moreover, after any strong earthquake the necessary restoration requires strengthening and, often, partial reconstruction, with a significant loss of the authenticity in respect to construction techniques. Therefore, it is necessary to have tools to implement a preventive policy, which takes into account the conservation requirements. Slight damage occurred due to previous earthquakes might suggest the possible collapse mechanism that the building would experience in the case of a strong event, but a reliable seismic assessment cannot be performed without quantitative models.

The seismic assessment of existing buildings is a complex task, basically for two different reasons: (1) the difficulty of interpreting and modelling the seismic response; and (2) the difficulty of acquiring as-built information on material parameters and structural details, due to their spatial variability in the buildings and the need of avoiding invasive testing.

In the last decades, earthquakes have proven that particular strengthening interventions carried out in the last century have revealed to be ineffective and, in some cases, even worsen to the seismic behaviour of the structure. Thus, proper methods of analysis and verification procedures are required for the seismic assessment and the design of interventions, with the aim of risk mitigation of cultural heritage.

Finally, it has to be stressed that, if carefully planned, the use and exploitation of cultural heritage constructions represents a sustainable approach for the conservation, because it underlies/undertakes as a continuous “health monitoring”, even in the cases in which some interventions and modifications are required. A detailed assessment through proper procedures and models allows to avoid invasive and useless interventions.

1.1 Cultural Heritage: The Origin and the Establishment of the Concept

The United Nations Educational, Scientific and Cultural Organisation (UNESCO) was constituted in London on November 16, 1945. Aimed at continuing the work begun decades before by the League of Nations, UNESCO articulated its commitment to the concept of a common cultural heritage and to the idea of strengthening and conserving this heritage through international collaboration and cooperation in its constitution [1]. In 1957 UNESCO was involved with organizing the First International Congress of Architects and Specialists of Historic Buildings, which took place in Paris and wherein a recommendation to create an "international assembly of architects and specialists of historical buildings" had met with approval. In May 1964 UNESCO's executive board adopted a resolution with an identical goal to that of the 1957 Paris congress and, in the same year, during the Second International in Venice, Italy, UNESCO put forward a resolution and draft status providing the basis for the establishment of an international non-governmental organization for monument and sites, named International Council on Monuments and Sites (ICOMOS), responsible for providing expertise in the form of consultants to UNESCO. The resolution was adopted along with twelve others, the first of which became the International Charter for the Conservation and Restoration of Monuments and Sites, known as the Venice Charter. In June 1965 the Venice Charter was ratified and the ICOMOS was officially founded in Warsaw, Poland. From its foundation, ICOMOS has stablished more than twenty-five International Scientific Committees on various themes and issues related with cultural heritage, which undertake research, develop conservation theory, guidelines and charters and foster training for better heritage conservation practice [2].

The Venice Charter is the first text wherein the concept of heritage is defined. In its introductory section it can be read that "Imbued with a message from the past, the historic monuments of generations of people remain to the present day as living witnesses of their age-old traditions. People are becoming more and more conscious of the unity of human values and regard ancient monuments as a common heritage. The common responsibility to safeguard them for future generations is recognized. It is our duty to hand them on in the full richness of their authenticity" [3]. In other words, heritage as concept can be defined as the collection of things which relates people to who they are, where they have come from, and why they are the way they are. According to [4], the documents following the Venice Charter focus on two different issues: (1) the definition of the general principles for the identification of new fields of conservation (addressed in the 1971 UNESCO Convention on the safeguarding of wetlands and in the Charter of the Council of Europe in 1972, wherein as a limited and fragile resource the soil is proposed as heritage); and (2) the attempt to integrate the principles of safeguarding with the control systems of the territory and of the economic and social development [4]. In the 1972 UNESCO Convention on the Protection of World, Cultural and Natural Heritage [5], the expression "cultural heritage" is used to refer monuments and sites of "exceptional universal value from the point of view of history art and science", a line followed later in the 1987 Charter for the Conservation of Historic Towns and Urban Areas [6], known as Washington Charter, where the need to protect historic cities is clearly stated. It is worth adding that the concepts of tangible and intangible values as the object of protection were recognized for the first time in this document. Another worthy highlighting document on this issue is the 1979 Burra Charter [7], where it is stated that the conservation of the cultural significance of a site, due to its aesthetic, historic, scientific or social value, must be safeguarded and protected. Despite its great influence, cultural heritage has not often had the recognition that it deserves. In fact, throughout history there have been many theories on the treatment and protection of cultural heritage, particularly to buildings, some of those have been considerate and respectful, whereas others have been destructive and oblivious [8].

1.2 The Safeguard of Cultural Built Heritage

Safeguarding any heritage asset, particularly heritage valued constructions, requires method, strategy and planning. The cultural built heritage includes and encloses the historical, ideological, architectural, artistic and material identity of a city and consequently any conservation, restoration or rehabilitation intervention must respect, as much as possible, the authenticity and compatibility with the original. Knowledge on past urban renewal and renovation processes are the basis of the definition of a methodology and strategy, keeping in perspective that every case has its singularities and necessary adaptations. The need for survey, through building appraisal and inspection is a decisive and guiding stage for the success of the intervention of any singular or collective regeneration process.

It is based on these concepts that the discussion presented in this chapter is developed, starting with a brief overview on the appraisal, inspection and monitoring of heritage valued construction, which is followed by the presentation of two different but complementary approaches. The first is dedicated to the conservation of restoration process of the Tower of the University of Coimbra, in Portugal, wherein non-structural and structural diagnosis and interventions were prepared and undertaken in the scope of the acknowledgement of University of Coimbra, the uptown ("Alta") and Sofia as World Heritage Sites [9]. The second is a comparison between possible seismic assessment procedures, applied to two different types of structures: an ottoman palace, the Hassan Bey's Mansion in Rhodes (Greece), and a mosque, the Great Mosque of Algiers (Algeria).

2 Survey Appraisal, Inspection and Monitoring of Heritage Valued Constructions

The survey is the starting point to assess the condition and identify defects of the constructions. Survey actions are often inadequate and unfruitful, because they are not based on a true knowledge of the building stock, from the type of materials used, construction techniques, possible systematic vulnerability features, etc. A poor survey can have a negative effect on the way the building is retrofitted and maintained, compromising its future well-being. Another aspect to take into account is the scale of appraisal and inspection pursued. This is, choosing the most adequate approach for inspection, appraisal and diagnosis is a complex task that can determine the success or the total failure of the survey purpose.

ICOMOS establishes guidelines on several levels [10]. On the survey and diagnosis level, the need of complete understanding of the structural and material characteristics of the construction is clearly stated. It recommends, as essential to collect historical information on the structure, techniques and construction methods used, subsequent alterations, present conservation state, etc. It further states that the diagnosis should be based on historical information and on qualitative and quantitative approaches and therefore, prior to any decision on intervention, it is indispensable to determine the causes of damages and degradations, and only then to evaluate the safety level of the construction based on its present knowledge. As outlined by [9], the rational approach for the survey stage must keep guided by the following general principles:

- each traditional building has different and singular aspects that make them unique, leading to slightly different survey needs, from case to case. The survey strategy to be adopted must be the most adaptable and sensitive to the building features;
- the selection of the means of inspection, appraisal and recording must be adaptable to the nature of the building, physical and in situ limitation of survey actions and available resources;
- the survey actions should be based on the general scope and most important and critical aims of the project. Any repair, maintenance, refurbishment action or intervention strategy should reflect the technical and financial effort made in the survey phase;
- the survey is a multidisciplinary task. The contribution of surveyor teams (engineers, architects, historians, archaeologists, etc.) with expertise opinion is very valuable. The greater challenge is to coordinate these specialists and their objectives;
- the surveying stage, through inspection, appraisal, diagnosis and recording tasks could attain very high level of complexity. Nevertheless, the focus on the project overview and in its general understanding must be always kept;
- the use of other sources of information, such as the documentary information is also very valuable and should be considered.

Fig. 1 Survey framework of heritage valued constructions

The surveying task is essentially a combination of complementary tasks: recording, diagnosis, inspection and testing. As depicted in Fig. 1, generally the survey process should involve three essential steps: preparation phase; field work and off-site work. In each one of these phases, several processes are carried out: organizing activities, research, analysis, recording and reporting are some of the major procedures.

3 The Tower of the University of Coimbra

With more than six centuries of history, the University of Coimbra was included in the World Heritage List of UNESCO in 2013. The area inscribed has about 36 hectares and comprises 31 groups of buildings with different ages, considered of major relevance to the history and the memory of the University [11]. Among those, the Tower of the Royal Palace, depicted in Fig. 2, is the most well known and one of the *ex-libris* of the old town, receiving more than 300 thousand annual visitors.

The 34-meter-high Tower, also known as Tower of the University of Coimbra, was planned during the reign of João V and it is considered one of the most original examples of Portuguese eighteenth century Baroque architecture. Started to build in 1728 and finished in 1733, about 22 years before the great Lisbon earthquake, during its history it suffered no more than a few and limited non-structural maintenance and restoration interventions.

Along the ten years of preparation of the dossier for UNESCO, two main challenges were identified: (1) to improve the conservation state of the buildings

Fig. 2 Tower of the University of Coimbra and Via Latina

located within the inscription area, assuring simultaneously their adequate and up-to-date response to the university everyday activities, and the preservation of their integrity and authenticity; and (2) to contribute for a needed methodological approach, as a sustainable example, more than just as an administrative acknowledgment, inspiring learning and research activities, and motivating the community for the preservation and valorisation of this heritage valued asset of national interest [12, 13].

The restoration project of the Tower followed these guidelines and was carried out by an internal multidisciplinary team of Engineers, Architects, Restoration experts, Archaeologists and a large number of other expertise contributions, with the scientific support of several professors and research groups [14]. The technical works were carried out by specialized companies (chosen through international public trends), under the supervision of University technical teams. The terms of these public trends included specific clauses on the need of compatibility between the efficient execution and ongoing of the restoration works, and the project of the pedagogical work site (presented in Sect. 3.5).

3.1 The Conservation and Restoration Project of the Tower

The general aim of the intervention is to preserve an architectural heritage element of symbolic meaning not only for the University of Coimbra but also for the city, through the strong physical presence it has in the landscape and for its socio-cultural meaning. Simultaneously, from the standpoint of a sustainable intervention over a cultural heritage asset, this project aims at reintroducing the visits to the Tower, which had to be suspended due to its poor condition. The project and the conservation work itself required a specific approach and was supported by a broad number of preliminary activities [11]:

- historical and architectural research, aiming at cross-referencing historical data of the 5-year construction of the Tower. This continuous process allowed to get a better understanding of the existing structure and to consolidate the criteria used to justify the intervention;
- the graphical base obtained both from the architectural and photogrammetric survey, as well as the mapping of the defects, provided information on features and dimensions essential for the restoration project;
- the analysis of the structural behaviour attained through a numerical model provided important data on the structural integrity of the Tower;
- the prior testing of cleaning methods defined a series of references for the execution of the project and the intervention, assuring the suitability of the solutions adopted.

As discussed in the introductory section of this chapter, the existent set along with the ethical principles inherent to the intervention in this kind of heritage led to a minimum action, mostly concerning a preventive maintenance and conservation. Moreover, the characteristics of the several materials implied coherent and sustainable methodologies and strategies of intervention, both in the preliminary works and in the several stages of intervention. The main purpose of maintaining all the original materials, establishing the physical and aesthetical balance of the architectural whole, is to assure that from the design to the execution, safeguarding the authenticity of the tower for future generations is kept [11].

3.2 State of Conservation

Despite the presence of several defects, the stones materials were in an acceptable state of conservation. The overall surface presented a heterogeneous colouring caused by different factors, namely, biological colonisation, films and dark crusts, oxidation spots of metallic elements and the orange patina resulting from the aging process of limestone. On the terrace, the abutment rail had several embedding spots that were causing fractures in the cornices. In addition, several floor slabs were identified as damaged or broken.

In the interior of the Tower, the plasters were degraded, both by the action of humidity and nitre, and the layers of whitewash were detached. The stone slabs of the stairs were fractured and cracked due to erosion and use. In the most fragile areas, there were also some situations of loss of material. Finally, in the gap of the clock weights, the surface was damaged and parts of the coating plaster was missing, leaving the ceramic bricks at sight.

3.3 *Material, Mechanical and Modal Characterisation*

In order to rapidly assess the mechanical conditions of the Tower, a numerical model was constructed and calibrated on the basis of a series of ambient vibration measurements, which were used to identify the structural modal shapes and natural frequencies.

3.3.1 Construction of a Finite Element Model (FEM)

Taking advantage of the already referred architectural and photogrammetric survey of the structure, the numerical model was built using 4 node tetrahedral finite elements into the software ADINA. Since both the type of foundations and the characteristics of the foundation soil were unknown, it was assumed that all displacements of the base nodes are restricted in the definition of the support conditions of the model. Moreover, the horizontal displacements of the shared walls between the tower and adjacent buildings were considered restrained in the normal direction. Regarding the mechanical properties of the materials, although the Tower is composed of two leaf masonry, an inner leaf of ceramic bricks plastered and whitewashed painted and an outer leaf of faced limestone masonry blocks, the Tower walls were assumed as homogeneous in the analysis by taking an equivalent Young's Modulus and an equivalent shear Modulus. As described in [15], such values were calibrated resorting to a dynamic identification procedure.

3.3.2 Modal Identification

The measurements of the dynamic behaviour of the Tower were performed using a frequency analyser to record the data acquired from eight accelerometers, four of them fixed and the remaining four movables, in time frames of 45 min under ambient noise vibration condition. The model analysis was subsequently performed by means of peak picking and frequency domain decomposing (FDD) techniques, implemented in the ARTeMIS Extractor software [16], from which natural frequencies as well as modal damping and shapes were estimated. The measurement plan included two stages: the first, performed only at the level of the bells and top terrace of the tower; and the second, on twenty points located at different levels of the Tower strategically selected on the basis of the comparative analysis between the measurements performed in the first stage and the results of the initial FEM [15]. The goal of the measurements consisted in identifying the first five natural frequencies and corresponding vibration mode shapes (presented in Fig. 3), with the purpose of calibrating the FEM.

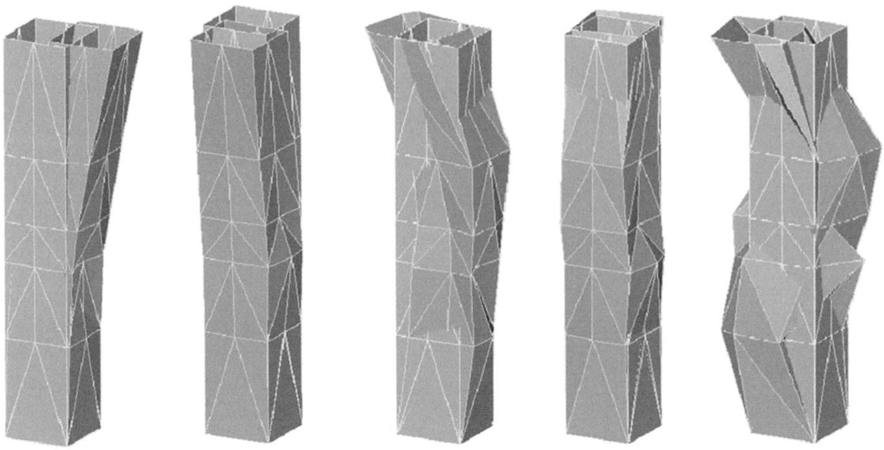

Mode 1	Mode 2	Mode 3	Mode 4	Mode 5
$f_1 = 2.133$ Hz	$f_2 = 2.473$ Hz	$f_3 = 6.557$ Hz	$f_4 = 8.255$ Hz	$f_5 = 9.709$ Hz

Fig. 3 First five vibration mode shapes and corresponding natural frequencies measured [15]

3.3.3 Calibration of the Numerical Model

Taking into account the uncertainties existing in the definition of some key parameters, a two-step calibration methodology, involving (1) the correction of the support conditions, through the analysis of several models with different support conditions; and (2) the values of the equivalent Young's and shear Modulus by trial-and-error, was iteratively carried out until the first five natural frequencies present a suitable coincidence. Having calibrated the model, values of 5.5 and 0.34 GPa were found for the Young's and shear Modulus respectively. It is worth noting that these values are in good agreement with other published studies, namely with [17], where a value of Young's Modulus up to 5 GPa and a shear Modulus of 0.5 GPa were assumed in the modal identification of a 48-meter-high masonry tower built in the fifteenth century. Figure 4 depicts the mesh and supports of the calibrated FEM of the Tower.

Finally, Fig. 5 shows the first five vibration mode shapes obtained with the calibrated model and their corresponding frequencies.

As revealed from the comparison between the results presented in Figs. 3, 5 and Table 1, the approximation obtained between the measured and the numerical vibration modes and frequencies reveals a good agreement.

From the results shown in Table 1, it can be concluded that there is no evidence that the structural integrity of the Tower is compromised. However, as is further discussed in [15], if inadmissibly lower values of the Young's Modulus and/or shear Modulus had been obtained or if a different level of accuracy had been registered for one or more modal frequencies, one could have been concluded that the integrity of the Tower might be affected [15].

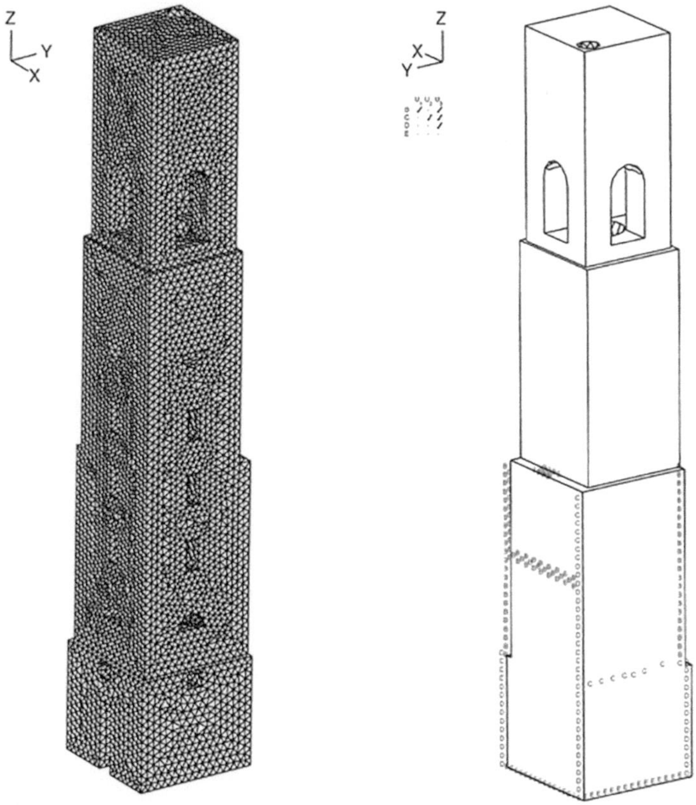

Fig. 4 Mesh and supports of the calibrated finite element model (adapted from [15])

Since the planned restoration and rehabilitation works include the correction of minor structural defects, such as cracking and small movement and displacement of stone leaf facing blocks (see Sect. 3.4.1), it was decided to carry out a re-assessment using the same methodology, correcting and adjusting the numerical model, if necessary, even though no need for deeper strengthening operations [15].

3.4 Catalogue of the Surveyed Information and Description of the Conservation Works

A more precise analysis of the existent defects was performed after the installation of the scaffolds. The Tower garland was one of the areas that showed more fractures and cracks that had remained unperceived until then. Therefore, the existent architectonical and photogrammetric survey was redone. Throughout the conservation works, mapping information on three important features was regularly

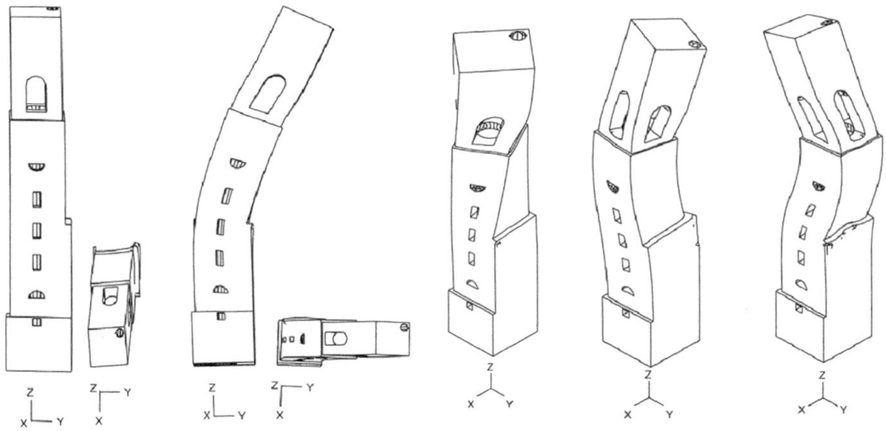

Mode 1 $f_1 = 2.152$ Hz | **Mode 2** $f_2 = 2.403$ Hz | **Mode 3** $f_3 = 6.581$ Hz | **Mode 4** $f_4 = 8.160$ Hz | **Mode 5** $f_5 = 9.583$ Hz

Fig. 5 Vibration mode shapes and corresponding frequencies obtained from the calibrated finite element model (adapted from [15])

Table 1 Comparison between the measured and the numerical natural frequencies [15]

Mode	Measured frequency (Hz)	Numerical frequency (Hz)	Error (%)
1	2.133	2.152	−0.89
2	2.473	2.403	2.83
3	6.557	6.581	−0.37
4	8.255	8.160	1.15
5	9.709	9.583	1.30

updated: (1) revision of the pre-existent defects survey; (2) record of all the tests performed during the appraisal works; and (3) record of all the conservation treatments of restoration performed during the restoration works. All the intervened elements during the works were also registered through general and detailed photographs before, during and after the actions of conservation and restoration.

3.4.1 Conservation and Restoration Works

The conservation and restoration works took place from March 2010 to August 2010. Besides the main action of cleaning and treatment of all surfaces, assumed as ordinary planned maintenance, other reactive maintenance actions aiming at restoring the stability conditions and the cohesion of the architectural elements, as well as of ornamentation elements that presented signs of instability and risk of imminent fall, were also taken. The suppression or mitigation of the action of agents responsible for the degradation of materials and the attainment of better conditions

Fig. 6 Conservation works performed

to resist the action of external atmospheric agents were other maintenance actions carried out [11]. It is worth highlighting that the specific nature of this intervention entailed the adoption of individual strategies and methodologies, not only during the preparation works, but also during the accomplishment of the different tasks. Figure 6 schematises the conservation works performed, which are herein grouped and described in the following paragraphs.

Disinfection and Elimination of the Microbial Colonisation

The vegetation had developed mostly in the superior part of the Tower, with more impact in the upper two-thirds. The number and the species found was variable according to the façade elevation and orientation and to the exposure to the atmospheric agents. To revert this process, a herbicide was applied by spraying directly on the vegetation growth, with particular incidence on the new leaves and shoots. The first application was performed without packaging the upper vegetation, and a second one, after a short period of time, with black plastic packaging completely closed with nylon and/or tape, avoiding this way the vegetation from direct light and increasing the efficiency of the product. The removal of the existing vegetation was carried out mechanically and manually. Finally, a biocide was applied in order to eliminate the microbial colonisation.

Treatment of the Stone Material

The cleansing actions performed sought a balanced and continuous chromatic reading, avoiding the removal or change the time patina. The methods used for cleansing were selected in function of the location and type of dirt or chromatic change, and the results achieved in the previous tests. The exterior cleansing was carried out resorting to different methods/techniques of water seepage, namely brushing with soft nylon brushes, cleaning spray complemented with interspersed soft brushing, water jet cleaning machine at low pressure and mechanical cleansing of films and old mortar. The interior cleansing was performed mainly with mechanical methods, namely spatulas and rotary abrasive devices.

Before the intervention, most of the joints were fully or partially open, allowing the proliferation of plants in their interior. Moreover, in areas of direct access, namely at the ground level of the Tower and in the bells' area, the joints were filled with incompatible and inappropriate materials, such as Portland cement. During the opening process, all corroded steel and non-functional elements were removed and, after that, the joints were cleaned with compressed air. When biological colonisation was still present inside the joint, it was removed resorting to a wet process.

The voids resulted both from the opening of joints and from the cleaning of stone surfaces were filled with traditional lime mortar. All the mapped cracked and fractured elements were consolidated, as well as the loose stone elements and fragments which were assembled resorting to a resin. Stainless steel and fiberglass bolts were used to ensure stability in cases of excessive volume or weight, also in cases where there was greater fragility as a result of the adhesive resins used. Following the mentioned actions, the areas of fractures and cracks were filled with fine-grained micro-plastering mortar.

After the execution of all conservation and restoration treatments, including the last application of biocide, a water repellent product was used aiming at reducing the capacity of water absorption of the stone surface and extending the efficiency of the final biocide treatment, allowing however water vapor permeability and increasing the durability of the treatments.

Treatment of the Metallic Elements

On the roof, it were identified: the metallic elements with no structural function, namely the fitting elements of the metallic railing; the non-structural elements, resulting from an existent old mechanism; the elements without any defined current function; and the structural elements of stone block laying and fixing. All metallic elements were identified and mapped by category and treatment action.

The methodology followed for the treatment of the metallic elements, with few exceptions, consisted on manual and mechanical removal by using pliers, drills, chisels and mallet. The non-structural metallic elements detached and with no identified function, were removed.

3.4.2 A Final Note About the Conservation Works Performed

Concluded the works and dismantled the scaffolds, the impact of the actions performed was notorious. The new Tower that arose from the cleansing operation and the chromatic contrast with the previous image is clear. The conservation work performed showed stone material that does not show a chromatic heterogeneity, allowing a better perception of the sculptures, many of which were imperceptible due to the existent strong biological colonisation. The maintenance of the natural aging patina of the stone was assured by tackling the defects causes of degradation and strong visual impact.

3.5 The Pedagogical Restoration Work Site Initiative "Tower-PSite"

Since the early stage of the project, it was understood that the restoration works of the Tower of the University of Coimbra could be an exceptional opportunity to test a pedagogical work site based on the permanent information and interaction with different public-targets, in order to promote, on one hand, the relevance of a responsible restoration process on the preservation of our collective memory and built heritage and, on the other hand, to promote the awareness of general public to both the technical issues and the philosophical concerns of this kind of work. Additionally, several secondary goals were also identified, namely, the increase on the scientific and technical discussion about restoration and the promotion of a public and academic recognition of the multidisciplinary of the knowledge areas involved. Other positive effects resulting from this initiative were the increase of external visibility, both national and international, through media and web, as well as the increasing credibility of the protection strategies proposed to UNESCO within the nomination to the World Heritage List [14].

Regarding the target public, four main groups were identified and subdivided into two specific categories: tourists (structured tourism; family tourism), technical and scientific public (professors and post-graduation students; professionals and researchers), general public (locals; undergraduate school community) and foreign non-visiting public (national; international). In this regards, it should be noted that Coimbra town has about 143.000 inhabitants, where students represent more than 25%, and that University of Coimbra is visited by more than 300.000 tourists a year.

To fulfil the goals mentioned above and get close to target public, four types of activities have been organized: Multi-level information outdoors; Website and follow up "newspaper"; Guided visits; and Seminars. Figure 7 establishes a holistic matching between these activities and target public.

Each one of these activities are individually detailed in the next paragraphs.

Multi-level Information Outdoors

As already referred, the Tower is visited by about 300 thousand tourists every year, who expect to observe it as the *ex-libris* of the University. For this reason, as can be seen in Fig. 8, canvas covering of the scaffolds has been adopted with real size photo of the tower on all surfaces and, at the ground floor level, the bay that protects the work site was transformed into an outdoor with multilevel information, with a studied design and a hierarchy of written and graphic information [14].

Website and Follow Up "Newspaper"

A local "newspaper" (wall or outdoor "newspaper" at the work site) and a website were created in order to provide periodical information about the progress of the restoration works. This task required a significant volume of work of several experts, namely designers, technicians, translators, etc. Unfortunately, after the two first months, this activity had to be cancelled due to a clear lack of human resources.

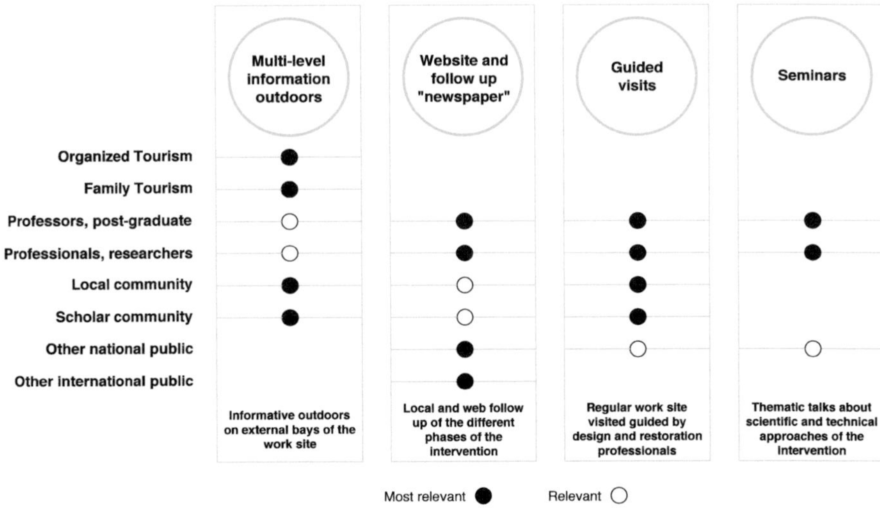

Fig. 7 Relationship between activities and target public in "Tower-PSite" project

Fig. 8 Hierarchical organization of the information presented on the outdoors

Guided Visits

The most visible activity was the guided visits, oriented by different technicians and researchers, such as engineers, architects, historians, archaeologists, etc., every week, up and down the 33 m high scaffold, contacting closely to specialized workers, being part of everyday site discussion and activity. Slight site adaptation had to be made in order to guaranty safety and space circulation conditions.

Seminars

For a more detailed approach, not only in technical terms but also from a scientific and philosophical point of view, four thematic seminars were organized with the collaboration of the Science Museum of the University of Coimbra, with four complementary perspectives: (1) history and identity; (2) architecture and performance; (3) restoration and integrity; and (4) safety and longevity.

4 Seismic Assessment and Preservation of Historical Structures

International standards (Eurocode 8-Part 3 [18], ASCE/SEI 41/13 [19]) adopt the evaluation of the seismic risk to existing buildings the Performance-Based Assessment (PBA), which considers several Performance Levels (PLs) that must be fulfilled in the occurrence of corresponding earthquake hazard levels (defined by the return period). The need to check the achievement of PLs that are close to structural collapse strongly recommends the use of static nonlinear models and displacement-based procedures for the assessment, because the use of linear analysis with the behaviour factor approach is not reliable enough.

The specific case of cultural heritage assets is treated in some recommendation documents [10, 20, 21], which are not only aimed to seismic vulnerability but consider all possible causes of damage and deterioration, with the aim of making a diagnosis and designing a rehabilitation intervention. They point out the complex configuration of this kind of structures, also due to the relevant transformations that have usually occurred over the time, as well as the difficulty of adopting a proper modelling strategy. All these recommendations stress the importance of the qualitative approach, founded on the historical analysis, the accurate investigation of structural details and the interpretation of seismic behaviour, on the basis of observed damage on the building (due to previous events, if any) or on similar structures.

As already mentioned in Sect. 1, it is worth noting that a preliminary assessment is usually sufficient for the diagnosis in many critical situations, such as material deterioration or soil settlements. On the contrary, the evaluation of seismic vulnerability without the support of calculations is overambitious, because the qualitative approach can only suggest which is the expected seismic behaviour and the historical analysis is not sufficient to prove the building safety. This is the reason why the Italian Guidelines for the seismic assessment of cultural heritage [22] clearly states that it is not possible to avoid a quantitative calculation of the structural safety, even if models have to be based on an accurate knowledge and the results can be adjusted by taking into account qualitative evaluations.

The PERPETUATE project [23], funded by the European Commission, has developed guidelines that are coherent with the latter cited recommendations but frame the problem of the seismic assessment of cultural heritage assets and design of interventions within the PBA approach, outlined by the international standards for current buildings. The aim is to define, even for the complex case of old masonry structures, an assessment procedure repeatable and verifiable, which leads to the quantitative evaluation of safety levels, taking also properly into account historical and qualitative information.

In case of historical buildings PLs have to be linked also to cultural relevance concepts: thus, the use and safety of people, the conservation of the building and the conservation of artistic assets (if present) have been considered in an integrated approach. Since pushover analysis is considered the standard tool for the PBA, detailed acceptance criteria are proposed for the identification of target PLs on the pushover curve, by considering the displacement u as Engineering Demand Parameter (EDP) and defining proper thresholds.

Specific PLs are introduced in PERPETUATE taking into account three different groups of requirements (n = U, B, A): *use and human life* (U); *building conservation* (B); *artistic asset conservation* (A). The seismic input is defined by the hazard curve, obtained through a Probabilistic Seismic Hazard Analysis (PSHA), which gives the selected Intensity Measure (IM) as a function of the annual probability of occurrence (or the return period). Possible IMs are: peak ground acceleration (PGA), spectral acceleration for a given period, maximum spectral displacement, Arias intensity, Housner intensity [24]. In the standard case of nonlinear static analysis, the seismic demand is represented by an Acceleration-Displacement Response Spectrum (ADRS), which must be completely defined, for the specific site of the building under investigation, as a function of the assumed IM.

Figure 9 summarizes the basic principles and steps of PBA according to PERPETUATE guidelines, where the displacement-based approach is adopted as the standard method for vulnerability assessment of cultural heritage and design of preventive interventions. In the following the attention is focused only on the use of static nonlinear analysis (pushover), while PERPETUATE procedure also considers the use of Incremental Dynamic Analysis (IDA) [25].

The outcome of the assessment is IM_{PL}, which is the maximum value of the intensity measure that is compatible with the fulfilment of each target PL: it is computed by nonlinear static procedures with overdamped spectra [26]. Thus, through the hazard curve, it is possible to evaluate the annual rate of exceedance λ_{PL} of the earthquake correspondent to this performance (or its return period $T_{R,PL} \approx 1/\lambda_{PL}$). These values are compared with the target earthquake hazard levels $\bar{T}_{R,PL} \approx 1/\bar{\lambda}_{PL}$, defined for the assessment as a function of asset characteristics, in terms of safety and conservation requirements.

This general methodological path has been particularized in PERPETUATE guidelines for different architectural assets; a classification is proposed [27], which is related to the different types of seismic behaviour, considering both building morphology (architectural shape and proportions) and technology (masonry type, horizontal diaphragms, effectiveness of wall-to-wall and floor-to-wall connections). It consists of six architectonic classes: (A) box-type buildings; (B) assets studied by independent macroelements; (C) slender structures studied by monodimensional models; (D) arched structures; (E) massive structures; (F) blocky structures subjected to rocking. Different modelling strategies can be adopted for describing the seismic behaviour of each kind of asset. Moreover, the problem of seismic local

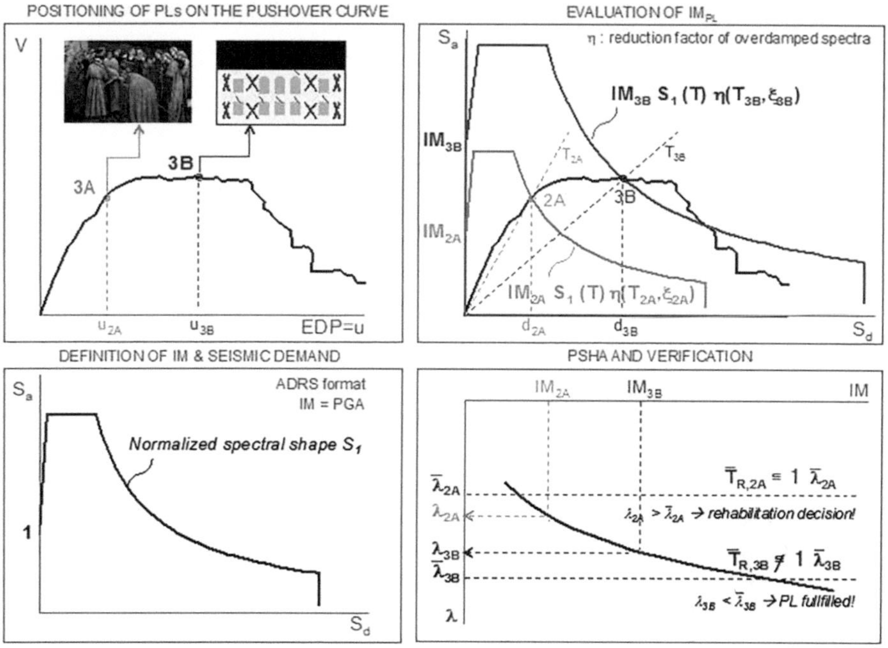

Fig. 9 PBA of architectonic and artistic assets according to PERPETUATE guidelines [23]

mechanisms is treated, which has to be taken into account in all the above-mentioned architectural assets classes, in order to assess the vulnerability of single elements that are not described by the structural models used for the assessment at global scale. The seismic assessment considers also the presence of artistic assets that has to be preserved; three different classes have been introduced: (P) artistic structural elements (e.g. carved stone column); (Q) artistic assets strictly connected to structural elements (e.g. frescoes, mosaics, stuccoes); (R) artistic assets that are independent elements (e.g. pinnacles, spires, merlons).

The application of PBA is particularized for each class by analysing also the use of different modelling strategies and the proper approach to describe the seismic behaviour of the asset. For example, it is necessary to evaluate if the seismic behaviour of the building can be represented by a single model or by a set of different models. The former is the case of assets made by a single element (such as those belonging to classes C, D and F) or by many macroelements (masonry walls, horizontal diaphragms etc.) that can be represented by a global model (such as those of class A, which presents the so-called “box-type” behaviour). On the contrary, the need to consider different models is characteristic of complex assets, made by macroelements that behave quite independently; in this case the assessment requires to develop more than one model, even of different types (it is typical for assets of class B), and the result of the analyses in each macroelement must be then properly blended, in order to define the seismic assessment of the whole asset.

4.1 PBA Procedure of Complex Architectonic Assets

In this chapter the attention is focused on the PBA of complex architectonic assets belonging to classes A—*assets subjected to prevailing in-plane damage* (e.g. palaces, castles, …) and B—*assets subjected to prevailing out-of-plane damage* (e.g. churches, mosques, …); the global assessment, in terms of compatible Intensity Measure ($IM_{PL,G}$), also implies the verification of possible local mechanisms. Despite this, for the sake of brevity these latter are not explicitly treated, while more details on this issue are illustrated in [28].

In case of Classes A and B, the PBA is faced by applying two alternative modelling approaches (Fig. 10):

- buildings characterized by box-behaviour: in this case a 3D model of the whole building is possible (global scale approach);
- buildings made by a set of N_m macroelements, which exhibit an almost independent behaviour: each macroelement is modelled independently (macroelement scale approach) and the seismic load needs to be assigned by a proper redistribution; the assessment of whole asset is then made through a proper combination of results achieved in each macroelement.

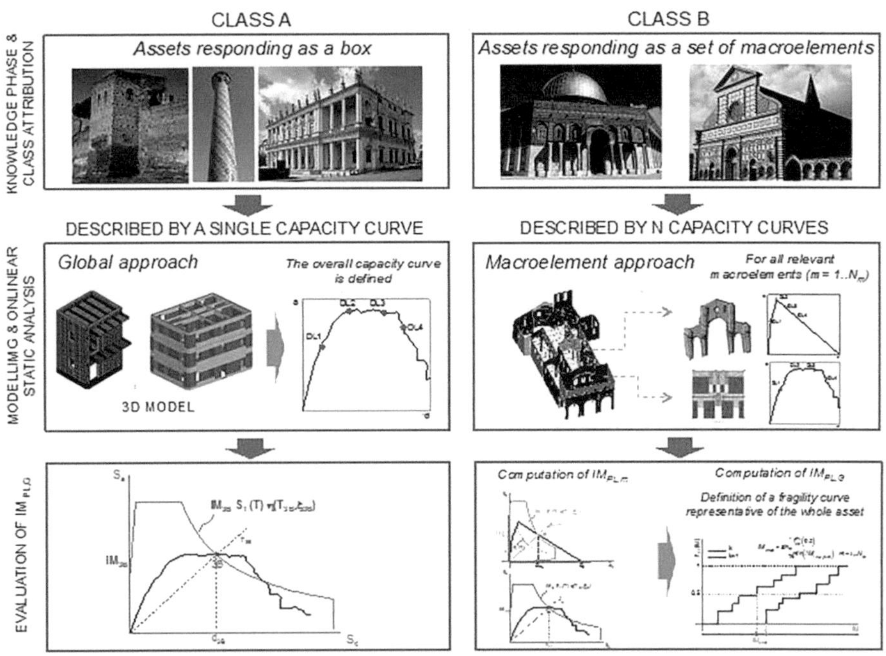

Fig. 10 Basics of PBA for assets of *Classes A* and *B*

The global scale approach is typical of buildings of class A but can be sometimes adopted also for architectonic assets of class B, when macroelements are well connected and there is a horizontal diaphragms which is able to redistribute inertial actions among them. The macroelement scale approach is necessary for most of structures of class B, but also for very few buildings of class A, when horizontal diaphragms are very flexible and/or internal walls are sparse.

One of the critical issues in the PBA is the availability of reliable criteria to define the PLs on the pushover curve. To this aim, a multiscale approach has been proposed that takes into account the asset response at different scales: structural elements scale (local damage), elements scale (damage in macroelements) and global scale (pushover curve). It aims firstly to define proper Damage Levels (DL_k, k = 1.4) on the pushover curve, which may be correlated by proper criteria to the PLs [23]. In case of Class A, its application implies to perform checks at these different scales by considering the evolution of various variables; at the end, the displacement on the overall pushover curve corresponding to a certain DL is defined as the minimum among the displacements corresponding to the attainment of those conditions.

In the case of Class B, once evaluated the $IM_{PL,m}$ for each macroelement that composes the asset, it is necessary to define the intensity measure representative of the whole response ($IM_{PL,G}$). Also in this case a multiscale approach is proposed, aimed to define a fragility curve of the whole assets by combining the contribution offered by each macroelement. In particular, it is computed as:

$$P_{PL}(IM) = \sum_{m=1}^{N_m} \rho_m H(IM - IM_{PL,m}) \tag{1}$$

where: H is the Heaviside function (0 if $IM < IM_{PL,m}$; 1 otherwise); ρ_m is the weight that has to be assigned to each macroelement. Finally, the value of $IM_{PL,G}$ is obtained as the minimum of the following two conditions: (1) the lower value of IM for which the fragility curve has $P_{PL}(IM) \geq 0.5$; (2) the value of IM for which the fragility curve of the performance level (k + 1) is greater than 0.

4.2 *Examples of Application: The Hassan Bey's Mansion in Rhodes and the Great Mosque of Algiers*

The procedure illustrated in previous sections is applied to two assets, the Hassan Bey's Mansion in Rhodes and the Great Mosque in Algiers, which belong to Classes A and B, respectively. Only the PBA of the global response is considered, by focusing herein the attention to some specific aspects of the procedure: (1) the selection of the proper modelling strategy; (2) the definition of performance levels on the capacity curve; (3) the analogies and differences in applying the proposed multiscale approach to such different classes. Moreover, the effect of increasing the stiffness of diaphragms as a possible strengthening intervention is discussed for both assets. More detailed information and results on these two buildings may be founded in [29] and [30] where: in the case of Hassan Bey's Mansion, the use of sensitivity analysis for planning the investigation tests and the effect of uncertainties are also illustrated; while in the case of Great Mosque, an in depth discussion is present on the integrate use of different modelling strategies and the definition of the mechanical properties.

4.2.1 Choice of the Modelling Strategy

The Hassan Bey's Mansion is a typical Ottoman mansion located in Rhodes (Greece), built at the end of the eighteenth century, which has undergone many changes during the nineteenth century. It consists of two storeys and an attic at the South-East corner, with overall dimensions 17.75 m by 15.50 m. The plan is quite regular; the wall thickness varies between 0.35 and 0.60 m at the ground floor, while it is thinner (about 0.27 m) at the upper levels (first storey and attic). The building is a masonry structure formed by sandstones and lime mortar: a rubble masonry characterizes the ground floor, while a cut stone masonry the other levels (ashlar masonry). Diaphragms are made by timber floors (with a single boarding), while the building is covered by wooden ceiling (and the attic by wooden roof and French tiles). Actually the building is not in use and characterized by a very bad maintenance state: thus, the PBA carried out refers to the original state of the building, where "original" means before the ongoing deterioration, in order to provide information on the original safety level of the structure.

The Great Mosque, also known as El Jedid Mosque, is located in Algeria's capital city. It was built in 1097 under the direction of Sultan Ali Yusuf (1106–1142), and it is the oldest mosque in Algiers as well as one of the few remaining of Almoravid architecture. Its architectural features and layout, with naves perpendicular to the *qibla* wall, and its rectangular courtyard, bordered on both its narrower sides by a *riwaq* (gallery), were destined to become a model of much religious architecture, particularly in al-Aqsa Maghreb mosques in Algeria. The building is almost square in plan, measuring approximately 40 by 50 m. The interior is a series of hallways, passages and rooms, with the common theme of pillars and archways throughout the building based on a 9 by 11 grid.

According to the architectonic asset classification proposed in PERPETUATE [27] and on basis of the specific features and the expected seismic behaviour of these assets, Hassan Bey's Mansion belongs to Class A—*Assets subjected to prevailing in-plane damage* while the Great Mosque to Class B—*Assets subjected to prevailing out of plane damage*. For this latter such assumption is supported by the fact that the building is characterized by a large hall partitioned by a set of orthogonal system of arcades, without any intermediate horizontal diaphragms, except the wooden roof that is not enough stiff to guarantee a "box-behaviour". Following this classification, the modelling strategies illustrated in Fig. 11 have been adopted.

In particular, in the case of Hassan Bey's Mansion a global 3D model has been assumed by adopting a Structural Element Model (SEM) based on the equivalent frame approach by using the software Tremuri [31]. The choice of such approach is justified by the quite regular opening pattern; moreover, the use of a software able to simulate the presence of flexible floors (modelled as orthotropic membrane finite elements) is essential for the simulation of the original state of the building. Moreover, a distinctive feature of the building is the presence of many infilled openings consequent to the various transformations that occurred during the centuries. In the following, results presented refer to a model in which they have been considered as windows (thus assuming the infill material as not able to interact effectively with the original masonry panels of the building), while in [29] this uncertainty has been analytically treated by the logic tree approach.

On the contrary, in the case of Great Mosque, the most suitable modelling strategy is different for each type of macroelement that constitutes the building in two orthogonal directions, that is: (1) the system of internal arcades; (2) the four external walls; (3) the portico (forward the NW façade). In particular, while the external walls and the portico have been modelled through the equivalent frame approach, for the arcade system a Macro Block Model (MBM) by using the MB-PERPETUATE software [32] has been adopted (Fig. 11). Indeed, in the examined case, the a priori identification of the kinematism to be analysed by the limit analysis has been supported by the combined use also of a detailed finite element model (Fig. 12). In particular, the latter has been performed by using ANSYS software and by assuming the constitutive laws proposed in [33, 34] to describe the nonlinear response of masonry material. Further details on the models and mechanical properties adopted are illustrated in [30].

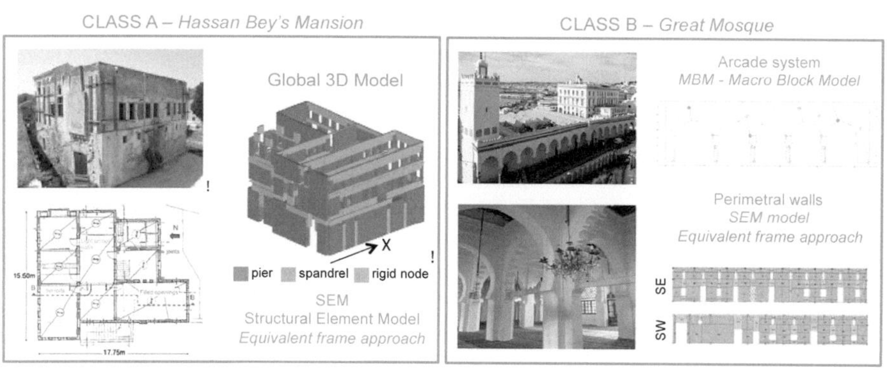

Fig. 11 Modelling strategy adopted in case of Hassan Bey's Mansion (belonging to *Class A*) and Great Mosque (belonging to *Class B*)

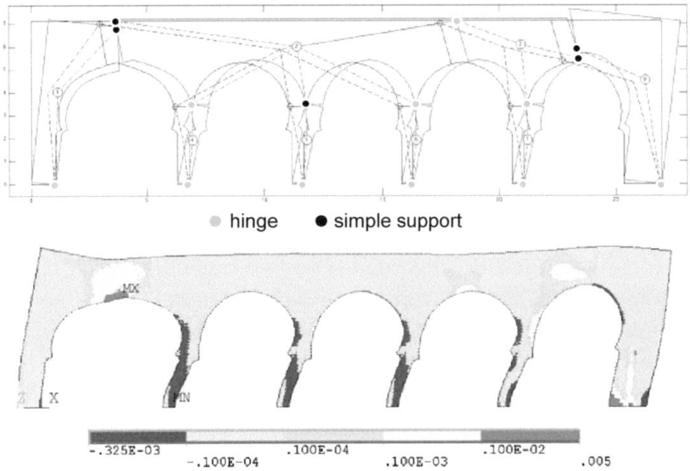

Fig. 12 Kinematism analysed for the Y5 arcade through the MBM model and inelastic strain perpendicular to bed joints, obtained by means of the CCLM model, from [30]

4.2.2 Nonlinear Analyses and Definition of Performance Levels

Once selected the most suitable modelling strategies, the PBA proceeds with the execution of nonlinear static and kinematic analyses in case of SEM and MBM models, respectively. As aforementioned, one of the most critical issues in PBA is the adoption of proper criteria to define the performance levels on the pushover curves. Firstly, it is necessary to specify the PLs selected for the examined buildings. For the Great Mosque the considered PLs are: 2U—*Immediate occupancy*, 3U—*Life Safety* and 3B—*Significant but restorable damage*; on the contrary, only the PL 3B is assumed for the Hassan Bey's Mansion. Indeed, in the case of Great Mosque also the verification with respect to the preservation of an artistic asset has been considered: it consists in a *mihrâb* constituted by an arched niche decorated by two spiral column on the both sides, some stuccos and small decorated tiles attached to South-East (SE) wall.

In particular, the position of PLs has been assumed to be coincident with the corresponding damage levels (DL). These latter have been computed on basis on the multiscale approach proposed in [23] in case of SEM models and on basis of the criteria proposed in [28] in case of MBM ones.

For the Great Mosque, PLs have been defined for each macroelement. In particular, Fig. 15b) illustrates their position in case of two arcades representative of the recurring systems in X and Y directions: performance level 2U corresponds to the intersection between the elastic branch and that from the incremental kinematic analysis; while, PLs 3U/3B (assumed to be coincident) correspond to a displacement capacity equal to $0.25d_0$, where d_0 is the displacement in which the capacity curve is zero. It is worth noting that the initial branch of the pushover curve (that correspond to a period equal to 0.55 and 0.6 s in case of Y5 and X11 arcades,

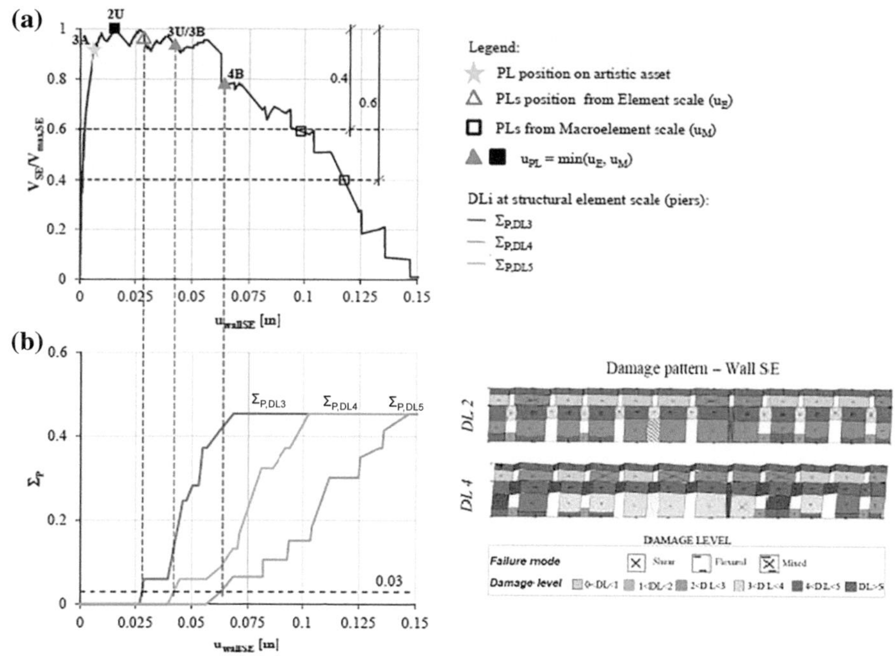

Fig. 13 Definition of PLs on the pushover curve of SE wall of Great Mosque according to the multiscale approach (by the strips is indicated the pier which the *mihrâb* is connected to) (adapted from [30])

respectively) has been calibrated on basis of results coming from the detailed finite element model. Figure 13 depicts the application of the multiscale approach for the SE perimetral wall, in which the variables monitored are: the cumulative rate of piers (Σ_P) and spandrels (Σ_S) that reached a certain damage level at local scale (where the summation is extended to the elements present in each macroelement); fixed rates of the base shear of the macroelement examined. In this case, checks at structural element scale tend to prevail.

The application of the multiscale approach in the case of Hassan Bey's Mansion has been extended by monitoring the reaching of fixed values of the interstorey drift in each wall (see Fig. 14 for those oriented in X direction) and fixed rates of the overall base shear; moreover, at element scale, the summation has been extended to all the elements present in the building. Finally, Fig. 15a) shows the final position of DLs (assumed as reference to define the corresponding PLs) on the overall pushover curves for X and Y directions, deriving from the minimum among checks performed at three different scales. Checks performed at macroelement scale tend to prevail in this case: this is mainly due to the fact that in the original state, the seismic response of Hassan Bey's mansion is strongly affected by the presence of flexible diaphragms that do not allow the distribution of actions among the walls (as evident from Fig. 14).

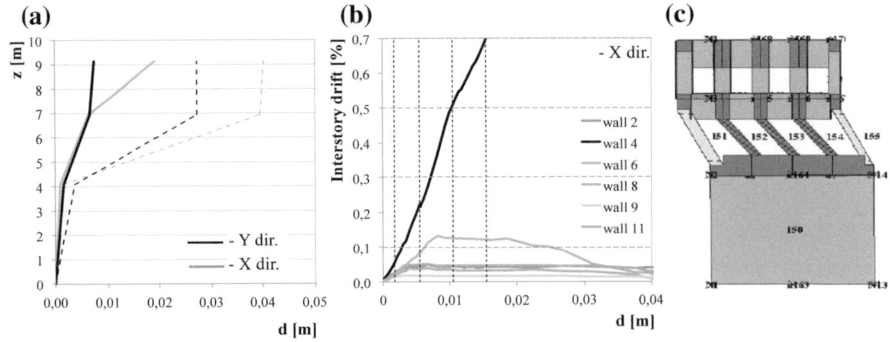

Fig. 14 Role of checks at macroelement scale (in terms of interstorey drift) in case of Hassan Bey's Mansion: **a** profile of the deformed shape in height at DL3 (*continuous line*: mean value; *dotted line*: maximum value occurred), **b** evolution of interstorey drift at first level in case of -X dir. (*vertical lines* correspond to the DLs coming from the multiscale approach, *horizontal lines* indicate the thresholds assumed as reference at macroelement scale), **c** damage pattern of Wall 4 (see Fig. 10 for the legend) (adapted from [29])

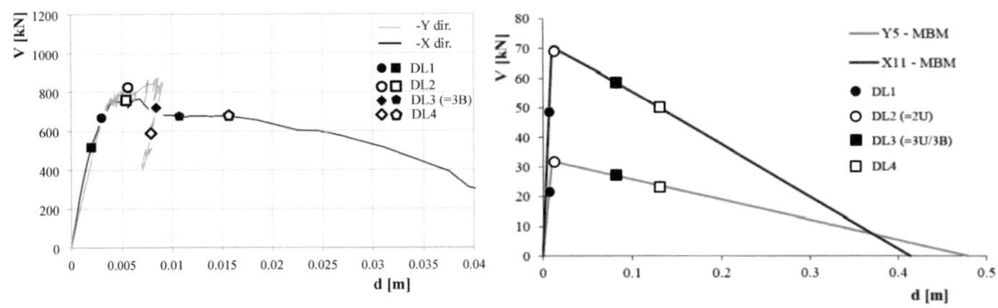

Fig. 15 Definition of PLs on the pushover curves of: **a** Hassan Bey's Mansion (*circles* indicated the DLs in Y direction, while *square* for the X direction), **b** two arcade systems of Great Mosque

4.2.3 Performance Based Assessment and Computation of the Maximum IM Compatible with the Fulfillment of Performance Levels

Once the pushover curves have been obtained and the PLs fixed on them, the PBA consists of computing the value of $IM_{kn,G}$. In both cases, the Peak Ground Acceleration (PGA) has been assumed as reference IM, being the two assets quite rigid. In particular, the computation of $IM_{kn,G}$ is based on the use of overdamped spectra [26], while the conversion of the pushover curve (representative of the MDOF system) in the capacity curve (equivalent SDOF) is made: (1) through the participation coefficient (Γ) and the participation mass (m^*), according to the proposal originally illustrated in [35], in the case of nonlinear static analyses (SEM model); (2) as explained in [28], in the case of nonlinear kinematic analyses (MBM model).

In the case of the Great Mosque, the computation of $IM_{kn,G}$ at global scale passes from that of each single macroelement ($IM_{kn,m}$). In particular, Fig. 16 shows the construction of the global fragility curves according to (1).

Table 2 summarizes the resulting values of $IM_{kn,G}$ for two examined assets, where the reference target values of the seismic demand are also reported (in terms of PGA), which have been computed on basis of the probabilistic seismic hazard analysis illustrated in [36]. The return periods assumed as reference $\bar{T}_{kn}$ reflect the importance coefficients assumed for the two assets, equal to 1 in the case of requirement related to the building conservation (B) but equal to 1.2 in the case of that related to the use and human life (U) in the case of the Great Mosque (due to its condition of usc, frequent and subjected to possible crowding). As evident from Table 2, both assets show some deficiencies in fulfilling the required PLs: very strong in the case of Hassan Bey's Mansion in both directions and in particular in Y direction in the case of the Great Mosque.

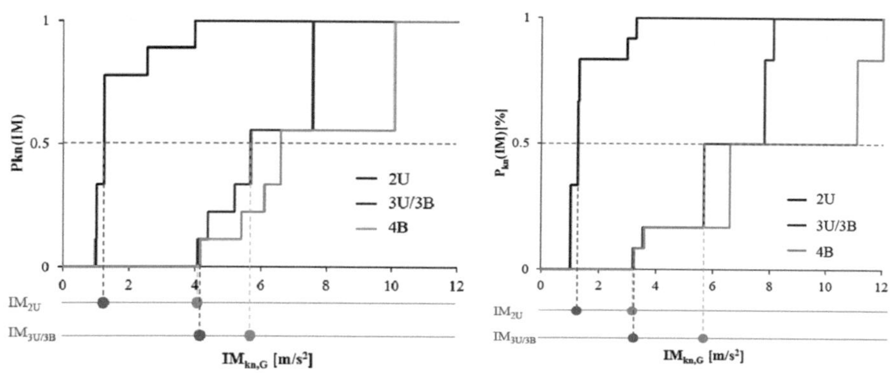

Fig. 16 Great Mosque case study: fragility curves representative of the seismic behaviour of the whole asset in *X* (*left*) and *Y* (*right*) directions and computation of $IM_{kn,G}$ [30]

Table 2 $IM_{kn,G}$ values and target seismic demand for two examined case studies

Case study	$\overline{PGA}$ (m/s^2) ($\bar{T}_{kn}$[*years*])			$IM_{kn,G}$ (m/s^2) (T_{kn}[years])			
	2U	*3U*	*3B*	2U		3U/3B	
				X	*Y*	*X*	*Y*
Hassan Bey's Mansion	–	–	1.78 (475)	–	–	0.55 (95)	0.71 (119)
Great Mosque	1.96 (120)	3.8 (570)	3.55 (475)	1.10 (55)	1.23 (63)	4.16 (692)	3.23 (383)

4.3 Preventive Strategies by Strengthening Interventions

The PBA in the original state of the two examined assets highlighted the need of strengthening interventions. In the following the effect of a possible intervention consisting in the stiffening of diaphragms is illustrated. In both cases it could be achieved by adopting some solutions still based on the conservation of timber floors (e.g. based on a double boarding), thus more compatible in terms of preservation and also more effective for the seismic response, because these solutions are not associated to a significant increase of masses.

While in the case of Hassan Bey's Mansion such intervention only affects the capability of floors to redistribute the actions among walls, in the case of the Great Mosque it modifies more significantly the behaviour, that now involves the independent response of each wall/arcade while in the strengthened configuration consists of a "box-type" structure, passing from Class B to Class A. The change in the class implies the modelling strategy has to be updated, requiring the adoption of a global 3D model. Among the different possible choices, the SEM approach has been considered due to its quite limited computational effort. However, in order to provide a reliable response not only for ordinary walls but also for the arcade system, in this latter case it has been necessary to calibrate: (1) the geometry of the equivalent frame idealization; (2) the mechanical parameters of masonry to be adopted in order to correctly simulate the damage response. To this aim, results achieved through the MBM and finite element models constituted as essential supporting tool. Figure 17 illustrates by way of example the complete 3D SEM model and a sketch aimed to clarify the rules adopted in the equivalent frame idealization of arcade systems.

Figure 18 shows the resulting pushover curves for the Great Mosque in X and Y directions and the final position of the PLs that have to be checked (defined on basis of the application of the multiscale approach aforementioned). In terms of PBA and computation of $IM_{kn,G}$, the intervention revealed to be quite effective leading to the fulfilment of all PLs, corresponding to a value of 2.65 and 3.96 m/s^2 in Y direction (the most critical one) for 2U and 3U/3B, respectively.

Fig. 17 3D SEM model of the Great Mosque and rules adopted for the equivalent frame idealization of the arcade system [30]

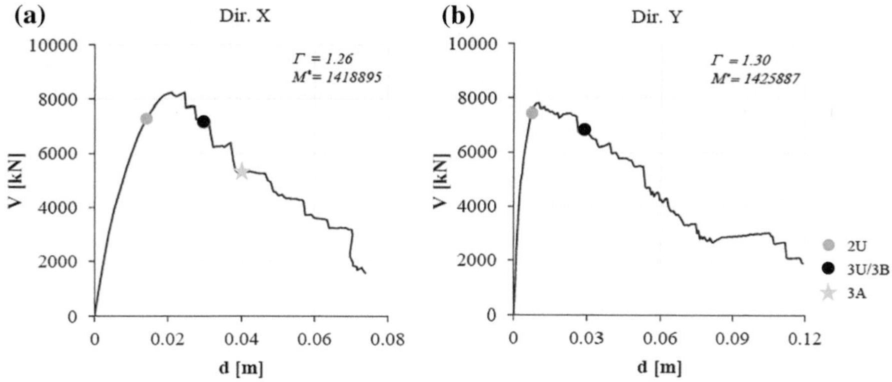

Fig. 18 Pushover curves obtained on the 3D model of Great Mosque and position of performance levels (adapted from [30])

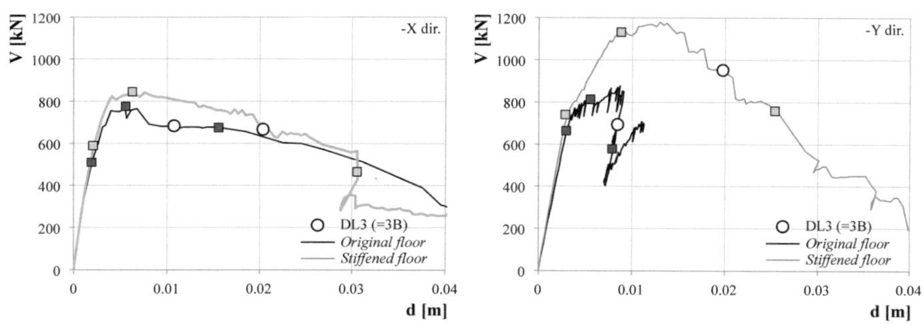

Fig. 19 Effect of floor stiffening in case of the Hassan Bey's Mansion on the positioning of damage levels on the pushover curve (adapted from [29])

Figure 19 shows the effect of diaphragm stiffening in terms of pushover curves and position of PLs in the case of Hassan Bey's Mansion. As evident, the *Y* direction is greatly affected in terms of both base shear and global ductility by the effect of the improved actions redistribution among walls. This is highlighted also by the damage pattern (Fig. 11), from where it is apparent that the damage is distributed among the different walls and not concentrated only in some of them.

Although in the case of *Y* direction the beneficial effect of such intervention is more evident than in *X*, it is interesting to note that in this latter case it affects the DLs position on the pushover curve (Fig. 19). In fact, more rigid floors tend to produce a more homogeneous behaviour limiting the occurrence of very high interstorey drift values in some single walls, this latter condition being very critical for the premature attainment of DL3 and DL4 in the case of flexible floors (see Fig. 20 and also Fig. 6). Indeed, the multiscale approach adopted revealed to be quite effective in capturing the effects on modification of such types of local behaviours. Despite this, in terms of final outcome of the PBA (values of $IM_{kn,G}$), in

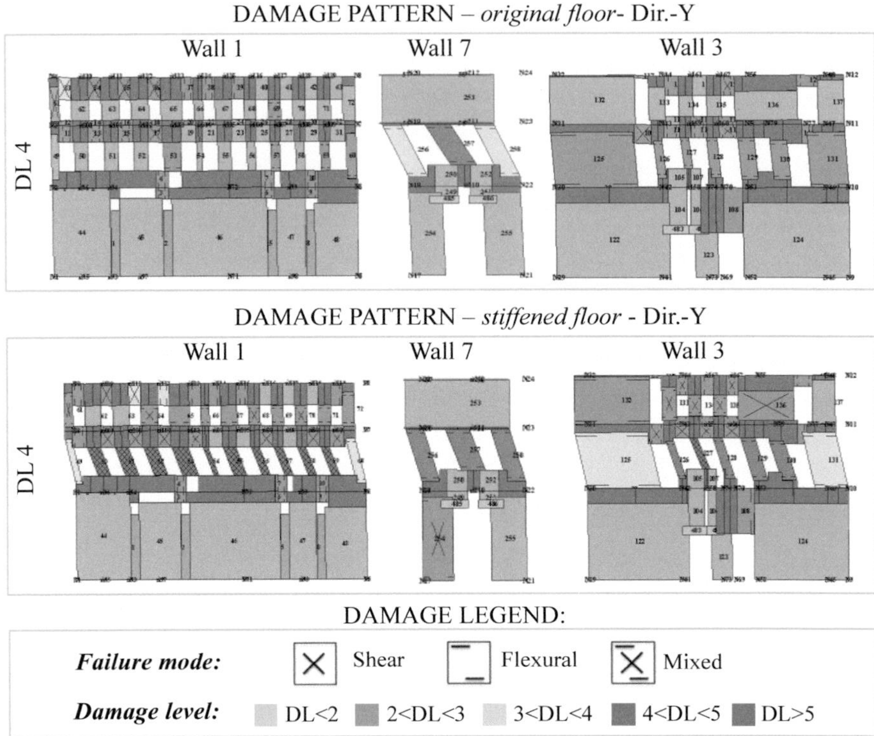

Fig. 20 Effect of floor stiffening in case of the Hassan Bey's Mansion on the damage pattern and the overall response (adapted from [29])

the case of Hassan Bey's Mansion such intervention proved to be not decisive. Indeed, the building is characterized by some strong structural deficiencies (like as the presence of very thin walls, numerous openings or of flue that strongly reduce the seismic capacity of walls), which require a more invasive strengthening.

5 Conclusions

The preservation of cultural heritage assets should consider both the monitoring of slow processes (material deterioration, anthropic transformations and climate change effects) and the risk associated with natural hazard (rare catastrophic events such earthquakes, floods, fire and cascading events).

With reference to the first aspect, the case study of the Tower of the University of Coimbra has proved the importance of the following steps:

- Diagnosis and appraisal—The importance and influence of the survey and appraisal must not be underestimated. It is the natural point of interest at which

all interested and involved parties will focus and discuss, identifying the needs of the building, understanding the buildings and finally putting forward solutions, demands, decisions and good practices;
- Structural assessment tools—A finite element model of the Tower of the University of Coimbra was developed and calibrated on the basis of vibration measurements and modal analysis using modal extraction techniques in the frequency domain. As is fully discussed in [15], despite the lack of information in relation to parameters with direct influence on the structural behaviour of the Tower, such as the supporting conditions and the mechanical characteristics of the materials, the analysis of the results obtained allows to conclude that the numerical model satisfactorily reproduces the dynamic response of the Tower;
- Pedagogical and scientific use—The excellent receptivity of all publics, the final positive evaluation and the reduced costs of the initiative lead to the conclusion that the process should be studied and organised in order to expand its implementation, not only in the University of Coimbra, but also in all the places were built heritage is a relevant resource and should be closer to populations and carefully protected and promoted.

Regarding the second aspect, related to the vulnerability to natural hazards, the relevant problem of seismic assessment has been deepened by the description of the PBA procedure for cultural heritage assets developed within the PERPETUATE project [23] pointing out as follows:

- Numerical analysis—The procedure has been applied to two different ancient masonry structures, a building and a mosque. These two structures highlighted how the choice of the most reliable modelling strategy needs to properly consider the specific configuration and the behaviour expected for the asset under examination. In some cases the combined use of different modelling strategies reveals very effective to manage the model uncertainties;
- PERPETUATE procedure—The parallel description of the different steps of PERPETUATE procedure on two case studies highlights its capability to treat the problem of seismic assessment within a general common framework. A distinctive feature of ancient masonry structures is the absence of rigid horizontal diaphragms; to this end, the proposed models appears to be able to describe the actual behaviour of these structures and the multiscale approach, differently formulated for "box-type" and macroelement structures, turns out to be able to define the displacement thresholds correspondent to each Performance Level.
- Strengthening interventions—Such displacement thresholds revealed to be quite effective also in capturing modifications in the behaviour induced by strengthening interventions not so evident in terms of the overall pushover curve (like as the case of the Hassan Bey's Mansion).

Acknowledgements It is indispensable an extensive acknowledgment to all the teams, all the institutions, all the supporters of the restoration process of the Tower of the University of Coimbra, that contribute with their work, their knowledge, with their advice, with their financial support, with their determination to the results presented and discussed in this chapter. Moreover, the PERPETUATE project was funded by Seventh Framework Programme of the European Commission, ENV.2009.3.2.1.1-Environment, Grant Agreement No. 244229 (www.perpetuate.eu).

References

1. Titchen SM. On the construction of outstanding universal value: UNESCO's World Heritage Convention (Convention concerning the Protection of the World Cultural and Natural Heritage, 1972) and the identification and assessment of cultural places for inclusion in the Wo 1995; Australian National University.
2. ICOMOS Hrsg. The international council on monuments and sites. Herit Risk. 2015;6–7.
3. International Council of Monuments and Sites (ICOMOS). International Charter for the Conservation and Restoration of Monuments and Sites (The Venice Charter 1964). Int Congr Archit Tech Hist Monum. 1964;1–4.
4. Vecco M. A definition of cultural heritage: from the tangible to the intangible. J Cult Herit. 2010;11(3):321–4.
5. UNESCO. Convention concerning the protection of the world cultural and natural heritage. In: General Conference at its Seventeenth Session 1972;1:135–145.
6. International Council of Monuments and Sites (ICOMOS). International charter for the conservation of historic towns and urban areas (The Washington Charter). 1987;1–3.
7. ICOMOS Australia. The Australian ICOMOS charter for the conservation of places of cultural significance, Burra, Australia 1979.
8. Goodwin C, Tonks G, Ingham J. Identifying heritage value in URM buildings. J Struct Eng Soc N Z. 2009;22(2):16–28.
9. Vicente R, Ferreira TM, Mendes da Silva JAR. Supporting urban regeneration and building refurbishment. Strategies for building appraisal and inspection of old building stock in city centres. J Cult Herit. 2015;16(1):1–14.
10. ICOMOS/ISCARSAH Committee. Recommendations for the analysis, conservation and structural restoration of architectural heritage. In: ICOMOS international committee for analysis and restoration of structures of architectural heritage 2005.
11. AAVV. University of Coimbra—Alta e Sofia, vol. I, II, III, IV, V, VI, VII, VIII. University of Coimbra, Coimbra (http://issuu.com/unescouc) 2012.
12. Mendes da Silva JAR, Lopes N, Marques C. O processo de candidatura a património mundial da Universidade de Coimbra: desafios e estratégias de gestão e salvaguarda. V Congreso Latinoamericano REHABEND 2014 about "Patología de la Construcción, Tecnología de la Rehabilitación y Gestión del Patrimonio".
13. Mendes da Silva JAR, Lopes N. O contributo dos "planos diretores dos edifícios para a gestão de longo prazo de conjuntos classificados". O caso da Universidade de Coimbra. V Congreso Latinoamericano REHABEND 2014 about "Patología de la Construcción, Tecnología de la Rehabilitación y Gestión del Patrimonio".
14. Mendes da Silva JAR. Full and pedagogical access to a restoration site: the tower of the University of Coimbra. ReUSO—III Congreso Internacional sobre Documentación, Conservación y Reutilización del Patrimonio Arquitectónico y Paisajístico 2015;2156–2163.
15. Júlio ENBS, da Silva Rebelo CA, Dias-da-Costa DASG. Structural assessment of the tower of the University of Coimbra by modal identification. Eng Struct. 2008;30(12):3468–77.
16. Structural Vibration Solution. ARTeMIS Extractor Pro. Release 5.3. Aalborg, Denmark, 2012.

17. Beneditini F, Gentile C. Ambient vibration testing and operational modal analysis of a masonry tower. In: Proceedings of the 2nd international modal analysis conference 2007;285–292.
18. CEN. Eurocode 8: design of structures for earthquake resistance. Part 3: assessment and retrofitting of buildings, Brussels, Belgium 2005.
19. ASCE/SEI 41/06. Seismic rehabilitation of existing buildings. Reston: American Society of Civil Engineers; 2007.
20. ISO 13822. Bases for design of structures: assessment of existing structures. 2nd ed. Switzerland: ISO International Standard; 2010.
21. CIB 335. Guide for the structural rehabilitation of heritage buildings. CIB Commission W023 – Wall Structures, ISBN 978-90-6363-066-9, 2010.
22. Recommendations P.C.M. 9/2/2011. Seismic assessment and risk mitigation of cultural heritage according to Italian Technical Code for Constructions (NTC 2008). G.U. n. 47 of 26-2-2011, Suppl. Ord. n. 54 (in Italian) 2011.
23. Lagomarsino S, Cattari S. PERPETUATE guidelines for seismic performance-based assessment of cultural heritage masonry structures. Bull Earthq Eng. 2015;13(1):13–47.
24. Douglas J, Seyedi DM, Ulrich T, Modaressi H, Foerster E, Pitilakis K, Pitilakis D, Karatzetzou A, Gazetas G, Garini E, Loli M. Evaluation of seismic hazard for the assessment of historical elements at risk: description of input and selection of intensity measures. Bull Earthq Eng. 2015;13(1):49–65.
25. Lagomarsino S, Cattari S. Seismic performance of historical masonry structures through pushover and nonlinear dynamic analyses (Chapter 11). In: Ansal A, editor. Perspectives on European earthquake engineering and seismology, geotechnical, geological and earthquake engineering, vol. 39. New York: Springer; 2015. p. 265–92.
26. Freeman SA. The capacity spectrum method as a tool for seismic design. In: Proceedings of 11th European conference of earthquake engineering. Paris, France; 1998.
27. Lagomarsino S, Abbas N, Calderini C, Cattari S, Rossi M, Ginanni Corradini R, Marghella G, Mattolin F, Piovanello V. Classification of cultural heritage assets and seismic damage variables for the identification of performance levels. In: Proceedings of structural repairs and maintenance of heritage architecture conference. WIT Press; 2011. pp. 697–708. ISSN 1743-3509.
28. Lagomarsino S. Seismic assessment of rocking masonry structures. Bull Earthq Eng. 2015;13 (1):97–128.
29. Cattari S, Lagomarsino S, Karatzetzou A, Pitilakis D. Vulnerability assessment of Hassan Bey's Mansion in Rhodes. Bull Earthq Eng. 2015;13(1):347–68.
30. Rossi M, Cattari S, Lagomarsino S. Vulnerability assessment of Great Mosque of Algiers. Bull Earthq Eng. 2015;13(1):369–88.
31. Lagomarsino S, Penna A, Galasco A, Cattari S. TREMURI program: an equivalent frame model for the nonlinear seismic analysis of masonry buildings. Eng Struct. 2013;56:1787–99.
32. Lagomarsino S, Ottonelli D. A Macro-Block program for the seismic assessment (MB-PERPETUATE). PERPETUATE (EC-FP7 project), Deliverable D29, www.perpetu-ate.eu.
33. Gambarotta L, Lagomarsino S. Damage models for the seismic response of brick masonry shear walls. Part II: the continuum model and its applications. Earthq Eng Struct Dyn. 1997;26:441–62.
34. Calderini C, Lagomarsino S. Continuum model for in-plane anisotropic inelastic behavior of masonry. J Struct Eng ASCE. 2008;134(2):209–20.
35. Fajfar P. A non linear analysis method for performance-based seismic design. Earthq Spectra. 2000;16(3):573–91.
36. Faouzi G, Nasser L. Scalar and vector probabilistic seismic hazard analysis: application for Algiers City. J Seismol. 2014;18:319–30.

Strengthening of Stone and Brick Masonry Buildings

Francesca da Porto, Maria Rosa Valluzzi, Marco Munari, Claudio Modena, António Arêde and Alexandre A. Costa

1 Introduction

The design of structural interventions on existing masonry buildings is mainly aimed at the identification and interpretation of the possible resistant mechanisms of the structure and their role in the global structural response, and of potential weaknesses: this approach allows understanding if it is necessary, and possible, to intervene for improving the safety level of the structure [1, 2].

In this context, "conventional" design activities, such as the choice of materials and techniques for the intervention, their design and verification, are involved in a later phase of the project. Indeed, investigations and analyses, that are part of the knowledge process, are of basic importance for the correct set up of the design of

F. da Porto (✉) · M. Munari · C. Modena
Department of Civil, Architectural and Environmental Engineering,
University of Padova, 35131 Padua, Italy
e-mail: francesca.daporto@unipd.it

M.R. Valluzzi
Department of Cultural Heritage, University of Padova, 35139 Padua, Italy
e-mail: mariarosa.valluzzi@unipd.it

A. Arêde
CONSTRUCT-LESE, Civil Engineering Department, Faculty of Engineering,
University of Porto, 4200-465 Porto, Portugal
e-mail: aarede@fe.up.pt

A.A. Costa
Department of Civil Engineering, School of Engineering,
Polytechnic Institute of Porto, 4200-072 Porto, Portugal
e-mail: alc@isep.ipp.pt

A.A. Costa
CONSTRUCT-LESE, Faculty of Engineering (FEUP), University of Porto,
4200-465 Porto, Portugal

A. Costa et al. (eds.), *Strengthening and Retrofitting of Existing Structures*,
Building Pathology and Rehabilitation 9, https://doi.org/10.1007/978-981-10-5858-5_3

structural interventions [3]. However, as already mentioned in Chapter "Structural Strengthening and Retrofit; Motivations, Concepts and Approaches", these aspects are not within the scope of this book, and in particular of this chapter, that will focus on strengthening approaches for existing masonry buildings.

In the following sections, after a brief introduction on the possible methodologies for the assessment of the structural behaviour of existing masonry buildings, traditional and innovative approaches for their strengthening and retrofitting will be illustrated. As stated in Sect. 2.3 of Chapter "Structural Strengthening and Retrofit; Motivations, Concepts and Approaches" [4], considering that earthquakes are one of the main causes of damage for existing masonry buildings, some parts of the following text will focus on seismic protection; those considerations are, in any case, useful for every type of action and damage.

Dealing with these topics, the appropriate division of roles between traditional and innovative materials and techniques is to be found: technology in itself is neither good nor bad and there is not a technique that is inherently more suitable than another one. In particular, in projects that have significant impact on the global behaviour, it seems that the role of traditional materials and techniques is irreplaceable, as they will obviously cause the least possible changes to the structural behaviour, valuing the original constructive features [5]. Technological innovation and use of "advanced" materials seem to find the right and, in many cases, potentially decisive role where they allow, thanks to properties that traditional materials do not have (e.g. unattainable ratios between weight or volume and resistance; penetration or adherence maintaining adequate porosity and breathability), executing local and highly targeted interventions on specific resistant elements and mechanisms, with minimal impact on resistance parameters, on which it is neither possible, nor necessary, to intervene.

Therefore, the development of innovative materials and systems, that ensure effectiveness, economic application and maintenance, low-intrusiveness towards the existing structure, respect of authenticity and of the original structural concept, collaboration between old and new materials and elements, high structural performance and improvement of structural behaviour, is of basic importance and has proceeded continuously during the recent years. Some examples are represented by the use of nanotechnology for mortars and grouts, taking also into account the substrate they are applied to; the use of Steel Reinforced Polymers/Steel Reinforced Grout (SRP/SRG) materials; the definition of new application methods and design rules for Fibre Reinforced Polymers (FRP); the development of compatible techniques for improving horizontal diaphragms and connections; the advance of innovative dissipative devices for anchors and tying, etc. It is also important to underline that it is nowadays possible to rely on new and integrated methodologies, like early warning techniques for intelligent interventions and advanced monitoring techniques for knowledge-based assessment, that allow progressive implementation and evaluation of interventions [6, 7].

On the other hand, in some cases, the research for new materials and products and new techniques for repairing and strengthening masonry buildings did not bring out real success. It is evident that many intervention techniques that were proposed in terms of "technological innovation" after past seismic events (e.g., injection of resins, reinforced injections, jacketing of masonry walls, indiscriminate replacement of wooden floors and roofs, and vaults, with heavy reinforced concrete structures) have not positively exceeded the test of time and of subsequent earthquakes [8–10]. Conversely, it was highlighted the need for revaluating traditional materials and techniques, and for a consequent recovery of the so-called "rules of art".

The issues of strengthening existing masonry buildings and of improving the general safety level can be dealt with starting from the consideration that efficient protection can only be achieved on the basis of the 'minimum intervention' approach. This requires that the potentials of existing materials and components are as much as possible exploited in terms of strength and that candidate interventions are validated and optimized under specific, real life conditions.

An important contribution on these tasks was given by the NIKER Project (New Integrated Knowledge based approaches to the protection of cultural heritage from Earthquake-induced Risk), which produced two important sets of guidelines on materials and techniques for interventions, and on the integrated knowledge-based methodology for the seismic protection of existing masonry buildings. A structured catalogue of interventions, related to the various structural elements, structural materials, and possible failure mechanisms, has been also developed [11].

2 General Considerations and Methodological Aspects

As already highlighted in Sect. 2.3 of Chapter "Structural Strengthening and Retrofit; Motivations, Concepts and Approaches" [4], common masonry buildings were built following "codes of practice" and traditional rules, according to material types (masonry materials and arrangement, multi-leaf walls, etc.), structural types (plan and elevation irregularities, isolated or clustered buildings, etc.), and construction details (in particular, poor connections) which, in some cases, yield to significant structural deficiencies, particularly in the case of seismic actions. This has been in many cases surveyed and shown for churches [12, 13] and monumental buildings [6, 14], but it has been evidenced since the past, until the most recent earthquake events, also for common masonry buildings [15–18]. In addition, the continuous changes experienced by existing masonry buildings over time and the deterioration phenomena produce many uncertainties in the knowledge of the structural layout and of the mechanical properties of the constituting materials, that are reflected in the understanding and assessment of their structural behaviour and safety conditions [19].

The observation and study of structural failure modes of original and repaired structures, also evidenced that existing masonry buildings are constructions whose real structural behaviour cannot be assessed through "standard" methods, where "clustered buildings" in historic centres represent an exemplary case [20]. In general, in the case of existing masonry buildings, although traditional elastic analysis can give useful indications for a preliminary overall interpretation of the structural behaviour, non-linear static and dynamic analyses, cited in Sect. 3.5 of Chapter "Structural Strengthening and Retrofit; Motivations, Concepts and Approaches" [4], are generally preferred for evaluating the ultimate conditions and assessing the structural safety [21]. However, the sophisticated procedures of analysis used for modern buildings, applied to existing buildings, can lead to inadequate results. Hence, the interpretation of the actual building behaviour on the basis of such procedures can be little reliable, or even misleading. Indeed, this kind of analyses represent the global behaviour of a structure, but, in the majority of the cases, in existing masonry constructions, structural elements are not able to manage an effective load redistribution, the building will not develop an overall behaviour and the response will be dominated by the activation of a series of local mechanisms [22, 23].

Indeed, in these buildings, the absence of systematic connections between intersecting walls and between walls and horizontal diaphragms, the presence of deformable floors, may cause kinematic mechanisms related to the loss of equilibrium of individual structural portions rather than to states of stress exceeding the ultimate capacity of materials [24, 25]. In this framework, new approaches and tools for the analysis and evaluation of the safety levels are made available. The structural models, on which they rest upon, represent in a more articulated and flexible way how existing masonry buildings reach limit conditions. These procedures calculate the values of horizontal static-equivalent forces (i.e., the values of the mass multiplier) that trigger specific mechanisms of local failure. The latter generally consist of out-of-plane overturning of structural macro-elements, composed by single walls or by wall sub-assemblages (such as intersecting walls, walls and floors or roof portions, etc.), but also in-plane mechanisms have been proposed.

This limit analysis approach depends on few geometric and mechanical parameters and therefore it does not require an extremely accurate survey and time-consuming computation. In addition, these models can easily process the inevitable uncertainty of the prediction, by the use of appropriate numerical techniques that take into account the lack of sufficient information in calibrating probabilistic methods (this problem makes often illusory the precision of complex linear or nonlinear behavioural models [26]).

It is also worth noting that the limit analysis of local macro-elements points out the need for local strengthening interventions, essentially aimed at the improvement of connections between local portions of the buildings, and also allow comparative evaluations prior and after their execution. Conversely, global methods of analysis, when they are not properly used, could not only lead to misleading interpretation of the actual building behaviour, but also force the execution of invasive interventions, which permanently change aspect and structural behaviour of the buildings [27].

Therefore, the conventional concept of structural safety 'verification', attained and satisfied by means of the methods proposed by the modern codes for new buildings, is replaced by the softer concept of 'evaluation', intended as the *positive evaluation on the ratio between the attained safety level […] and the reference safety level* [28]. This concept is in agreement with the idea of 'structural improvement', which describes an intervention that reasonably increases the safety level of an existing structure, without necessarily bringing it to same safety level required for a new construction. Hence, the specific problem of structural safety and intervention for existing structures can be dealt with within the framework of “evaluation-improvement” [29].

3 Traditional and Innovative Structural Interventions

As stated before, the improvement of building performance should not disregard the beneficial effects of traditional methodologies, although a good balance with the adopting of innovative techniques and advanced materials can be followed [30]. Indeed, the observation of the performance of existing masonry buildings confirmed limits and consequences of some types of strengthening interventions, but also effectiveness and advantages of new methods that will be described in the following sections.

In general terms, the main aim is the conservation of both materials and structural features, therefore interventions should avoid, as much as possible, significant alterations to the original structure [31]. An overall discussion on structural interventions for repairing and strengthening existing masonry buildings can be found in Modena et al. [32, 33].

The intervention strategy has to be defined in the framework of the approaches defined in Sect. 4 of Chapter “Structural Strengthening and Retrofit; Motivations, Concepts and Approaches” [4], choosing the most appropriate technique, among the less invasive ones, and those with the greatest compatibility with the original structure. Several research programs and experimental validations of the various technological solutions for vertical masonry elements and horizontal diaphragms, diversified according to the construction material, type of construction, type of regional environmental conditions, etc., were carried out by developing and applying special testing procedures and by numerical parametrical assessment. In the following paragraphs a general introduction of the most important techniques used to strengthen existing masonry buildings is presented.

3.1 Interventions to Improve Connections

As a pre-requisite to obtain a satisfactory global behaviour of the structure, it is necessary to improve the connections between intersecting masonry walls and between masonry walls and horizontal diaphragms (floors and roofs) [27, 34].

This goal may be obtained by: (i) inserting ties in a proper configuration (number and positions) so that a balanced behaviour on the two main directions can be achieved (Fig. 1); (ii) positioning confining rings (Fig. 2) or tie-beams (Fig. 3), with due care to avoid stress concentration at corners.

In particular, reinforced masonry or steel tie-beams should be preferred to reinforced concrete ones, especially when masonry walls are not sufficient to bear

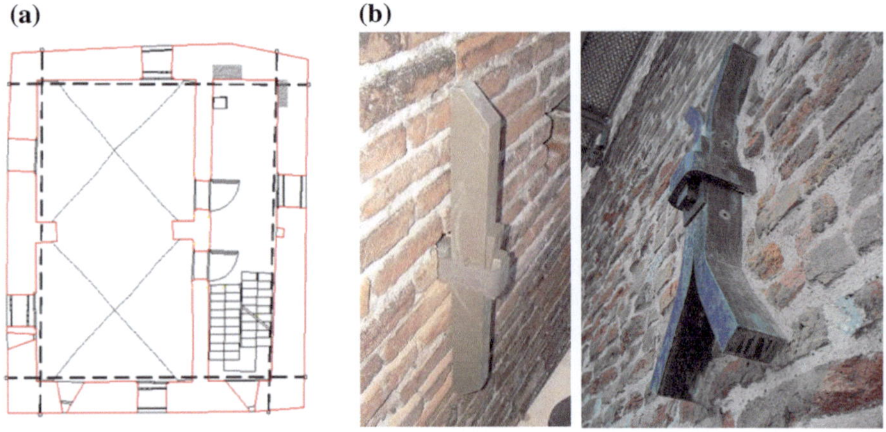

Fig. 1 Positioning of stainless steel ties: **a** tie positioning, plan [31], **b** view of external anchors

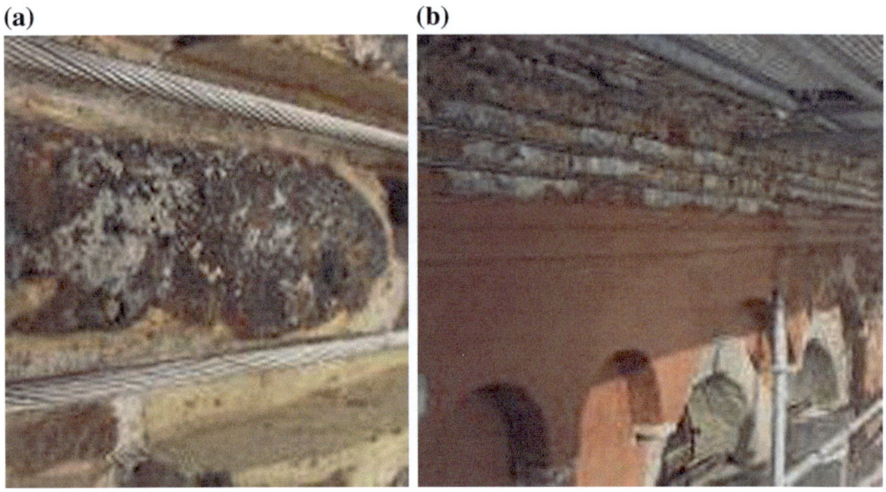

Fig. 2 External confining stainless steel cables: **a** detail of cable insertion between mortar joints [31], **b** view of positioned cables, façade

the additional loads uniformly, as in the case of multi-leaf walls. Indeed, severely damaged buildings after earthquakes often showed incomplete (i.e. not-closed ring) or scarcely reinforced concrete tie-beams. Such a loss of continuity (both for the structural element and/or steel reinforcement) makes the intervention not only ineffective but dangerous, due to the increase of mass caused by concrete elements and their difficult collaboration with masonry. Moreover, avoiding breaches in the wall (that is instead the way how concrete beams are usually built), which contributes positively for keeping the continuity of masonry sections (and, consequently, improves their stress bearing capacity), and using light strengthening

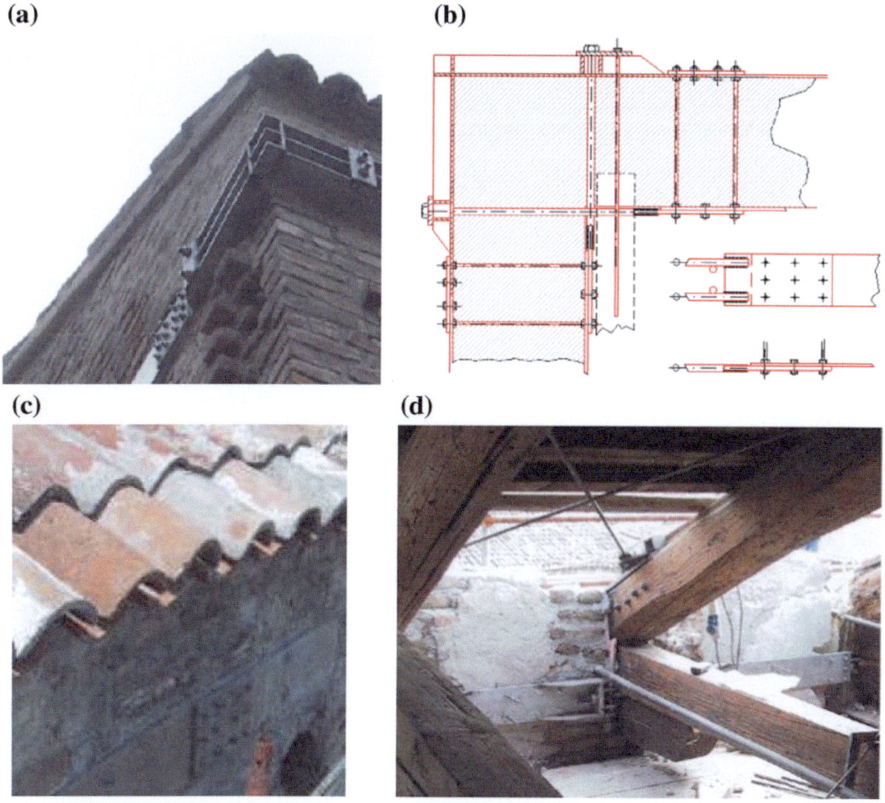

Fig. 3 Stainless steel tie-beam: **a** view of corner [31], **b** detail of connections inside thickness of wall; view of metallic belt from **c** outside and **d** inside

solutions, should be considered as general rules for a good practice and proper choice of interventions [30].

Tying of walls, in concurrence with enough (but not necessarily too high) floor stiffness, strongly improves box-like behaviour of buildings by preventing out-of-plane collapses in favour of in-plane response [35]. The combination with confining rings or metallic belts installed from both sides and connected through the section (Fig. 3b), especially when applied at the top of the building, may represent a low intrusive and light solution, which is also able to improve the interaction with the roof.

Effective connections between floors and walls are fundamental to permit a global behaviour of the building, as they mainly provide a better load redistribution, according to the layout of walls, and exerts a restraining action towards the overturning of perimetral walls [36]. Several solutions are possible; in case of wooden floors, a satisfactory connection is provided by fasteners anchored on the external face of the wall (Fig. 4), so that it is also possible to take advantage of the existing timber beams acting as ties for the building. Local strengthening of masonry in the

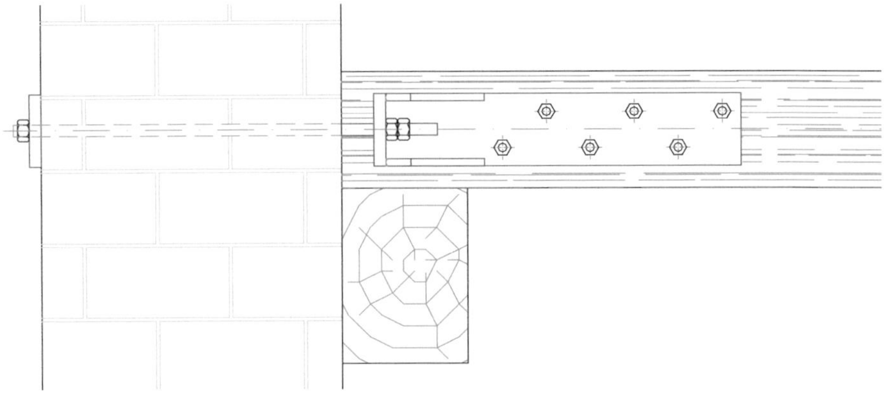

Fig. 4 Strengthening interventions, diffuse wooden floor connections to masonry wall

zone supporting the forces applied by the anchoring devices may be required; to this aim, bed joints repointing or injections can be implemented on regular brickwork or irregular and poor stone masonry, respectively [37].

Several innovative intervention techniques for connections (wall-to-wall; floor-to-wall; roof-to-wall) and different strengthening elements able to dissipate seismic energy have been recently developed, e.g.: connection between vertical masonry walls by steel anchors, in series with dissipative anchoring devices; dovetail halved joints of roof frames strengthened with damping and/or reinforcing elements; coupled injected anchors placed in polyester tubular sleeves and inserted in pre-cored holes acting as double shear connections between masonry walls and timber beams of floors [38, 39].

3.2 *Interventions to Increase the Strength of Masonry*

Interventions, aiming at increasing the masonry strength, may be used to re-establish the original mechanical properties lost because of material decay and/or to improve the performance of low quality masonry. Techniques should be employed with cautiousness and, to the possible extent, comply with restoration principles [40], among which the compatibility—at mechanical and chemical–physical level, of reinforcing materials with the existing ones is fundamental [25, 41, 42].

Several solutions are available, some transmitted and validated by tradition and history, others introduced with the advent of modern technologies over time and spread progressively on existing structures [43].

What mainly concerns stone masonry and fired-brick is given in the following. Some techniques may be more suitable for one or the other type of masonry, according to the peculiar morphology of the wall or the arrangement of the

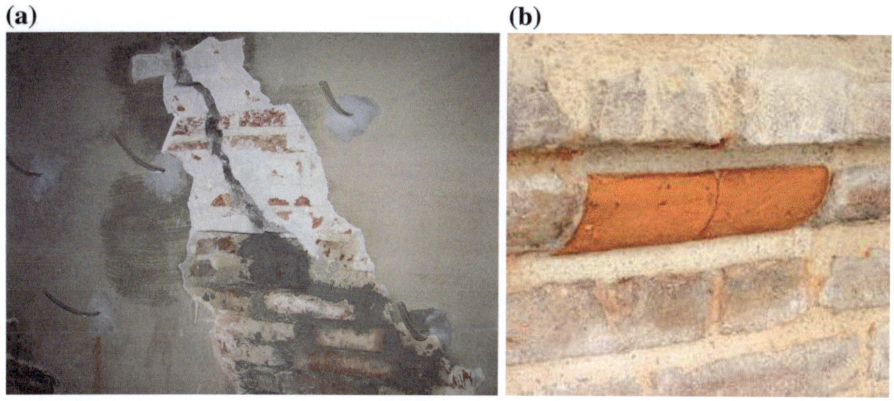

Fig. 5 Example of "scuci-cuci" interventions: **a** rebuilding in cracked portion of wall (combination with injections), **b** replacement of damaged bricks [31]

constituent materials (irregular or regular texture, multi-leaf or monolithic sections), and the specific structural need.

The "scuci-cuci" techniques (Fig. 5), i.e., the partial rebuilding, aims at restoring the wall continuity along cracks (substitution of damaged elements with new ones, re-establishment of the structural continuity) and to recover locally heavily damaged parts of masonry walls. It can be applied to both brick and stone masonry, paying attention at using materials that are similar, for geometry, dimensions, stiffness and resistance, to those employed in the damaged wall, to improve collaboration among new and existing elements. A further advantage can be achieved with adequate connections between new and existing portions—orthogonal and in the plane of masonry, as well as with combinations of injections—in case of presence of voids as in rubble masonry, to obtain greater homogeneity and compactness.

The injection of grout admixtures has been proposed since a long time, particularly for stone masonry walls [40, 45,46]. This typology, often represented by two- or three-leaf walls with scarcely connected sections, is highly vulnerable, under both vertical and horizontal actions, and exposed to brittle out-of-plane collapses. The technique consists in the injection of grouts through a regular pattern of drilled holes (Fig. 6a), aiming at increasing the connection between the leaves of masonry walls and filling the voids existing in the inner masonry core. To accomplish compatibility requirements and improve durability, no cement-based mortar grouts should be preferred; particularly, natural hydraulic lime or pozzolana binders are suitable to consolidate regular and irregular three-leaf stone masonry walls [47–50]. In fact, the use of high-strength materials is not profitable since the main advantage of the intervention is given by voids' filling with a bonding material for which binders similar with the existing ones can be better exploited [47, 51]. Furthermore, grout injections do not increase significantly the stiffness of walls, as oppositely occurring

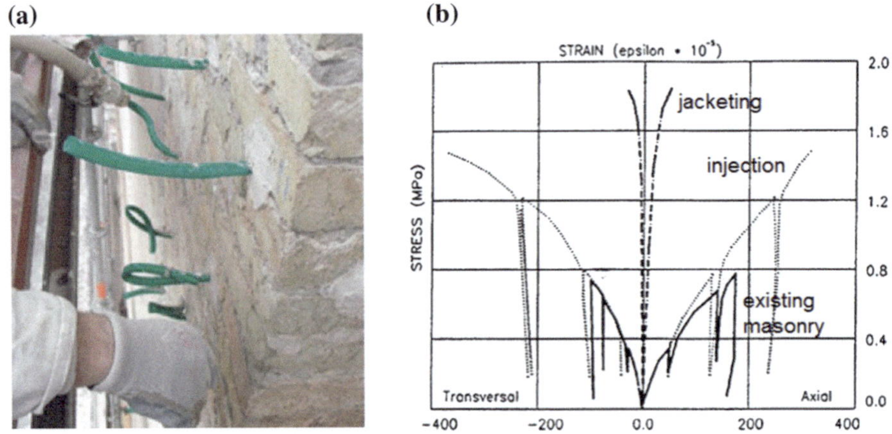

Fig. 6 **a** Strengthening interventions with grout injections [31], **b** comparison of structural behaviour between injections and jacketing [49]

with jacketing, thus also resulting in a mechanically compatible technique (Fig. 6b) [52]. A recent review can be found in Quelhas et al. [53].

Several compression and shear–compression tests, different types of numerical analyses have contributed to the understanding of the mechanical behaviour of injected stone masonry walls, demonstrating the effectiveness of the proposed technique [54, 55]. Both monotonic compression laboratory tests and cyclic shear–compression tests carried out on three-leaf rubble masonry panels resulted in a good increase of strength (about two times between injected and not-injected walls) (Fig. 7a, b) and a reasonable stiffness increase, due to higher compactness of the wall (from about 2 to about 4, for compression and shear behaviour, respectively). However, the best results consist of the changing of brittle mechanism on a more compact behaviour of the section (increased displacement capacity at maximum load, better distribution of damage), fostered by the bonding action of the grout.

Diagonal compression tests carried on-site with panels resting after the severe damage occurred in L'Aquila (as in Casarin et al. [57], Fig. 8a) also showed an increase ratio in strength similar to the one obtained in laboratory, whereas the shear modulus seems more scattering, varying of 5–10 times (Fig. 8b). The original constituents, especially the mortars, were fully characterized from the mechanic, petrographic, textural, mineralogical and chemical point of view, in order to choose the most suitable restoration products; six grouts were selected to consolidate 21 panels representative of typical constructions of the area [58].

The behaviour of multi-leaf masonry can be also improved with "diatoni", i.e., masonry units placed orthogonally to the wall plane (the headers). The insertion of small-sized ties across the wall, also proved to be effective in reducing transversal deformations under axial loads (see Valluzzi et al. [47]), and noticeably reduces local problems under out-of-plane actions, as assessed through shaking table tests

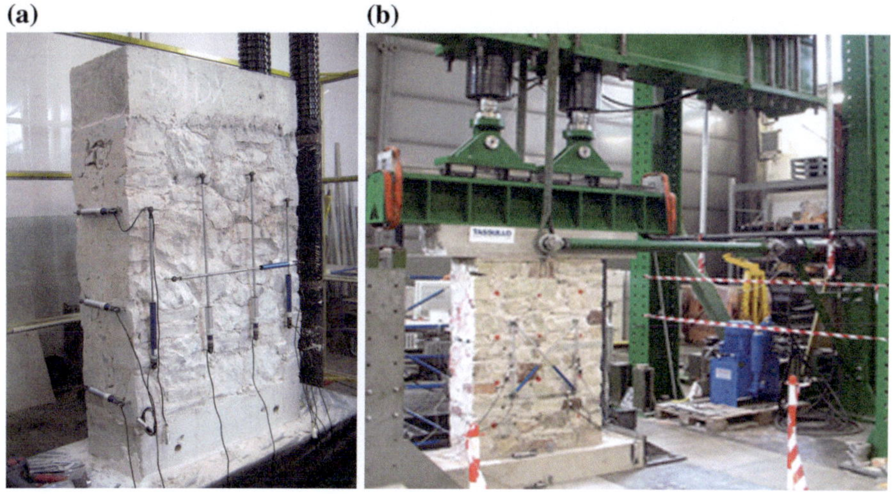

Fig. 7 Consolidation of rubble stone masonry walls with hydraulic lime-based grouting. Experimental tests in laboratory: **a** compression tests [56], **b** shear–compression tests [54]

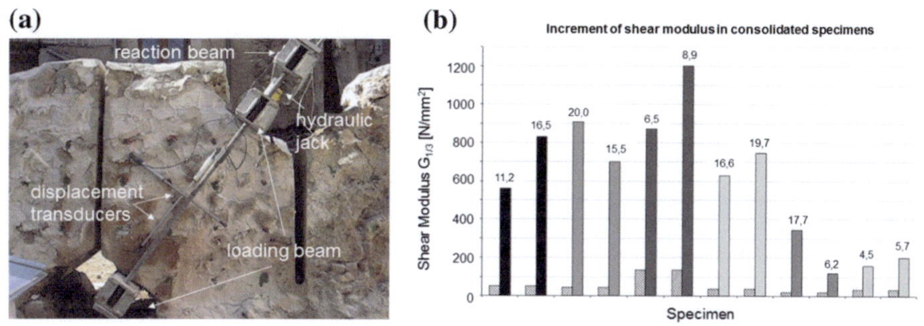

Fig. 8 Consolidation of rubble stone masonry walls with hydraulic lime-based grouting. On site experimental tests: **a** diagonal compression setup, **b** variation of shear modulus for various types of grouts selected for injection on irregular stone masonry walls [57]

(Fig. 9, [59]). In both cases, the combination of grout injections and transverse ties results in the largest overall wall strength [47, 59, 60, 61].

Several wide research programs based on shaking table testing of substructures and reduced scale building models, made with stone masonry, before and after strengthening with injections, transversal tying, or traditional timber lacing techniques have been carried out [62, 63]. Tests confirmed the effectiveness of injections, provided that proper connections among components (both in vertical and horizontal directions) are included, and floors present a good diaphragm action (Fig. 10, [62, 64]). The overall stiffness between injected and not-injected models is not increased and, therefore the dynamic performance does not change [65] either.

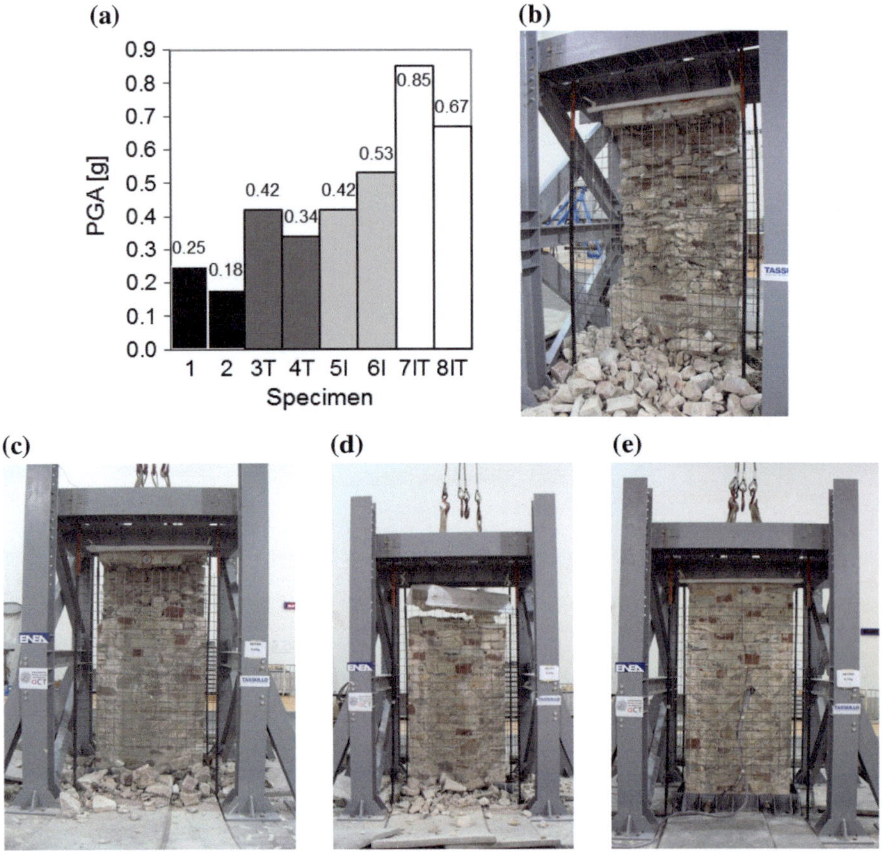

Fig. 9 Shaking table tests on three-leaf rubble stone masonry walls: **a** maximum PGA sustained by panels—examples of wall failures, **b** unreinforced (*panels 1, 2*), **c** with steel ties (3T, 4T), **d** with grout injections (5I, 6I), **e** with combined steel ties and grout injections (7IT, 8IT) [59]

However, higher values of seismic input can be reached thanks to improved monolithic performance achieved due to the connection effect provided by the injection [63].

In order to increase the strength in the out-of-plane direction the use of various types of reinforced plasters has also proved to be effective on stone masonry walls [66, 67]. One of the drawbacks of a common solution of reinforced plaster relies on the use of cement-based mortar to protect the steel reinforcing mesh from corrosion problems. However, improvements in terms of strength, displacement ductility and energy dissipation can be significant (respectively, up to 5, 2 and 11 times those of non-intervened specimens [68], Fig. 11), if that solution is introduced with other previously mentioned techniques, such as the connections between walls and floor/roof.

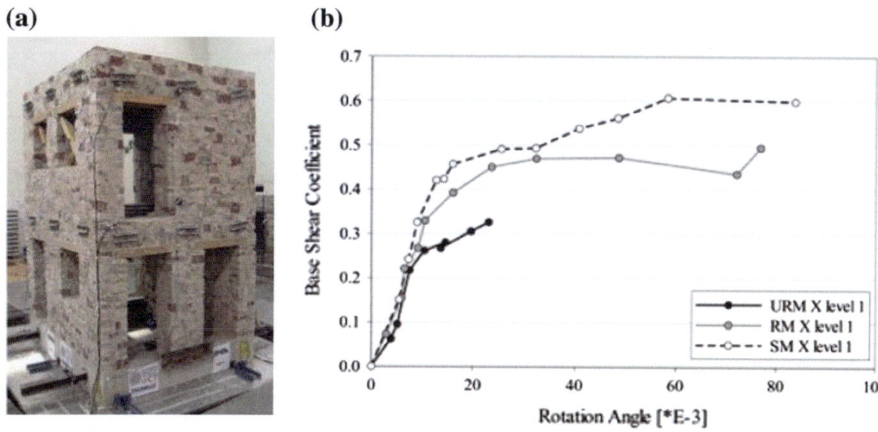

Fig. 10 Shaking table tests on injected three-leaf masonry wall [62]: **a** view of building model, **b** comparison of base shear trends among models (*URM* unreinforced, *RM* repaired and *SM* strengthened)

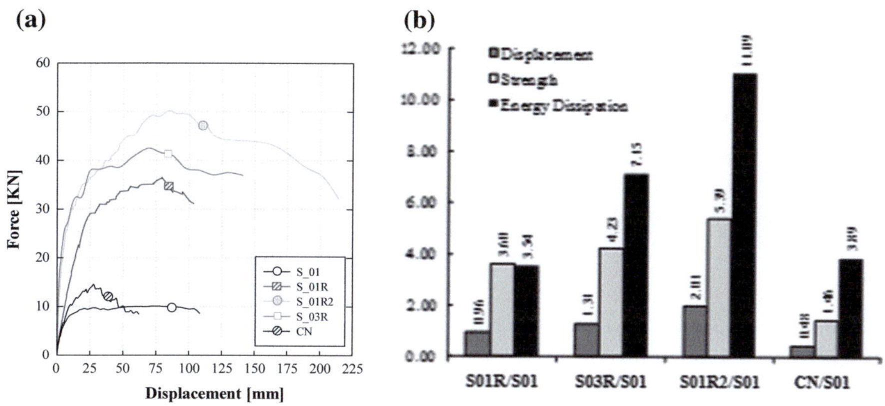

Fig. 11 Experimental results on original (non-strengthened: S_01 and CN) and strengthened (S_01R, S_01R2 and S_03R) specimens with reinforced plaster, under out-of-plane cyclic tests [68]: **a** lateral out-of-plane force-displacement envelopes, **b** global comparison in terms of ratios relative to non-strengthened specimen S_01

Aiming at minimizing the problem of durability and compatibility, other approaches were developed resorting to compatible reinforcing mesh types, for instance materialized with commercial polypropylene grids, and lime-based mortar [69]. In order to avoid the presence of steel connecting elements in the wall, the use of zinc screws and polypropylene connectors (as used on ETICS system) attaching the mesh to stone elements is also a possibility [69], Fig. 12. The improvements (relative to non-strengthened specimens) are important, namely in terms of displacement capacity, energy dissipation and damage distribution (Fig. 12).

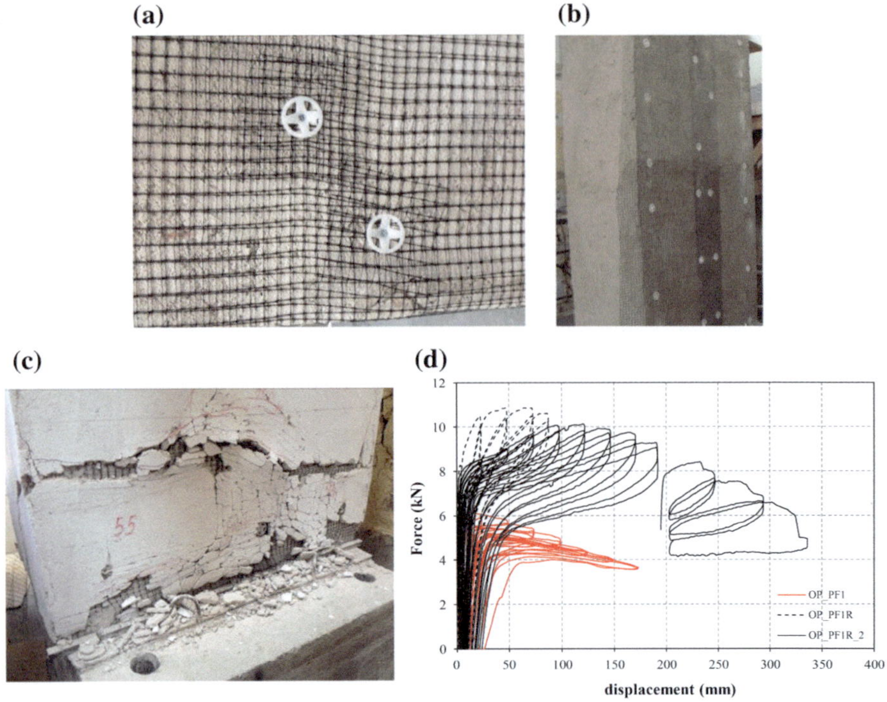

Fig. 12 Application of strengthening compatible solutions [69]: **a** mesh and connectors' detail, **b** global wall view prior to lime-based mortar application, **c** distributed cracking pattern, **d** experimental results, where PF1R refers to the retrofitted specimen (two loading stages)

The improvement in terms of strength is not meaningful, comparatively to the use of cement based steel reinforced plaster, because low strength mesh was adopted and no additional measures were taken to improve connections between wall leaves.

Reinforced grids have been also increasingly adopted with composite materials applied mainly to brick masonry, due to the more regular surface, in comparison with common irregular stonework, formerly using Fibre Reinforced Polymers (FRP) [70–72] or Steel Reinforced Polymers (SRP) [73, 74]. Other more compatible and removable solutions have been progressively proposed, e.g. Textile Reinforced Mortar (TRM), Fibre Reinforced Cementitious Matrix (FRCM) or Steel Reinforced Grout (SRG), with the same purpose of using fibre (or steel wires) reinforced inorganic matrices for jacketing the walls, and/or increase shear and flexural capacity [75–78], repair cracks or tying walls [79]. Indeed, this technique can provide relevant increase in strength (in the order of dozens of times), although the main advantages, particularly in seismic area, consist of the better distribution of cracks and the reduction of brittle detachments or collapse [80]. Nevertheless, the effectiveness of the intervention strongly depends of the bonding properties at the

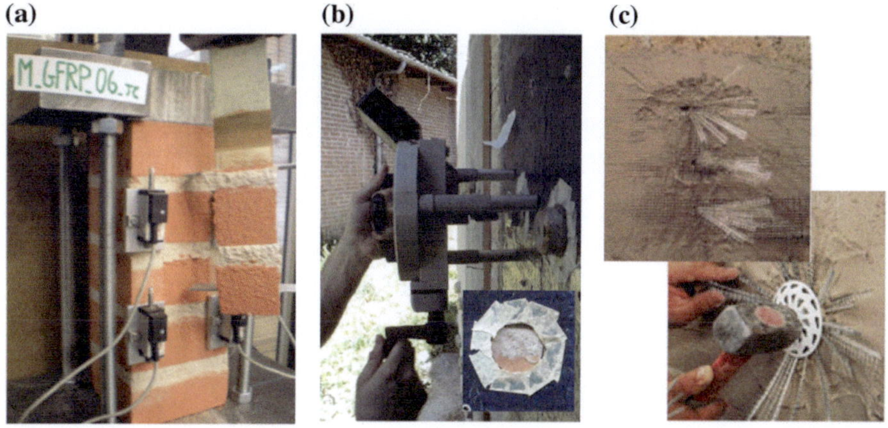

Fig. 13 Strengthening with composites: **a** delamination of glass FRP sheet in laboratory single-lap shear tests [91], **b** salt crystallization affecting carbon FRP-to-masonry bond measured by on site pull-off tests [88], **c** anchoring of SRG plaster with spikes made of steel wires [92]

composite-to-masonry interface, that can be affected by several aspects (e.g., the surface roughness and preparation, the presence of mortar joints, or severe environmental conditions) which can reduce durability (Fig. 13) [69, 82, 83, 84, 85, 86, 87, 88, 89, 90].

Deep repointing is a suitable technique to repair deteriorated bed joints, adaptable both to stone and to brick masonry. Reconstruction of bed joints, 70–80 mm deep in stone masonry about 50 cm thick, can provide significant increase in shear strength and improved behaviour, especially if combined with grout injections [93]; this intervention requires particular care in the execution phase, to ensure proper collaboration between old and new mortar. Bed joints reinforcement is a technique particularly suitable for masonry with regular courses (as commonly found in brickwork), since it consists of the insertion of steel bars in the joints [94, 95]. Laboratory tests and numerical models show the usability of composite materials as FRP small bars instead of steel, to ensure low intrusiveness and to control creep deformations (Fig. 14).

3.3 Interventions to Improve Diaphragm Action of Floors and Roofs

The role of diaphragms in the structural behaviour of existing masonry buildings is fundamental, especially when dealing with horizontal loads (e.g. wind, earthquake, etc.), as they contribute to transfer lateral actions to the walls parallel to the direction of the external force. Interventions able to increase the in-plane stiffness of existing floors contribute in redistributing horizontal actions on the basis of the

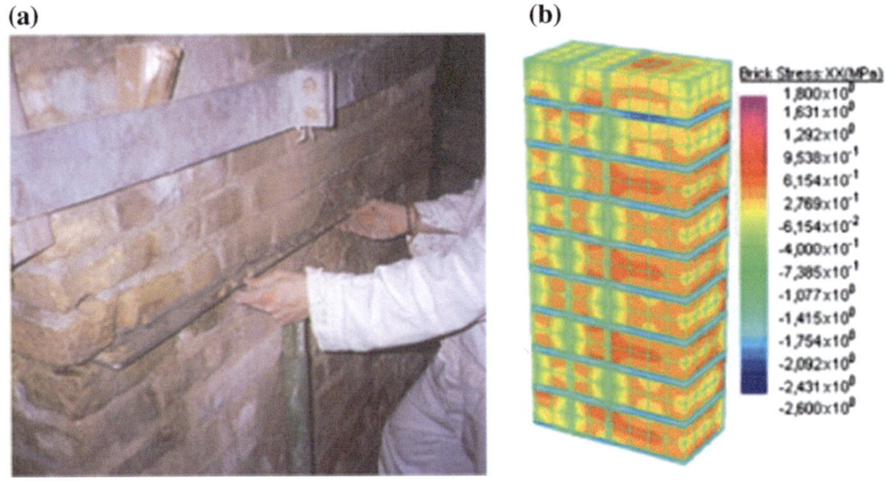

Fig. 14 Reinforced mortar repointing: **a** on-site strengthening intervention using stainless steel elements [95], **b** FE modelling of masonry wall strengthened with CFRP thin strips [96]

masonry wall stiffness, depending on the achieved diaphragm action of the floor [27, 36].

Several techniques have been developed in order to improve the diaphragm action, while respecting traditional construction practice and avoiding excessive mass increase at the floor level. As shown in Fig. 15, these consist in: (i) additional wooden planking (in orthogonal direction or inclined at 45°, using tongue-and-groove joints with nails or screws as connectors); (ii) application of diagonal metallic belts or composite materials strips fixed at the extrados of wooden floors; (iii) application of wooden diagonal nets [97–99].

Nailing an additional layer of wooden planks at the floor or roof extrados (outer surface) represents a light intervention, that does not modifies the overall structural element, but improves the diaphragm action [102]. The use of metallic bracing ties (strips or bars) can also improve the stiffening effect (Fig. 16).

Nevertheless, connections between floor/roof structures and walls is crucial to provide proper box-like behaviour under dynamic actions, as highlighted by shaking table tests performed on two-storey double-leaf stone masonry buildings with roof strengthened with plywood layers anchored with steel bars and intermediate floor reinforced with a collaborating concrete slab [103], according to Piazza et al. [104]. Also, the effectiveness of moderate in-plane stiffening of wooden diaphragms (adding a second layer of diagonal planks to the original one) and of improving wall-to-diaphragm connections (by means of reinforced masonry and steel tie-beams, at roof and intermediate floor levels, respectively) on the overall dynamic behaviour of masonry buildings, is evidenced by shaking table tests in Magenes et al. [105].

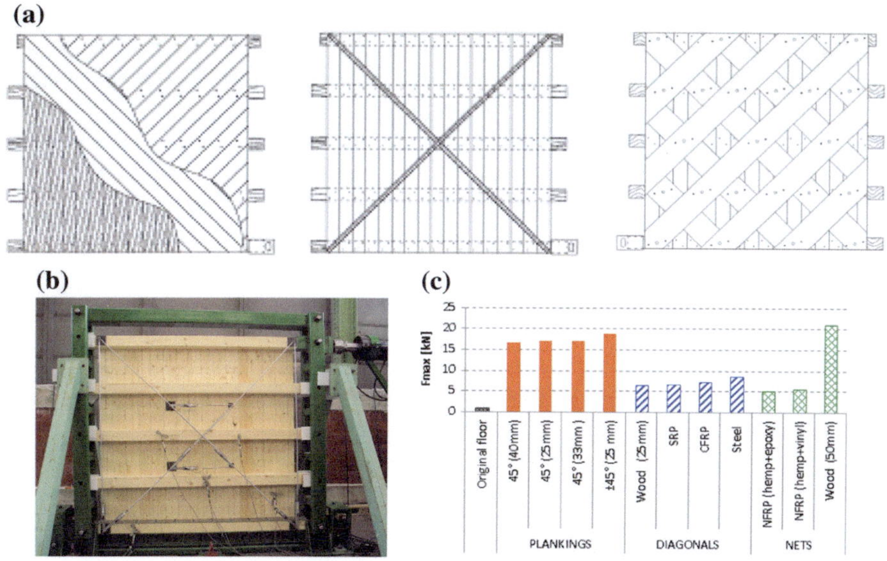

Fig. 15 **a** Floor strengthening interventions: double planking, steel diagonals with orthogonal planking, wooden diagonal nets' technique based on hardwood pin connections between boards and beams [100, 101], **b** in-plane monotonic and cyclic laboratory tests, **c** comparison of maximum lateral forces from cyclic tests [97] using different techniques

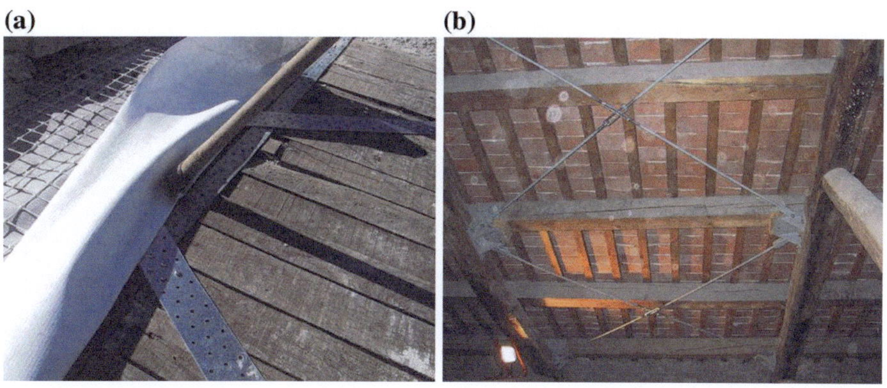

Fig. 16 In-plane stiffening of existing wooden roofs: **a** extrados metallic strips, **b** steel bracing bars at the intrados [32]

3.4 *Strengthening of Masonry Vaults and Reduction of Thrust*

Among structural components in masonry buildings, arches and vaults deserve particular attention, being a widespread type of horizontal diaphragm in European historical centres. Because of age or for accidental causes (such as earthquakes), these structures can suffer several types of damage. Strengthening materials and repair techniques are required to re-establish the performance of such components and to prevent their brittle collapse in future hazardous conditions.

A traditional intervention consists of inserting tie-rods, to compensate the thrust induced on bearing walls [106]. Tie-rods are generally inserted at the vault intrados, but can also be placed at the extrados, with some precautions. Simple techniques to evaluate the tie-rod force and their effectiveness on site have been implemented [107].

To absorb the thrust of vaults and arches, and to increase their lateral stiffness, buttresses and vertical masonry diaphragms (‘*frenelli*’) can be adopted, whilst jacketing the extrados using concrete or reinforced concrete may have a negative impact in terms of mass increase and removability in future.

Techniques based on the application of composite materials, e.g., carbon or glass FRPs, placed at the intrados or at the extrados of the structure (Fig. 17), were initially developed to control the brittle mechanism of collapse of unreinforced vaults and increase their maximum capacity under both vertical and horizontal loads. Several experimental works evidenced quite good performance of these techniques, namely: (i) the activation of new mechanisms to analyse for design and assessment [108]; (ii) the higher effectiveness of extrados applications in comparison with intrados ones (as in Valluzzi et al. [109], Barbieri et al. [110], Foraboschi [111], Baratta and Corbi [112]), unless connection pins or spikes are applied [113] and (iii) the better exploitation of less strong composites (e.g., GFRP in comparison with CFRP) by avoiding masonry damage [114]. More recently, the advanced trend of composites, SRP/SRG or TRM, also applied with basalt nets (BTRM—Basalt Textile Reinforced Mortar), have demonstrated the advantages of better compatibility, water permeability and removability. Static tests (Borri et al. [115]; Girardello et al. [116, 117]; Garmendia et al. [118]) and dynamic shaking table experiments (Giamundo et al. [119]) confirmed the relevant improvement of the ultimate capacity of the vault (Fig. 18).

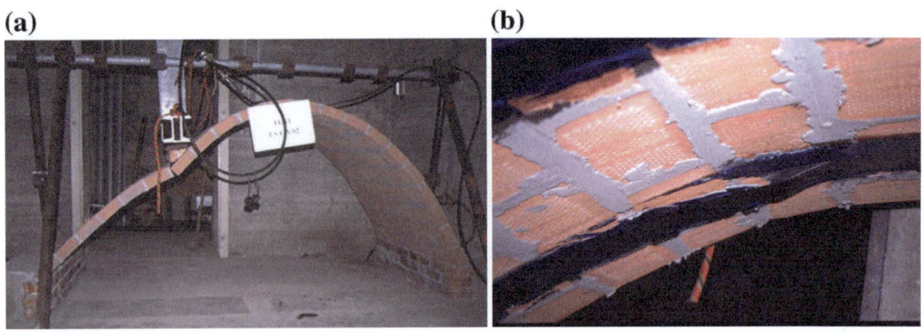

Fig. 17 Collapse mechanism of a vault strengthened using CFRP strips: **a** extrados reinforcement, **b** intrados reinforcement [108]

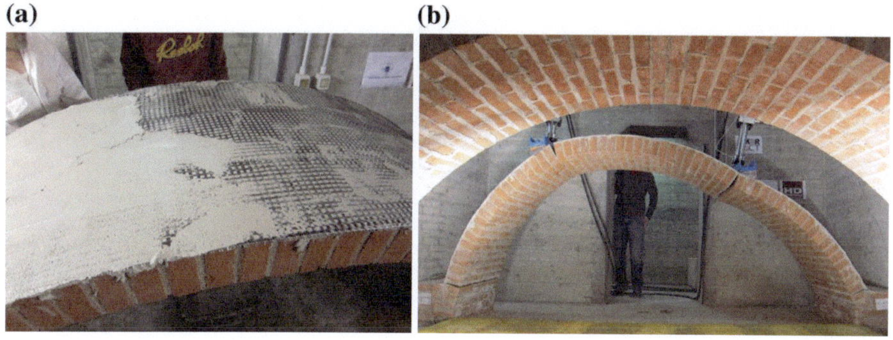

Fig. 18 Strengthening of masonry vault with BTRM: a application of basalt grid, b deformation of vault under testing [117]

4 Conclusions

The choice of suitable intervention techniques to retrofit or strengthen masonry components in existing buildings cannot disregard the importance of improving the collaboration among structural components in order to avoid or minimize the occurrence of brittle mechanisms that can be particularly severe in seismic zones (e.g., overturning of walls, collapse of vaults). Tying opposite walls or confining rings are effective to improve the buildings' box-like behaviour, provided effective connections exist between walls and with horizontal structures. A further step is then the opportunity to increase the performance of single components, according to the specific structural problem, but also evaluating, when needed, the possibility of integrating techniques and combining the capabilities of various intervention solutions. Both traditional techniques and modern/innovative materials available nowadays constitute a quite wide range of possibilities, as long as conservation principles are taken into account, thus orienting the choice towards compatible, removable and increasingly sustainable techniques. Although specific protocols for execution, design and control are still not available for many techniques, the experimental research (both on site and in laboratory) represents a fundamental support for their validation and practical use. In this framework, this chapter resumed the advance of scientific approach to balance tradition and innovation in the field of conservation of existing masonry building structures.

Acknowledgements Part of this work reports to research financially supported by Project POCI-01-0145-FEDER-007457—CONSTRUCT—Institute of R&D in Structures and Construction, funded by FEDER funds through COMPETE2020—Programa Operacional Competitividade e Internacionalização (POCI) and by national funds through Fundação para a Ciência e a Tecnologia (FCT).

References

1. ISO 13822. Bases for design of structures—assessment of existing structures. Annex I (Informative) Historic Structures. International Organization for Standardization; 2010.
2. Carocci CF, Tocci C, editors. Giuffrè A. Leggendo il libro delle antiche architetture. Aspetti statici del restauro. Saggi 1985–1997. Cangemi; 2010 (in Italian).
3. Valluzzi MR, Bondì A, da Porto F, Franchetti P, Modena C. Structural investigations and analyses for the conservation of the "Arsenale" of Venice. J Cult Herit. 2002;3(1):65–71.
4. Macchi G, Calvi GM, Sullivan TJ. Structural strengthening and retrofit; motivations, concepts and approaches. In: Costa A, Arêde A, Varum H, editors. Strengthening and retrofitting of existing structures. Berlin: Springer; 2017.
5. Modena C, Casarin F, da Porto F, Garbin E, Mazzon N, Munari M, Panizza M, Valluzzi MR. Structural interventions on historical masonry buildings: review of Eurocode 8 provisions in the light of the Italian experience. In: Proceedings of the workshop "Eurocode 8 perspectives from the Italian standpoint. Napoli, Italy; 2009.
6. Modena C, Casarin F, da Porto F, Munari M. L'Aquila 6th April 2009 earthquake: emergency and post-emergency activities on cultural heritage buildings. In: Garevski M, Ansal A, editors. Earthquake engineering in Europe, vol. 17; 2010. pp. 495–521.
7. Lorenzoni F, Casarin F, Modena C, Caldon M, Islami K, da Porto F. Structural health monitoring of the Roman Arena of Verona. Italy. J Civ Struct Health Monit. 2013;3(4):227–46.
8. Valluzzi MR, Munari M, Modena C, Binda L, Cardani G, Saisi A. Multilevel approach to the vulnerability analysis of historic buildings in seismic areas part 2: analytical interpretation of mechanisms for vulnerability analysis and structural improvement. Restor Build Monum. 2007;13(6):427–41.
9. Modena C, Valluzzi MR, da Porto F, Casarin F. Structural aspects of the conservation of historic stone masonry constructions in seismic areas. Int J Archit Herit Conserv Anal Restor. 2011;5(4–5):539–58.
10. Corradi M, Borri A, Vignoli A. Strengthening techniques tested on masonry structures struck by the Umbria–Marche earthquake of 1997–1998. Constr Build Mater. 2002;16:229–39.
11. NIKER 2010-2012, FP7-ENV-2009-1-GA244123: New integrated knowledge based approaches to the protection of cultural heritage from earthquake-induced risk; Coordinator: Università di Padova; www.niker.eu.
12. da Porto F, Silva B, Costa C, Modena C. Macro-scale analysis of damage to churches after earthquake in Abruzzo (Italy) on April 6, 2009. J Earthq Eng. 2012;16(6):739–58.
13. Sorrentino L, Liberatore L, Decanini LD, Liberatore D. The performance of churches in the 2012 Emilia earthquakes. Bull Earthq Eng. 2014;12(5):2299–331.
14. Cattari S, Degli Abbati S, Ferretti D, Lagomarsino S, Ottonelli D, Tralli A. Damage assessment of fortresses after the 2012 Emilia earthquake (Italy). Bull Earthq Eng. 2014;12 (5):2333–65.
15. Tomaževič M. The influence of rigidity of floors on the seismic behaviour of old stone masonry buildings. Eur Earthq Eng. 1991;5(3):28–41.
16. Sorrentino L, Liberatore L, Liberatore D, Masiani R. The behaviour of vernacular buildings in the 2012 Emilia earthquakes. Bull Earthq Eng. 2014;12(5):2367–82.
17. Decanini L, De Sortis A, Goretti A, et al. Performance of masonry buildings during the 2002 Molise, Italy, earthquake. Earthq Spectra. 2004;20:191–220.
18. Penna A, Morandi P, Rota M, Manzini CF, da Porto F, Magenes G. Performance of masonry buildings during the Emilia 2012 earthquake. Bull Earthq Eng. 2014;12(5):2255–73.
19. Binda L, Saisi A. Research on historic structures in seismic areas in Italy. Prog Struct Eng Mater. 2005;7:71–85.
20. da Porto F, Munari M, Prota A, Modena C. Analysis and repair of clustered buildings: case study of a block in the historic city centre of L'Aquila (Central Italy). Constr Build Mater. 2013;38:1221–37.

21. Magenes G, Bolognini D, Braggio C (eds). Metodi semplificati per l'analisi sismica non lineare di edifici in muratura. CNR-Gruppo Nazionale per la Difesa dai Terremoti; 2000 (in Italian).
22. Binda L, Cardani G, Saisi A, Valluzzi MR. Vulnerability analysis of the historical buildings in seismic area by a multilevel approach. Asian J Civ Eng (Build Hous). 2006;7(4):343–57.
23. Valluzzi MR, Munari M, Modena C, Binda L, Cardani G, Saisi A. Multilevel approach to the vulnerability analysis of historic buildings in seismic areas—part 2: analytical interpretation of mechanisms for the vulnerability analysis and the structural improvement. Int J Restor Build Monum. 2007;13(6):427–41.
24. Giuffrè A, Carocci C. Codice di pratica per la sicurezza e la conservazione del centro storico di Palermo. Laterza; 1999 (in Italian).
25. Binda L, Gambarotta L, Lagomarsino S, Modena C. A multilevel approach to the damage assessment and seismic improvement of masonry buildings in Italy. In: Bernardini A, editor. Seismic damage to masonry buildings. Kalamazoo: Balkema; 1999.
26. Bernardini A. Random and fuzzy sets in the modelling of uncertain engineering Systems. In: Elishakoff I, editor. Whys and Hows of uncertainty modelling: probability, fuzziness and anti-optimization. Berlin: Springer; 1999.
27. Tomaževič M, Weiss P. Seismic behaviour of plain and reinforced masonry buildings. ASCE J Struct Eng. 1994;120(2):323–38.
28. Linee Guida per la valutazione e la riduzione del rischio sismico del patrimonio culturale con riferimento alle Norme Tecniche sulle costruzioni (D.M. 14/01/08), Decreto P.C.M. 9/02/2011.
29. Modena C, da Porto F, Valluzzi MR. Conservazione del patrimonio architettonico e sicurezza strutturale in zona sismica: insegnamenti dalle recenti esperienze italiane. Materiali e Strutture, Problemi di Conservazione. 2012;I:1–2.
30. Valluzzi MR. Challenges and perspectives for the protection of masonry structures in historic centers: the role of innovative materials and techniques. RILEM Tech Lett. 2016;1:45–9.
31. Modena C, Valluzzi MR, da Porto F, Casarin F. Structural aspects of the conservation of historic masonry constructions in seismic areas: remedial measures and emergency actions. IJAH Int J Archit Herit. 2011;5(4–5):539–58.
32. Modena C, Valluzzi MR, da Porto F, Casarin F, Munari M, Mazzon N, Panizza M. Assessment and improvement of the seismic safety of historic constructions: research and applications in Italy. In Proceedings of the I Congreso Iberoamericano sobre construcciones históricas y estructuras de mampostería. Bucaramanga, Colombia; 2008.
33. Modena C, da Porto F, Valluzzi MR, Munari M. Criteria and technologies for the structural repair and strengthening of architectural heritage. Int J 3R's. 2013;4:606–21.
34. Tomaževič M, Lutman M. Seismic behaviour of masonry walls: modelling of hysteretic rules. ASCE J Struct Eng. 1996;122(9):1048–54.
35. Tomazevic M, Lutman M, Weiss P. Seismic upgrading of old brick masonry urban houses: tying of walls with steel ties. Earthq Spectra. 1996;12(3):599–622.
36. Tomaževič M. Earthquake-resistant design of masonry buildings. Series on innovation in structures and construction. In: Elnashai AS, Dowling PJ, editors. vol.1. London: Imperial College Press; 1999.
37. Modena C, Valluzzi MR, Folli TR, Binda L. Design choices and intervention techniques for repairing and strengthening of the Monza cathedral bell-tower. Constr Build Mater. 2002;16:385–95.
38. Paganoni S, D'Ayala D, Miccoli L, Hračov S, Urushadze S, Wünsche M, Adami C-E, Vintzileou E, Moreira S, Oliveira DV, James P, Cóias e Silva V. Connections and dissipative systems with early warning. In: Proceedings of the 8th international conference on structural analysis of historical constructions. Wroclaw, Poland; 2012.
39. Moreira S, Ramos LF, Oliveira DV, Lourenço PB. Design parameters for seismically retrofitted masonry-to-timber connections: injection anchors. Int J Archit Herit. 2016;10:2–3.

40. ICOMOS. Recommendations for the analysis, conservation and structural restoration of architectural heritage. International Scientific Committee for Analysis and Restoration of Structures of Architectural Heritage; 2001.
41. Binda L, Saisi A, Tedeschi C. Compatibility of materials used for repair of masonry buildings: research and applications. In: Kourkoulis SK, editor. Fracture and failure of natural building stones: applications in the restoration of ancient monuments. Berlin: Springer; 2006. p. 167–82.
42. Valluzzi MR. Strengthening of masonry structures with fibre reinforced plastics: from modern conception to historical building preservation. In: SAHC'08—VI conference on structural analysis of historical constructions, evaluating safety and significance, vol I; 2008. pp. 33–45.
43. Miccoli L, Muller U, Silva B, da Porto F, Hracov S, Pospisil S, Adami CE, Vintzileou E, Vasconcelos G, Poletti E. Overview of different strengthening techniques applied on walls used in historical structures. In: Proceedings of the 8th international conference on structural analysis of historical constructions. Wroclaw, Poland; 2012.
44. Binda L, Modena C, Baronio G. Strengthening of masonries by injection technique. In: Proceeding of 6th NaMC. Philadelphia; 1993. pp. 1–14.
45. Tomaževič M, Apih V. The strengthening of stone-masonry walls by injecting the masonry-friendly grouts. Eur Earthq Eng 1993;6(1):10–20.
46. Vintzileou E, Tassios TP. Three leaf stone masonry strengthened by injecting cement grouts. ASCE J Struct Eng. 1995;121:848–56.
47. Valluzzi MR, da Porto F, Modena C. Behavior and modeling of strengthened three-leaf stone masonry walls. RILEM Mater Struct. 2004;37(267):184–92.
48. Vintzileou E, Miltiadou-Fezans A. Mechanical properties of three-leaf stone masonry grouted with ternary or hydraulic lime-based grouts. Eng Struct. 2008;30(8):2265–76.
49. Oliveira DV, Silva RA, Garbin E, Lourenço PB. Strengthening of three-leaf stone masonry walls: an experimental research. Mater Struct. 2012;45(8):1259–76.
50. Kalagri A, Miltiadou-Fezans A, Vintzileou E. Design and evaluation of hydraulic lime grouts for the strengthening of stone masonry historic structures. Mater Struct. 2010;43:1135–46.
51. Silva B, Pappas A, Guedes JM, da Porto F, Modena C. Numerical analysis of the in-plane behaviour of three-leaf stone masonry panels consolidated with grout injection. Bull Earthq Eng. 2016. doi:10.1007/s10518-016-9969-5.
52. Modena C, Bettio C. Experimental characterization and modeling of injected and jacketed masonry walls. In Proceedings of the Italian-French symposium strengthening and repair of structures in seismic area. Nice; 1994.
53. Quelhas B, Cantini L, Guedes JM, da Porto F, Almeida C. Characterization and reinforcement of stone masonry walls. In: Costa A, Guedes JM, Varum H, editors. Structural rehabilitation of old buildings. Series: building pathology and rehabilitation, vol. 2, no. 5; 2014. pp. 131–56.
54. Silva B, Dalla Benetta M, da Porto F, Modena C. Experimental assessment of in-plane behaviour of three-leaf stone masonry walls. Constr Build Mater. 2014;53:149–61.
55. Vintzileou E. Testing historic masonry elements and/or building models. In: Ansal A, editor. Perspectives on European earthquake engineering and seismology. Geotechnical, geological and earthquake engineering, vol. 34; 2014. pp. 267–307.
56. Silva B, Dalla Benetta M, da Porto F, Valluzzi MR. Compression and sonic tests to assess effectiveness of grout injection on three-leaf stone masonry walls. Int J Archit Herit Conserv Anal Restor. 2014;8(3):408–35.
57. Casarin F, Dalla Benetta M, da Porto F, Valluzzi MR, Modena C. Masonry panels strengthened by grout injections after the 2009 Abruzzo earthquake. In: Proceedings of the 14th European conference on earthquake engineering. Ohrid, Republic of Macedonia; 2010.

58. Artioli G, Casarin F, da Porto F, Mazzoli C, Secco M, Valluzzi MR. Restoration of historic masonry structures damaged by the 2009 Abruzzo earthquake through cement- and polymer-free injection grouts. In: Proceedings of the 2nd historic mortars conference. Prague, Czech Republic; 2010.
59. Valluzzi MR, Mazzon N, Garbin E, Modena C. Experimental characterization of out-of-plane seismic response of strengthened three-leaf stone masonry walls by shaking table tests. XV Convegno L'Ingegneria Sismica in Italia. Padova, Italy; 2013.
60. Giaretton M, Valluzzi MR, Mazzon N, Modena C. Out-of-plane shake-table tests of strengthened multi-leaf stone masonry walls. Bull Earthq Eng. 2017. doi:10.1007/s10518-017-0125-7.
61. Adami CE, Vintzileou E, Mouzakis C, Karapitta L. Three-leaf stone masonry models with timber ties: the effect of timber ties and strengthening techniques on seismic response. In: SAHC. Wroclaw, Poland; 2012.
62. Mazzon N, Valluzzi MR, Aoki T, Garbin E, De Canio G, Ranieri N, Modena C. Shaking table tests on two multi-leaf stone masonry buildings. In: Proceedings of the 11th Canadian masonry symposium. Toronto (Canada); 2009.
63. Vintzileou E, Valluzzi MR, Mazzon N, Posipsil S, Miccoli L. Systemic improvement of overall seismic response. In: Proceedings of the 8th international conference on structural analysis of historical constructions. Wroclaw, Poland; 2012.
64. Mazzon N, Chavez Cano M, Valluzzi MR, Casarin F, Modena C. Shaking table tests on multi-leaf stone masonry structures: analyses of stiffness decay. In: SAHC2010 Structural analysis of historical constructions—Strengthening and retrofitting, Part 1, proceedings on 7th international conference; Shanghai, China; 2010. p. 647–52.
65. Aoki T, Mazzon N, Valluzzi M, Casarin F, Modena C. Dynamic identification and damage detection of multi-leaf stone masonry building by shaking table test. ASCE J Struct Eng. 2010;56:99–105.
66. Costa AA, Arede A, Costa A, Oliveira CS. Out-of-plane behaviour of existing stone masonry buildings: experimental evaluation. Bull Earthq Eng. 2012;10:93–111.
67. Costa AA, Arede A, Costa A, Guedes J, Silva B. Experimental testing, numerical modelling and seismic strengthening of traditional stone masonry: comprehensive study of a real Azorian pier. Bull Earthq Eng. 2012;10(1):135–59.
68. Costa AA, Arede A, Costa A, Oliveira CS. In situ cyclic tests on existing stone masonry walls and strengthening solutions. Dyn Earthq Eng Struct. 2011.
69. Arêde A, Gomes A, Marques D, Costa, AA. Unreinforced vs. strengthened stone masonry walls: experimental study of their out-of-plane behavior. In: SAHC 2016: 10th international conference on structural analysis of historical constructions, Sept 13–15, Leuven, Belgium; 2016.
70. Triantafillou TC. Strengthening of masonry structures using epoxy-bonded FRP laminates. ASCE J Compos Constr. 1998;2(2):96–104.
71. Shrive NG. The use of fibre reinforced polymers to improve seismic resistance of masonry. Constr Build Mater. 2006;20(4):269–77.
72. Valluzzi MR, Tinazzi D, Modena C. Shear behavior of masonry panels strengthened by FRP laminates. Constr Build Mater. 2002;16(7):409–16.
73. Borri A, Castori G, Corradi M. Shear behavior of masonry panels strengthened by high strength steel cords. Constr Build Mater. 2011;25(2):494–503.
74. de Felice G, De Santis G, Garmendia L, Ghiassi B, Larrinaga P, Lourenco P, Oliveira D, Paolacci F, Papanicolau CG. Mortar-based systems for externally-bonded strengthening of masonry. Mater Struct. 2014;47:2021.
75. Prota A, Marcari G, Fabbrocino G, Manfredi G, Aldea C. Experimental in-plane behavior of tuff masonry strengthened with cementitious matrix-grid composites. ASCE J Compos Constr. 2006;10(3):223–33.
76. Valluzzi MR, da Porto F, Garbin E, Panizza M. Out-of-plane behavior of infill masonry panels strengthened with composites materials. RILEM Mater Struct. 2014;47(12):2131–45.

77. Papanicolaou CG, Triantafillou TC, Karlos K, Papathanasiou M. Textile-reinforced mortar (TRM) versus FRP as strengthening material of URM walls: in-plane cyclic loading. Mater Struct. 2007;40:1081–97.
78. Corradi M, Borri A, Castori G, Sisti R. Shear strengthening of wall panels through jacketing with cement mortar reinforced by GFRP grids. Compos B Eng. 2014;64:33–42.
79. de Santis S, de Felice G. Traditional and innovative techniques for the seismic retrofitting of masonry buildings. In: SECED 2015, earthquake risk and engineering towards a resilience world, Cambridge UK.
80. Valluzzi MR, Modena C, De Felice GM. Current practice and open issues in strengthening historical buildings with composites., RILEM. Mater Struct. 2014;47(12):1971–85.
81. Capozucca R. Experimental FRP/SRP—historic masonry delamination. Compos Struct. 2010;92:891–903.
82. Garbin E, Panizza M, Valluzzi MR. Experimental assessment of bond behaviour of fibre-reinforced polymers on brick masonry. Struct Eng Int. 2010;20(4):392–9.
83. Valluzzi MR, Oliveira DV, Caratelli A, Castori G, Corradi M, de Felice G, Garbin E, Garcia D, Garmendia L, Grande E, Ianniruberto U, Kwiecień A, Leone M, Lignola GP, Lourenço PB, Malena M, Micelli F, Panizza M, Papanicolaou CG, Prota A, Sacco E, Triantafillou TC, Viskovic A, Zając B, Zuccarino G. Round robin test for composite-to-brick shear bond characterization. Rilem Mater Struct. 2012;45:1761–91.
84. Sciolti MS, Aiello MA, Frigione M. Influence of water on bond behavior between CFRP sheet and natural calcareous stones. Compos Part B. 2012;43(8):3239–50.
85. Ghiassi B, Marcari G, Oliveira DV, Lourenço PB. Water degrading effects on the bond behavior in FRP-strengthened masonry. Compos Part B. 2012;54:11–9.
86. Tedeschi C, Kwiecien A, Valluzzi MR, Zajac B, Garbin E, Binda L. Effect of thermal ageing and salt decay on bond between FRP and masonry. RILEM Mater Struct. 2014;47 (12):2051–65.
87. Cardani G, Valluzzi MR, Panizza M, Girardello P, Binda L. Influence of salt crystallization on composites-to-masonry bond evaluated on site by pull-off tests. In: MuRiCo4, 4th international conference on mechanics of masonry structures strengthened with composite materials—modeling, testing, design, control. Ravenna, Italy, 2014. Key Engineering Materials, vol. 624. Trans Tech Publications, Switzerland; 2015. pp. 338–45.
88. Cardani G, Valluzzi MR, Panizza M, Binda L. In-situ evaluation of composites-to-masonry bond in a natural aggressive environment. In: 12th NAMC North American masonry conference, Denver, Colorado, USA; 2015. (CDROM).
89. Carozzi FG, Poggi C. Mechanical properties and debonding strength of fabric reinforced cementitious matrix (FRCM) systems for masonry strengthening. Compos Part B Eng. 2015;70:215–30.
90. de Felice G, Aiello MA, Bellini A, Ceroni F, de Santis S, Garbin E, Leone M, Lignola GP, Malena M, Mazzotti C, Panizza M, Valluzzi MR. Experimental characterization of composite-to-brick masonry shear bond. Mater Struct. 2016;49:2581–96.
91. Panizza M, Garbin E, Valluzzi MR, Modena C. Experimental investigation on bond of FRP/SRP applied to masonry prisms. In: CICE 2012—conference on composites in civil engineering, Rome; 2012 (CD-ROM).
92. De Santis S, Casadei P, De Canio G, de Felice G, Malena M, Mongelli M, Roselli I. Seismic performance of masonry walls retrofitted with steel reinforced grout. Earthq Eng Struct Dyn. 2016;45:229–51.
93. Corradi M, Tedeschi C, Binda L, Borri A. Experimental evaluation of shear and compression strength of masonry wall before and after reinforcement: deep rejointing. Constr Build Mater. 2008;22:463–72.
94. D'Ayala D. The use of bed joint reinforcement to improve the performance of historic masonry buildings. In: Proceedings of the 5th international masonry conference. London; 1998.
95. Valluzzi MR, Binda L, Modena C. Mechanical behaviour of historic masonry structures strengthened by bed joints structural repointing. Constr Build Mater. 2005;19:63–73.

96. Valluzzi MR, Disarò M, Modena C. Bed joints reinforcement of masonry panels with CFRP bars. In: Bruno D, Spadea G, Swamy RN, editors. International conference on composites in construction, Rende (CS), Italy; 2003. p. 427–32.
97. Valluzzi MR, Garbin E, Dalla Benetta M, Modena C. Experimental assessment and modelling of in-plane behaviour of timber floors. In: Proceedings of the 6th international conference on structural analysis of historical constructions, vol. II. Bath, UK; 2008. pp. 755–62.
98. Valluzzi MR, Garbin E, Dalla Benetta M, Modena C. Experimental characterization of timber floors strengthened by in-plane improvement techniques. In: International conference on structural health assessment of timber structures, SHATIS'13, Trento, Italy. Advanced materials research, vol. 778. Trans Tech Pub; 2013. pp. 682–89.
99. Valluzzi MR, Garbin E, Dalla Benetta M, Modena C. In-plane strengthening of timber floors for the seismic improvement of masonry buildings. In: World conference on timber engineering, Riva del Garda, Italy; 2010 (CD-ROM).
100. Modena C, Valluzzi MR, Garbin E, da Porto F. A strengthening technique for timber floors using traditional materials. In: Structural analysis of historical constructions. Padova; 2004. p. 911–21.
101. Valluzzi MR, Garbin E, Modena C. Flexural strengthening of timber beams by traditional and innovative techniques. J Build Apprais. 2007;3(2):125–43.
102. Valluzzi MR, Girardello P, Francescato D, Pospisil S, Kral R, D'Ayala D, Paganoni S, Pelà L, Roca P, Banco J, El Harrouni K, Kukukdogan B, Parisi, MA. Optimization of design for floors, roofs and vaults. In: Proceedings of the 8th international conference on structural analysis of historical constructions. Wroclaw, Poland; 2012.
103. Senaldi I, Magenes G, Penna A, Galasco A, Rota M. The effect of stiffened floor and roof diaphragms on the experimental seismic response of a full scale unreinforced stone masonry building. J Earthq Eng. 2014;18(3):407–43.
104. Piazza M, Baldessari C, Tomasi R. The role of in-plane floor stiffness in the seismic behaviour of traditional buildings. In: Proceedings on 14th world conference on earthquake engineering. Beijing; 2008.
105. Magenes G, Penna A, Senaldi I, Rota M, Galasco A. Shaking table test of a strengthened full scale stone masonry building with flexible diaphragms. Int J Archit Herit. 2014;8(3):349–56.
106. Dillmann P, Bernardi P, Fluzin P. Use of iron for the building of medieval monuments. The Palais des Papes in Avignon and other French buildings. In: Archaeometallurgy in Europe, p. 1; 2003.
107. Cescatti E, da Porto F, Modena C, Casarin F. Ties in historical constructions: typical features and laboratory tests. In: Proceedings of structural analysis of historical constructions. Leuven, Belgium; 2016.
108. Panizza M, Valluzzi MR, Garbin E, Modena C. Bond mechanism of brick masonry vaults. In: Proceedings of the conference on structural faults + repair. Edinburgh, UK; 2008.
109. Valluzzi MR, Valdemarca M, Modena C. Behaviour of brick masonry vaults strengthened by FRP laminates. ASCE Int J Compos Constr. 2001;5(3):163–9.
110. Barbieri A, Borri A, Corradi M, Di Tommaso A. Dynamic behaviour of masonry vaults repaired with FRP: experimental analysis. In: Proceedings of the 6th international masonry conference. London; 2002.
111. Foraboschi P. Strengthening of masonry arches with fiber-reinforced polymer strips. J Compos Constr. 2004;8(3):191–202.
112. Baratta A, Corbi O. Stress analysis of masonry vaults and static efficacy of FRP repairs. Int J Solids Struct. 2007;44(1):8028–56.
113. Borri A, Casadei P, Castori G, Hammond J. Strengthening of brick masonry arches with externally bonded steel reinforced composites. J Compos Constr. 2009;6:468–75.
114. Briccoli Bati S, Rovero L, Tonietti U. Strengthening Masonry Arches with Composite Materials. ASCE J Compos Constr. 2007;11:1–33.

115. Borri A, Casadei P, Castori G, Ebaugh S. Research on composite strengthening of masonry arches. In: Proceedings of the 8th international symposium on fiber reinforced polymer reinforcement for concrete structures. Patras, Greece; 2007.
116. Girardello P, da Porto F, Dalla Benetta M, Valluzzi MR. Experimental behaviour of masonry vaults strengthened by innovative composite materials. In: 12th Canadian masonry symposium, Vancouver, British Columbia; 2013 (CD-ROM).
117. Girardello P, Pappas A, da Porto F, Valluzzi MR. Experimental testing and numerical modelling of masonry vaults. In: Rehabilitation and restoration of structures. In: Gettu R, Santhanam M, Menon A, Pillai RG, editors. Proceedings on international conference on rehabilitation and restoration of structures. Chennai, India; 2013. p. 117–27.
118. Garmendia L, Marcos I, Garbin E, Valluzzi MR. Strengthening of masonry arches with textile-reinforced mortar: experimental behaviour and analytical approaches. RILEM Mater Struct. 2014;47(12):2067–80.
119. Giamundo V, Lignola GP, Maddaloni G, da Porto F, Prota A, Manfredi G. Shaking table test on a full scale unreinforced and IMG-retrofitted clay brick masonry vault. Bull Earthq Eng. 2016;14(6):1663–93.

Seismic Retrofit of Adobe Constructions

Julio Vargas-Neumann, Cristina Oliveira, Dora Silveira and Humberto Varum

1 Introduction

Earth is one of the oldest and most widespread construction materials in use (Fig. 1). The unique properties and accessibility of this material justify its wide dissemination and continued use throughout the time. In fact, this kind of material has qualities such as low cost, local availability, recyclability, good thermal behaviour, and acoustic insulation—qualities which allow a more sustainable construction practice, with the preservation of the existing natural resources. In addition, this type of construction is associated with simple construction methods that require low energy consumption.

According to Houben and Guillaud [2], at the end of the twentieth century, approximately 30% of the world population lived in earthen buildings and, at

J. Vargas-Neumann
Civil Engineering Section, Department of Engineering, Pontifical Catholic University of Peru, San Miguel, Lima 32, Lima, Peru
e-mail: jhvargas@pucp.pe

C. Oliveira
Barreiro School of Technology, Polytechnic Institute of Setubal, 2839-001 Lavradio, Portugal
e-mail: cristina.oliveira@estbarreiro.ips.pt

C. Oliveira
CONSTRUCT-LESE, Faculty of Engineering (FEUP), University of Porto, 4200-465 Porto, Portugal

D. Silveira · H. Varum (✉)
CONSTRUCT-LESE, Civil Engineering Department, Faculty of Engineering, University of Porto, 4200-465 Porto, Portugal
e-mail: hvarum@fe.up.pt

D. Silveira
e-mail: dsilveira@fe.up.pt

A. Costa et al. (eds.), *Strengthening and Retrofitting of Existing Structures*, Building Pathology and Rehabilitation 9, https://doi.org/10.1007/978-981-10-5858-5_4

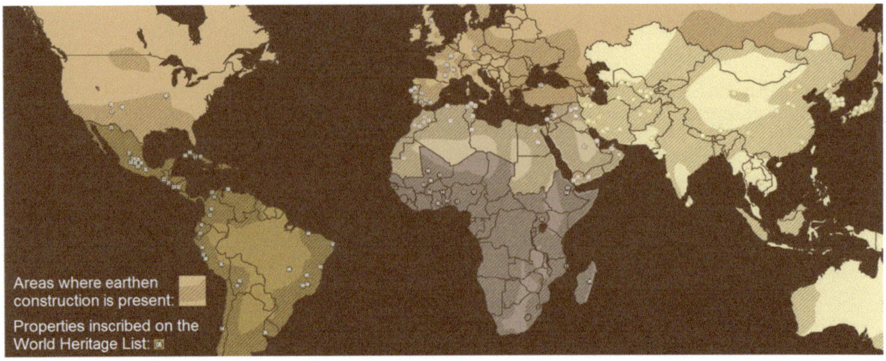

Fig. 1 Distribution of earthen construction in the world and properties inscribed on the World Heritage List (adapted from Gandreau and Delboy [1])

present, these figures should not be very different. Nowadays, earth is used mostly in two ways: (a) in traditional and vernacular earthen constructions, with popular and long-established techniques used in many developing countries—mainly in Africa, Latin America, and in some parts of Asia—usually by the poorest segments of the population; (b) in earthen buildings that combine the use of traditional techniques with more modern practices, mainly in developed countries such as the United States of America, some European countries, Australia, and New Zealand; in these countries, there is a growing interest in earth as a sustainable construction material and, as a result, numerous earthen buildings were built in the last decades [3, 4].

In earthen construction, a raw mixture of clay, silt, sand—and, sometimes, larger aggregates—is generally used. Other materials can also be added to improve the characteristics of the mixture, such as fibres and stabilisers (e.g. cement, lime) [3]. Rammed earth, adobe, cob, and compressed earth blocks (CEB) are among the main types of earthen construction methods still in use. Rammed earth consists of compacted earth in monolithic blocks that form the walls. Adobe is made with a mixture of earth moulded in prismatic blocks and sun-dried—frequently, straw is added to control drying cracks. Cob consists of a mixture of clay and a substantial amount of straw that is applied and moulded by hand in order to make monolithic layered walls. The compressed earth blocks (CEB)—a more recent earthen construction method, marked by the development of the CINVA-RAM press in the 1950s, in Colombia, and with great dissemination from the 1970s and 1980s [5]—are made using a mechanic press. In the production of these blocks, a small portion of lime or cement is added to the earth mixture for stabilization.

Even though earth is a construction material which has many advantages, it also presents some disadvantages that cannot be disregarded. In fact, earthen constructions are particularly vulnerable to the action of weathering agents, especially to the action of water and wind. In addition, this material is characterized by low tensile strength and fragile behaviour, and thus earthen constructions can display a deficient structural response, particularly when subjected to seismic loads. Even though the different types of earthen construction share some of the same characteristics and may show similar behaviour, there are specific problems associated with each of the different types of construction. In the present chapter, the focus will be on the structural behaviour of adobe construction. In effect, the seismic behaviour of adobe structures is typically characterized by brittle failure [6] and these constructions can suffer severe structural damage and eventually total collapse, with significant human losses.

Despite the growing interest in earthen construction as a sustainable alternative in developed countries, a large percentage of the existing earthen buildings are still associated with rural populations with low economic resources and, even though important research has been carried out in the last decades, the existing knowledge concerning earthen construction is still mainly empirical. Buildings are generally constructed and rehabilitated by non-specialized staff, normally without the use of appropriate behaviour improvement solutions. In addition, few countries have codes and standards for the rehabilitation and building with earth, and the existent documents are frequently incomplete [7]. It is also important to note that earthen construction continues to be used in regions with high seismicity, and thus there is an urgent need to develop means of improving the seismic behaviour of these structures. Furthermore, there is a vast and valuable architectural heritage stock which needs to be preserved. In 2012, a detailed inventory identified 150 properties partially or entirely built with earth which were inscribed in the World Heritage List (Fig. 1) [1]. It was also observed that, in these properties, adobe is the most commonly used earthen technique. Considering the total number of cultural properties inscribed on the World Heritage List in 2016 (814, according to the United Nations Educational, Scientific and Cultural Organization [8]), the properties partially or entirely built with earth identified in 2012 correspond to 18% of that total number. Of the 150 identified properties, 18 (i.e. 12%) are, in 2016, included in the List of World Heritage in Danger, corresponding to 49% of the cultural properties included in this list [8].

Considering the importance and vulnerabilities of the existing earthen constructions, one of the main questions asked by the technicians working with these constructions is how to effectively repair and strengthen damaged earthen structures. The utilization of traditional construction materials and techniques must be complemented with innovative technological tools, in order to provide adequate stability and control of displacements, which may reduce the structural vulnerability of this type of construction. Taking into account the location of these constructions and the low-income populations that generally build and live in them, the repair and strengthening techniques should be as inexpensive as possible, in order to be effectively applied by the populations. However, before discussing the possible repair and strengthening solutions, one must understand the behaviour of the materials individually and the performance of the structures with the materials working together.

2 Adobe Construction Materials and Structural Behaviour

2.1 Variability of the Mechanical Properties of Adobe Masonry

Adobe construction is made by assembling the adobe blocks together using an earth mortar for bonding. The adobe blocks used throughout the world have a high variability in composition, which is reflected in a significant heterogeneity in their mechanical properties [9–11].

A study conducted at the University of Aveiro, in Portugal, using lime adobes (i.e. adobes made with sandy soil and air lime binder) collected from different constructions in the region of Aveiro, showed that the mean unconfined compressive strength of the adobes, calculated per construction under analysis, ranged between 0.66 and 2.15 MPa, with coefficients of variation up to 50% (considering the variation of results obtained for each construction) [11]. The tensile strength corresponded to approximately 18% of the unconfined compressive strength, with even higher coefficients of variation (up to 65%). In a subsequent study, the modulus of elasticity of adobe specimens collected from existing constructions in the region was evaluated [12]. The mean modulus of elasticity, calculated considering all the specimens extracted from each adobe under analysis, varied between 7609 and 25,000 MPa, with a global mean value of 13,214 MPa. The coefficients of variation of the modulus of elasticity, determined considering the results obtained for the specimens of each adobe block, varied between 28 and 45%. The adobe blocks used in this study had mean dimensions of $0.44 \times 0.28 \times 0.12$ m^3 and specific weight of approximately 15 kN/m^3.

The heterogeneity in the mechanical properties of the adobe blocks and adobe masonry of existing constructions, in addition to being a result of the variability in the composition of the adobes and mortars, is also due to other factors. In traditional construction practices, there is no quality control of the materials and procedures used in the production of the adobes and earth mortars and in the construction of adobe structures, and so the earthen construction elements are not built uniformly.

The National Autonomous University of Mexico (UNAM) and the Pontifical Catholic University of Peru (PUCP) conducted, in the early 1980s, a very relevant international comparative study sponsored by the Organization of American States (OAS) [13]. This study concluded, by performing the same full-scale static experimental test in adobe walls from Mexico and Peru, made with adobes of similar strength, that there was a 300% difference between the shear strength of the adobe wall specimens from the two countries. In 1983, in order to investigate the reasons for this high variability of results, the United States Agency for International Development (USAID) supported a 1-year research project focused on the study of the seismic strength of adobe masonry, carried out by the Engineering Department of PUCP [14]. In this research project, it was possible to study the relationship of many physical variables with the indirect tensile strength of small walls. Some of the variables analysed were the granulometric distribution, Atterberg Limits, volumetric drying shrinkage, and density of the materials used in the adobes and mortars. Some of these variables—such as the volumetric drying shrinkage—showed a stronger correlation with the strength of the walls than others, but one of the main conclusions reached is that there are many possible causes for the heterogeneity in results obtained for adobe masonry [15].

More recently, another study was conducted at PUCP, focused on the influence of the thickness of joint mortars on the strength of small adobe masonry walls obtained in diagonal compression or indirect tensile tests. This study showed that changing the thickness of the joint mortars from 20 to 5 mm led to an increase in tensile strength of about 300%. This may be explained by the fact that the size of material imperfections is smaller in thin mortars than in thick mortars and so the tensile strength increases dramatically [16].

2.2 *Seismic Behaviour of Adobe Constructions*

Adobe masonry walls have low tensile strength and brittle behaviour and thus may show a poor structural response when subjected to cyclic horizontal forces, such as those induced by earthquakes. Figure 2 depicts the typical seismic damage observed in adobe constructions. This damage includes [17]: (a) diagonal in-plane cracking; (b) cracking near openings; (c) vertical cracking at the intersection of walls; (d) separation of the walls in the corners; (e) disintegration of the upper areas of the walls; (f) out-of-plane damage or collapse in gable-end and other walls; (g) separation between roof and walls; (h) bending damage at half of the walls height.

The most frequent collapse mode involves out-of-plane movement of the walls after the formation of vertical cracks in the intersection of the walls and the fall of gable and other freestanding walls through horizontal cracks, due to out-of-plane forces [18]. Subsequently to the collapse of walls, the roof may lose its support and collapse as well. It is important to note that the diagonal shear cracks due to in-plane forces weaken walls and thus leave them more vulnerable to out-of-plane forces.

There are several recent earthquakes that affected adobe buildings in a severe way, evidencing the vulnerability of this type of construction. Some examples are the El Salvador earthquakes, in 2001 [19], the Bam (Iran) earthquake, in 2003 [20], the Pisco (Peru) earthquake, in 2007 [21], the Maule (Chile) earthquake, in 2010 [22], and the Gorkha (Nepal) earthquake, in 2015 [23]. In the 2001 El Salvador earthquakes, for example, more than one million people lost their homes, with adobe houses being the most severely affected type of construction [19]. The 2010 Chile earthquake and the succeeding tsunami damaged approximately 370,000 buildings, of which about 37% were made with adobe [22]. In the region of Maule, in Curicó, in particular, approximately 90% of the existing adobe constructions were destroyed [22].

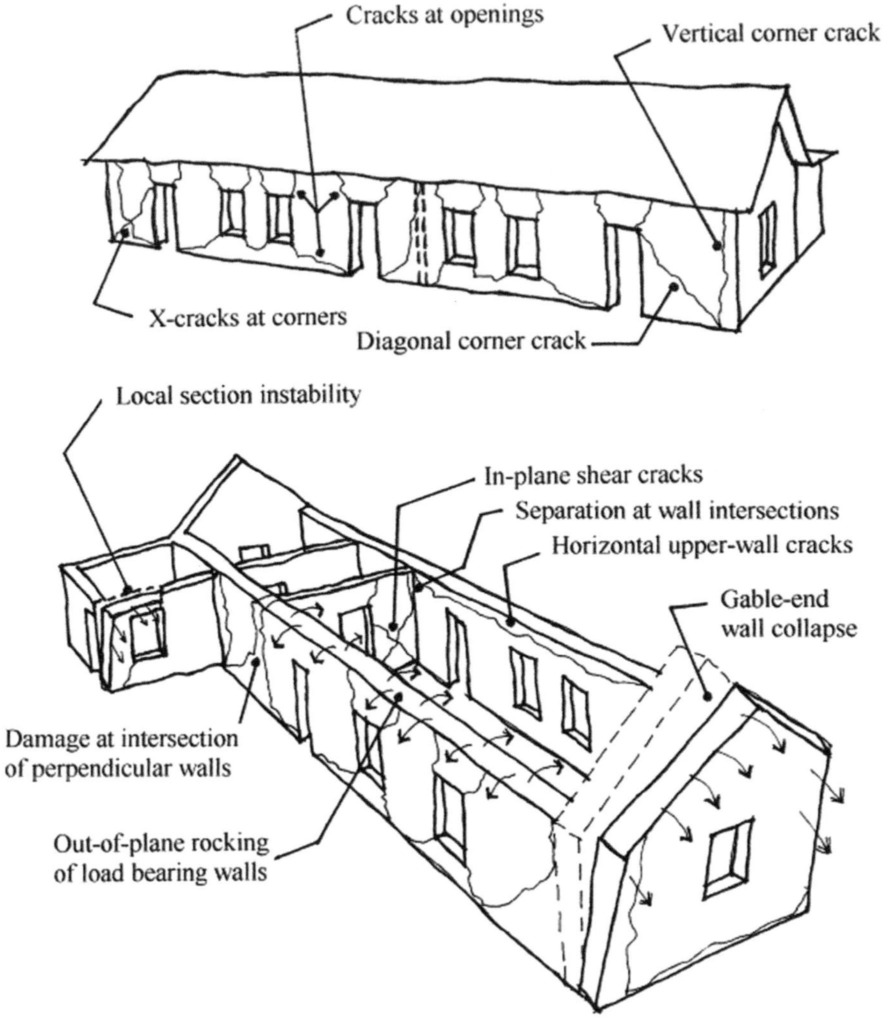

Fig. 2 Typical seismic damage in adobe constructions [17]

2.3 Research on the Mechanical and Structural Behaviour of Adobe Constructions

Important research work has been conducted on the characterization of the mechanical properties of adobe masonry and structural behaviour of adobe buildings. Some of the work that has been conducted is briefly described in the following paragraphs, focusing particularly on the research carried out at the Pontifical Catholic University of Peru (PUCP) and the University of Aveiro.

At PUCP, in an initial phase of the research on adobe construction, a series of tests were carried out to study the mechanical properties and behaviour of adobe masonry [24, 25]. Ten small square walls were tested in diagonal compression. The mean shear strength obtained was 0.026 MPa, and the mean modulus of rigidity (shear modulus) was 39.8 MPa. Adobe prisms with a height to thickness ratio (aspect ratio) of 4 were also tested in simple compression. A compressive strength varying between 0.80 and 1.20 MPa—depending on the age of the specimen—was obtained, and a modulus of elasticity of 170 MPa was computed.

At the University of Aveiro, ten adobe masonry walls, with dimensions of $1.26 \times 1.26 \times 0.29\ m^3$, were also tested in diagonal and perpendicular compression to the bed joints [26]. The walls were built using lime adobe blocks from a demolition in the region of Aveiro and lime mortar formulated in the laboratory with a composition similar to that traditionally used. From the tests, a mean shear strength of 0.026 MPa, mean modulus of rigidity (shear modulus) of 413 MPa, mean compressive strength of 0.331 MPa, and mean modulus of elasticity of 757 MPa were obtained.

At PUCP, at a later stage of the research, a full-scale double-T shaped adobe wall was submitted to a displacement controlled quasi-static in-plane cyclic test, with the objective of analysing the cyclic response of the wall and the damage pattern evolution caused by in-plane forces [27]. The main longitudinal wall (with a central window opening) was 3.06 m long, 1.93 m high, and 0.30 m thick; the transverse walls were 2.48 m long. With the transverse walls it was intended to: (a) simulate the influence of the connection between transversal walls found in typical buildings; (b) avoid rocking due to in-plane actions. In addition, a reinforced concrete beam was built at the top of the adobe wall to provide uniform distribution of the horizontal forces applied to the wall and to represent the loads corresponding to a roof composed by wooden beams, canes, straw, mud, and corrugated zinc sheet. The test consisted in applying a horizontal load at the top of the wall, in a series of increasing load cycles. Each displacement cycle was repeated twice. During the test, the cracks started at the corners of the windows and advanced diagonally up to the top and down to the base of the wall. During reversal loads, the cracks generated the X-shaped pattern that is typical in masonry walls subjected to in-plane cyclic forces. A maximum lateral load of 38 kN was reached in the test, with a corresponding shear strength of 41 kPa.

At the University of Aveiro, a double-T shaped wall was also built and tested [28, 29]. The wall was built using lime adobe blocks collected from a demolition in the region of Aveiro and mortar formulated with the composition traditionally used. The wall was subjected to a quasi-static in-plane cyclic test, with a cyclic horizontal demand of increasing amplitude applied 2.5 m above the base of the wall, until failure. The wall was 3.07 m high, 3.5 m long, and 0.29 m thick; the two transverse walls were 1.70 m long. A vertical uniform load was added to the top of the wall through an equivalent mass of 20 kN to simulate the common dead and live loads on typical adobe constructions. The maximum lateral force obtained in the cyclic test was 58.1 kN, for a drift of 0.03%, with a corresponding shear strength capacity of 57.3 kPa. The failure mode was brittle, as expected for adobe constructions.

At PUCP, an adobe model representing a typical Peruvian adobe construction was also tested. The model was built on a reinforced concrete foundation ring beam and subjected to a unidirectional dynamic test [30]. The model consisted of four walls 3.21 m long and 0.25 thick (with the exception of one wall, which had a thickness of 0.28 m, since it was plastered with mud stucco). The full-scale adobe model was subjected to three levels of unidirectional displacement signals, with the following maximum displacements at the base: 30, 80, and 120 mm, in order to represent the effects of a frequent, moderate, and severe earthquake, respectively. For the first two levels, typical vertical cracks occurred at the intersections of the walls causing their separation. Afterwards, X-shaped cracks were formed at the longitudinal walls, and cracks initiated at the transverse walls due to horizontal and vertical bending. Major damage was observed at the end of the second level of displacements, and total collapse occurred during the third level.

At the University of Aveiro, a full-scale adobe model was also built in the laboratory with adobes collected from an existing construction and mortar with traditional composition. The model was subjected to a monotonic test followed by a cyclic test, with the load imposed on the horizontal plane, until failure. The model had a rectangular geometry in plan view with dimensions of $3.00 \times 4.00\ m^2$, height of 2.35 m, and mean wall thickness of 0.35 m; it had three openings: one window and two doors located in the south, east, and west walls, respectively. The structure was built on top of a reinforced concrete footing. On the top of the walls, a reinforced concrete beam, with a weight of approximately 60 kN, was constructed to simulate the permanent loads that correspond to the weight of the roof and respective overloads and to allow the transference of the test forces to the walls. The pushover test led the structure to failure. A maximum shear strength at the base of 45.2 kPa was recorded, for a drift of 0.06%, and the structure displayed a maximum drift of 0.75%. The model suffered brittle failure, and the damage after the application of the cyclic load was typical of adobe structures subjected to seismic loads: in-plane cracking, X-shaped cracks, cracks near openings, and diagonal cracks in wall corners.

Other researchers have also conducted experimental work providing valuable information on the mechanical and structural behaviour of adobe construction (e.g. [9, 31–35]). In these studies, simple and diagonal compression, direct shear, and flexural tests were conducted on adobe wall panels, and adobe structural elements or house models were subjected to monotonic or cyclic horizontal loading tests and shaking table tests, with in-plane or out-of-plane loading.

All the research that has been conducted on the mechanical and structural behaviour of adobe construction reflects the increasing interest that this subject has raised in the scientific community but also highlights the pressing need for more knowledge on the behaviour of this type of construction. Recent and severe earthquakes have exposed the fragility of earthen buildings and prompted the development of research on how to effectively repair and strengthen these constructions.

3 Seismic Retrofit of Adobe Constructions

3.1 Introduction

As discussed in the previous section, adobe constructions may experience various types of damage when subjected to specific actions and are particularly vulnerable when under cyclic horizontal loads, such as those induced by earthquakes. It is possible to build new constructions with improved behaviour and also to enhance the behaviour of existing constructions, by strengthening them either before or after the damage has occurred. The structural design that defines the repair and strengthening solutions to be used in existing constructions must be based on the mechanical characteristics, behaviour, and defects of the structure and its materials, considering the different actions that may occur in the building lifetime.

The walls of adobe constructions are made with adobe bricks and earth mortars, while the roof can be made with different materials, such as timber and ceramic tiles or zinc sheets. The quality of this type of construction is highly dependent on historical, climatic, and ecological circumstances. The materials used are collected locally and the construction methods are usually based on the traditional and empirical knowledge gathered over time. The location of the constructions is a factor that also influences the type of actions to which they may be subjected. Regions of moderate to high seismic activity require special attention due to the seismic vulnerability of this type of construction. Rehabilitation in rural areas, often with a high degree of dispersion, is also different from rehabilitation conducted in urban areas. Additionally, the building use may also influence the type of rehabilitation and strengthening to be performed. Special use buildings, such as schools, hospitals, churches, or meeting places, that attract a great number of people, are evidently at greater risk. The rehabilitation and strengthening process must thus address all these different factors so that adequate solutions may be defined.

Three design criteria can be used in a process of rehabilitation and strengthening of adobe constructions. These can be based on the improvement of strength, stability, and structural behaviour or performance. The design based on improving the strength of the structures may, for example, help to define the geometry (thickness and/or number) of the walls to be used in the strengthening. The design based on improving the stability of the structures may help define the connections between structural elements (wall to wall, walls to floor, or walls to roof), the introduction of elements such as buttresses, as well as the geometry of the building. The design based on improvement of structural behaviour can be achieved by introducing specific reinforcements with high tensile capacity materials to tie potential cracked walls and also tie walls and roof, to control displacements and partial or total collapse, which is particularly important in seismic areas.

Although the great majority of knowledge on earthen construction still lies in traditional and empirical methods, long years of research on the subject conducted worldwide, with particular focus on the development of effective seismic strengthening solutions, have led to important results. In addition, different standards and codes have been or are being produced. Nevertheless, the need for more research on this topic and for the development of more complete standards and codes is recognized.

3.2 Research and Innovation

Important research has been conducted worldwide for the development of strengthening solutions for adobe construction, particularly in countries of significant seismic activity, since the effects of this activity can severely damage this type of construction. A research group from Pontifical Catholic University of Peru (PUCP), in particular, has been playing an important role in the acquisition of knowledge on the behaviour of adobe construction and development of seismic reinforcement solutions—part of the significant work developed by this research group in the last 45 years is reported in detail by Vargas et al. [36] and Blondet et al. [37]. In the early 1970s, the collapse of full-scale adobe house models with four walls was characterized through experimental tests using a tilt table, and different strengthening materials, such as wood and wire, were tested [38]. This research was then continued during this decade—new reinforcements using natural materials like cane or bamboo were tested on the same table, and it was concluded that the use of meshes inside of walls is an effective reinforcement solution [24]. From the 1980s until the present, many reinforcement solutions have been studied with full-scale shaking table tests [36]. Cane meshes and metallic and geosynthetic (polymer geogrid) meshes were evaluated for the seismic retrofit of existing adobe constructions [39]. The use of vertical canes with horizontal ropes and enveloping plastic meshes was compared and assessed [40]. The performance of adobe vaults with and without reinforcement was also investigated [41]. More recently, PUCP has conducted research with the aim of developing solutions to repair cracks in earthen walls damaged by earthquakes. The procedures for mud grout injection were assessed in dynamic and cyclic tests on full-scale models [42]. An experimental study for the development of a reinforcement solution using a mesh made with synthetic ropes tied with manual knots was also conducted [43].

The reinforcement of adobe walls was also investigated at the National University of Engineering, in Peru, using two different solutions [33]: wooden studs covered with a mat (*manta de Carrizo*); and wooden studs covered with a polyethylene layer. Both reinforcement solutions were applied to the corners of adobe wall specimens and embedded in cement-soil mortar. The wall specimens were then subjected to a monotonic in-plane test. The two types of reinforcement showed good results, increasing the lateral load and deformation capacity of the walls.

Another institution that has devoted significant attention to the seismic retrofit of adobe construction is the Getty Conservation Institute (GCI). In the 1990s, the GCI launched the Getty Seismic Adobe Project, with the aim of studying the seismic behaviour of historic adobe buildings and creating effective and low-impact seismic retrofit solutions. A significant part of the investigation focused on the shaking table testing of reduced-scale models of adobe walls and buildings. In the frame of this project, small-scale building models (1:5 scale) were tested at the Stanford University, in the USA, and large-scale building models (1:2 scale) were tested in the Institute of Earthquake Engineering and Engineering Seismology, in the Republic of Macedonia [44, 45]. In these tests, seismic retrofit solutions, such as vertical and horizontal straps, vertical centre-core rods, and bond beams, were evaluated—these solutions proved successful in preventing the collapse of the building models. More recently, the GCI, with the partnership of other institutions, created the Seismic Retrofitting Project. The aim of this project is to design and test retrofit solutions that can be easily implemented, making use of locally available materials and know-how [46].

Important research on adobe construction has also been conducted in Mexico. The use of synthetic meshes (geogrids) to strengthen adobe walls was investigated at the Autonomous University of Mexico State by conducting in-plane cyclic tests on full-scale adobe walls [47]. In this study, it was concluded that the use of synthetic meshes to strengthen adobe buildings is an adequate, accessible, and compatible solution. At the National Autonomous University of Mexico, walls were tested under cyclic lateral loading, and rural house models (1:2.5 scale), strengthened with different solutions, were tested on a shaking table [9]. All the reinforcement techniques used improved the strength of the models significantly, and the solution that proved most effective was a steel mesh applied to both surfaces of the walls.

In the last 10 years, a research group at the University of Aveiro, in Portugal, has also conducted research on the evaluation of the structural behaviour and safety assessment of existing adobe constructions and on the design of repair and strengthening solutions for these constructions [26, 29, 48–50]. The structural non-linear response of adobe elements has been investigated in a series of full-scale tests, in the laboratory and in situ, with constant vertical load combined with horizontal cyclic displacements. The behaviour of the adobe elements has been evaluated with and without specific strengthening solutions. A repair solution that consists in the injection of hydraulic lime gum into the cracks and a reinforcement solution made with a synthetic mesh applied to the wall surfaces proved very effective in improving the seismic performance of adobe structural elements [29].

The University of the Andes, in Colombia, has also devoted attention to the seismic retrofit of adobe construction. Adobe walls without reinforcement, with steel mesh reinforcement, and with wood reinforcement were tested under vertical loads combined with in-plane cyclic loads or under out-of-plane monotonic loads [51]. Adobe house models (1:5 scale) were tested on the shaking table, without and with reinforcement (steel mesh or wood reinforcement). In these tests, the strengthening solution with wood confining elements showed better seismic performance than the solution with steel mesh.

At the University of Technology, in Sydney, Australia, research for the development of low-cost, low-tech reinforcement solutions for adobe construction has also been carried out [52]. U-shaped adobe walls and a house model (1:2 scale) with different retrofit solutions were tested on a shaking table. The use of a system composed of stiff external vertical reinforcement (such as bamboo), external horizontal reinforcement (such as bamboo or wire), and a timber ring beam led to a great improvement in the structural behaviour of the specimens tested.

The research that has been conducted worldwide demonstrates that the strengthening techniques under consideration have the potential for improving the structural behaviour of adobe constructions by providing better structural continuity and confinement, thus reducing structural instability, which is fundamental under horizontal actions such as those induced by earthquakes. Additional research is needed, however, in order to better investigate possible strengthening techniques, so that these can become fully viable and applicable. New performance enhancement solutions should be developed with the purpose of reducing the seismic vulnerability of adobe structures, thus decreasing the risk associated with historical heritage and potential human losses.

3.3 Seismic Retrofit Solutions

3.3.1 Introduction

A building that, in the past, withstood some minor earthquakes and suffered many little visible cracks may experience unexpected and sudden collapse when subjected to a new earthquake. In fact, the effects earthquakes cause on structures are cumulative and therefore a structure that resisted before may not necessarily withstand a new earthquake without being deeply repaired. After the Third UN World Conference on Disaster Risk Reduction, held in 2015, in Japan, experts and participants concluded that it is wiser to rehabilitate and retrofit buildings before than after a future earthquake, during the mitigation time, to save lives and reduce damages [53]. There are three stages between the occurrences of two earthquakes: emergency, preparation, and mitigation. The last stage is, in fact, the preparation stage for a future earthquake (short, medium, and long term) and it is when the strengthening of structures should be considered, analysed, and applied.

As described in Sect. 2.2, there are several earthquake damage patterns of adobe construction which have to be addressed in different ways and must be taken into consideration in the design of effective strengthening solutions. It is also important to consider that the strengthening of adobe walls should not mainly aim to avoid cracking but rather to avoid the widening of cracks. The objective is to control the movement of the elements separated by fissures and thus prevent the collapse and loss of lives. Seismic forces are extremely strong when compared with the maximum strength of the structure. Therefore, high deformation is accepted with the inevitable cracking on the walls or other structural elements. The key factor is to guarantee a good connection between structural elements and to provide deformation capacity without collapse. Therefore, effective strengthening solutions tend to confine the adobe walls horizontally and vertically, using tension-resistant elements that form a mesh, applied internally or externally, and that can be made using different materials.

In the following subsections, a description of the steps involved in the repair stage of damaged adobe structures and a review of the research on different possible external mesh strengthening solutions to be applied in these structures are presented.

3.3.2 Repair Stage

An adequate assessment of the damage in the building should be made in order to evaluate what repair and strengthening solutions are possible and applicable. If the decision is to repair the adobe construction, the first thing to do must be to repair all the cracks. Widespread cracking is common after an earthquake, although cracking may also appear due to other reasons. Before beginning the repair stage, however, special attention should be paid to unstable walls or roofs. These elements should be shored up so that their fall can be prevented. In the following paragraphs, recommendations regarding the steps necessary to effectively repair damaged walls are provided [54].

In order to better evaluate the cracks, the existing coatings should be removed. There are different approaches to repair, depending on the size of the opening. For crack openings under 10 mm, the crack should be widened until 10 mm and filled with liquid mortar. If the crack opening is more than 20 mm or is accompanied by out-of-plane deformation, the adobe blocks on that area should be removed and the wall area reassembled using adobes in good condition and a new mortar. Cracks with openings between 10 and 20 mm should not be widened but only filled with liquid mortar.

In the case of horizontal cracks with detachment of the walls that lead to rotation of the wall along the crack, if the rotation is under 1:30 it is possible to straighten the wall, provided that there are no other horizontal cracks. Wooden shores should be placed against the walls by means of small boards supported by a strut on the ground. The lower horizontal crack must then be cleaned and injected with liquid mortar with 30% of humidity. The wall is pushed sequentially in the shores by banging on the floor struts, one by one. This way, the boards placed on the wall push gently and uniformly, forcing the wall to recover its original vertical position. If the rotation of the wall is greater than 1:30, the complete reconstruction of the wall should be considered.

In adobe houses, there are sometimes gable walls as a continuation of two opposite walls to facilitate the placement of the roof. As these elements are not braced by the remaining walls, during earthquakes they tend to rotate outwards causing a horizontal crack on their lower side. This is aggravated by the difference of vibration frequency between the gable and lower wall. In order to repair this type of damage, the roofs should be shored and the gable completely removed. A ring beam should be inserted and a new gable should be built above the ring beam. One possible solution for the new gables is to use 2 triangles of double-straw thatch

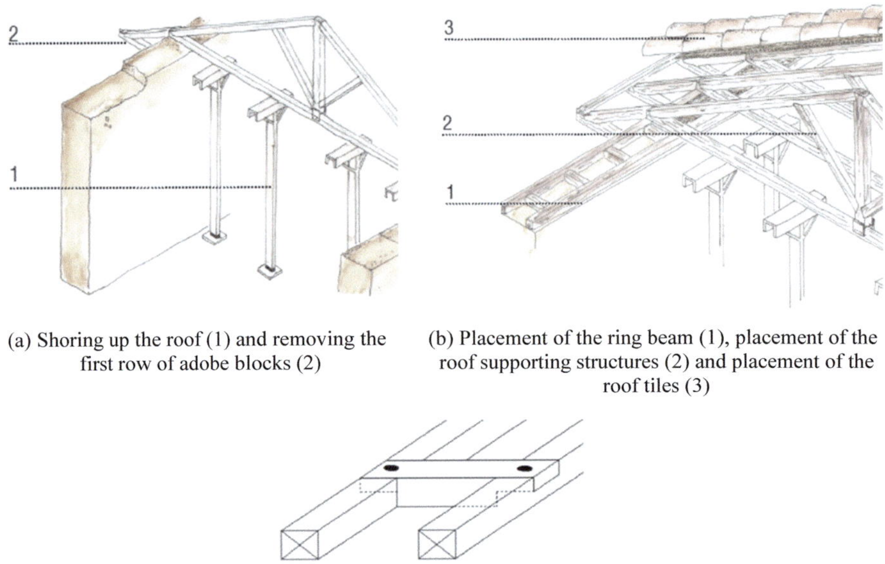

(a) Shoring up the roof (1) and removing the first row of adobe blocks (2)

(b) Placement of the ring beam (1), placement of the roof supporting structures (2) and placement of the roof tiles (3)

(c) Detail of the ring beam and its crosspieces

Fig. 3 Placement of a ring beam on an adobe building [54]

filled with clay (1 volume unit) and straw (6 volume units, for thermal reasons). The remaining cracks should be repaired as described earlier.

In order to ensure an adequate structural performance of the adobe building under seismic loads, the existence of a ring beam is desirable. This element connects the different walls together and leads to a homogenous distribution of loads from the roof to the walls, allowing them to collectively resist the imposed loads. Therefore, in the repair stage, a ring beam should be placed on top of the walls. A possible solution for this beam can be a timber element, consisting of two stringers joined by crosspieces at a distance corresponding to the wall thickness. In the repair stage, the roof should be shored up and the upper row of adobe blocks removed. The ring beam is then placed on the space formed, supported by the walls. Cracks are repaired, the ring beam is filled with an earth mixture similar to that used in the adobes and the roof can be placed adjusting the trusses with nails and ropes (Fig. 3).

3.3.3 Strengthening Stage: External Mesh

The reinforcement of adobe constructions using external compatible mesh—i.e. a grid of interconnected tension-resistant vertical and horizontal elements—has proved effective in many studies conducted in the last decades. In these studies, various types of mesh, using different materials, have been tested. In the following paragraphs, some of the studies carried out and the solutions developed are briefly described.

In the frame of the Getty Seismic Adobe Project, developed in the 1990s and previously presented in Sect. 3.2, many shaking table tests on reduced-scale adobe models were conducted, with the intention of investigating how to effectively strengthen historic adobe buildings [44, 45]. The external retrofit system tested consisted of horizontal and vertical nylon straps that formed a loop around the building or around individual walls. In several of the models tested, these straps were combined with other strengthening solutions, such as wood bond beams, wood diaphragms, steel centre-core rods, and local ties. The retrofitting techniques tested were able to reduce the tendency of the adobe models to collapse. The retrofit solution using vertical straps, in particular, was very effective in reducing the risk of out-of-plane collapse. Even though straps could not prevent the initiation of crack damage, they proved successful in preventing large displacements.

Between 1994 and 1997, at PUCP, different external retrofit methods were evaluated through seismic tests conducted on U-shaped walls and house models [55, 56]. One of the reinforcement solutions studied used non-continuous electro-welded wire grids placed in the corners of the walls and in the area where the walls meet the ring beam, embedded in a mixture of cement and thin sand. Figure 4 shows the damage on an adobe model with this retrofit solution, tested on the shaking table. Although the strengthening solution increased the global strength of the model, the cracks occurred abruptly causing brittle failure of the elements. This type of failure should be avoided as it puts human lives at risk. Thus, it was concluded that this retrofit solution would be effective in the case of a light or moderate earthquake but not in the case of a severe earthquake.

At the beginning of the twenty-first century, a new line of investigation was launched at PUCP, focused on the study of the use of different industrial materials to strengthen adobe buildings [36]. Different reinforcement solutions were tested on double-T shaped adobe walls subjected to cyclic tests [27]. One of the solutions tested with success was an external geosynthetic mesh wrapping the walls and

(a) (b)

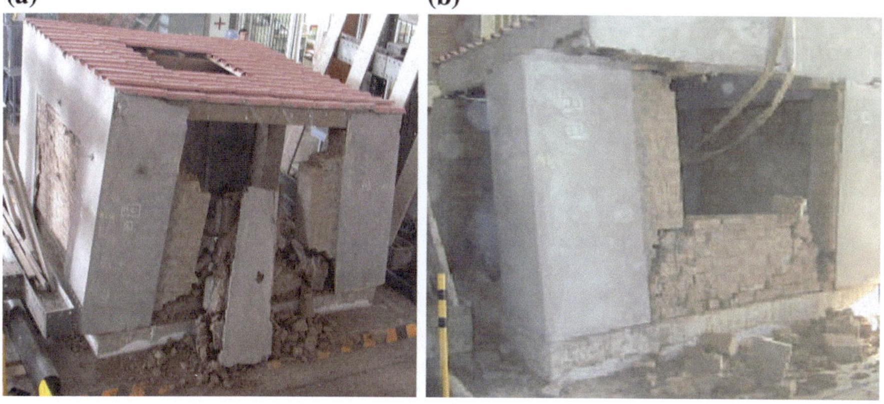

Fig. 4 **a** Sudden collapse of lintel, **b** Brittle collapse of the grids [36]

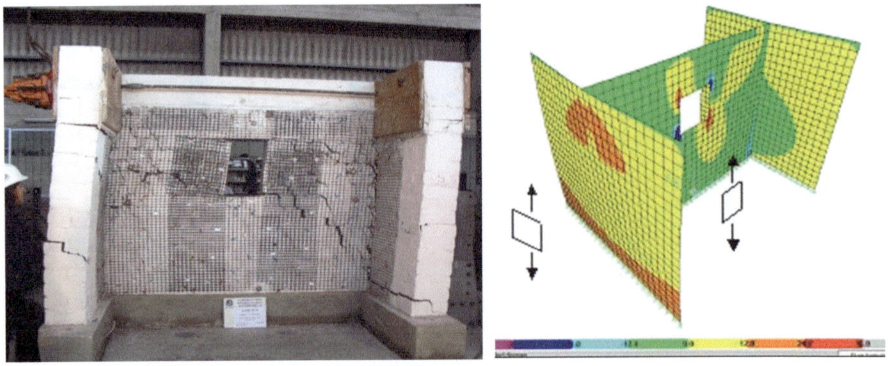

Fig. 5 Cyclic test and finite element model of an adobe wall strengthened with geogrid [27]

embedded in plaster (Fig. 5). In the frame of this project, two shaking table tests on adobe models were also carried out [40]. The first experimental test used a mesh made with natural cane and rope, while in the second test a geogrid was used. In the first model, canes were placed vertically, spaced by 0.40 m, and connected horizontally by ropes spaced 0.30 m apart (Fig. 6a). This grid was placed on each side of the walls, tied together with rope inserted through the walls. This strengthening solution provided better results than a solution using internal cane mesh. In the second model, the geogrid meshes were applied over the whole surface of the walls, both in the exterior and interior of the models (Fig. 6b). The results of the second test showed that the strengthening solution improved the structural behaviour of the model by controlling the displacements in the cracks, with better results than using the geogrid mesh as an internal reinforcement. At a later stage, more full-scale adobe models were tested on the shaking table, using different amounts of plastic mesh reinforcement [30]. In these tests, all the models showed an adequate seismic response.

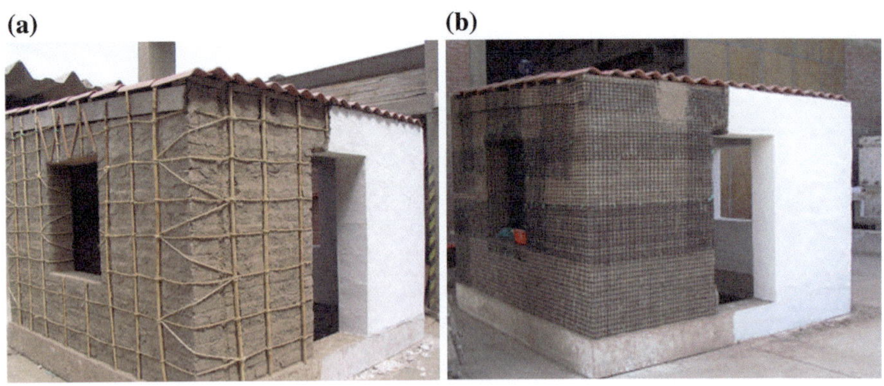

Fig. 6 **a** Strengthening using canes and rope, **b** strengthening using geogrid [40]

In order to provide adequate means and knowledge for populations to rehabilitate their own houses, it is desirable that the strengthening solutions be made with low-cost and readily available materials. Taking this into account, more recently, researchers from PUCP studied the materials available both in rural and urban areas in Peru. Different hardware and construction material stores were visited in order to select the materials to be used. A specific type of synthetic rope was found to be sold throughout the country as halyard, and so it was decided to study this rope in detail to understand its strength properties. Simultaneously, simple and efficient knots to unite the ropes, known by the local populations, were also studied.

For this type of reinforcement, the walls are wrapped with the synthetic ropes forming a mesh with vertical and horizontal elements, which confine the whole wall (Fig. 7). It is recommended that the diameter of these ropes be 5/32″ (≈4 mm). The distance between vertical layers should be less than 40 mm. Horizontal layers should be placed at the middle of the height of adobe blocks, every two rows. At the intersection of the horizontal and vertical layers, a perforation in the wall is made in order to pass the connecting rope—with a diameter of 1/8″ (≈3 mm)—transversally through the wall. The transversal rope is then connected to the inside and outside mesh with simple knots. The knots selected to be used in this type of reinforcement are knots used by fishermen and shepherds, known from the Andean population, and have several advantages: they are simple and fast to tie, consume little material, and fit in any direction.

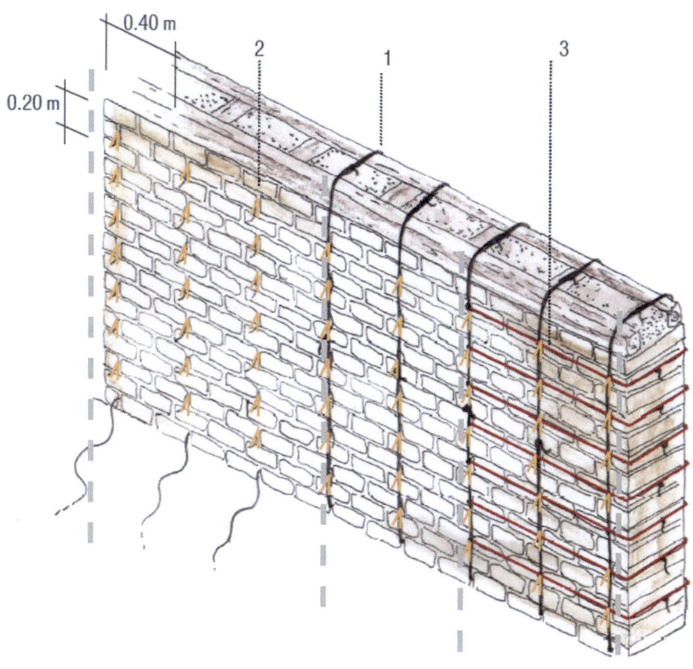

Fig. 7 Strengthening with synthetic ropes (adapted from MVCS [54]): *1* vertical ropes, *2* transversal ropes, *3* horizontal ropes

An adobe model using this kind of rope as reinforcement was constructed and tested [43]. The original model, unreinforced, was initially tested in the shaking table for a moderate earthquake. The module was then repaired using liquid clay injected into the cracks and strengthened with the rope system (Fig. 8). The ropes were applied by making knots around all walls, tied with manual tension (about 150 N–200 N), with a distance of about 0.25 m between layers of rope. The model was then tested again, this time considering a severe earthquake. After the occurrence of cracks, the ropes were able to control in a satisfactory way the movement between the different elements. The wooden ring beam on top of the walls allowed a very efficient connection between walls and roof.

A manual entitled "Fichas para la reparación de viviendas de adobe" was printed by the Ministry of Housing of Peru [54] based on the ongoing research on adobe construction. This manual includes the strengthening solution with synthetic ropes that was just described, developed at PUCP [43].

A research group at the University of Aveiro, in collaboration with other institutions, has also been developing work to study the seismic behaviour and retrofit of adobe construction. With this objective, a full-scale double-T shaped adobe wall was built in the laboratory, using traditional techniques and adobe blocks from a demolition site in the region of Aveiro (as previously described in Sect. 2.3) [29]. An in-plane cyclic test was carried out on the wall, and the damage was then repaired by pressure-injecting hydraulic lime gum into the cracks (Fig. 9a). The original plaster was removed, and a polymer mesh was applied to the surface of the wall. The mesh was fixed to the wall with angle beads and angle profiles in PVC, using highly resistant nylon thread (Fig. 9b). In order to evaluate the efficiency of the strengthening technique, an in-plane cyclic test was performed again, following the same procedure used in the first test (Fig. 9c).

The maximum shear strength of the retrofitted wall was approximately 70.7 kPa, with a corresponding force of 71.8 kN, and the shear strength obtained at a drift of

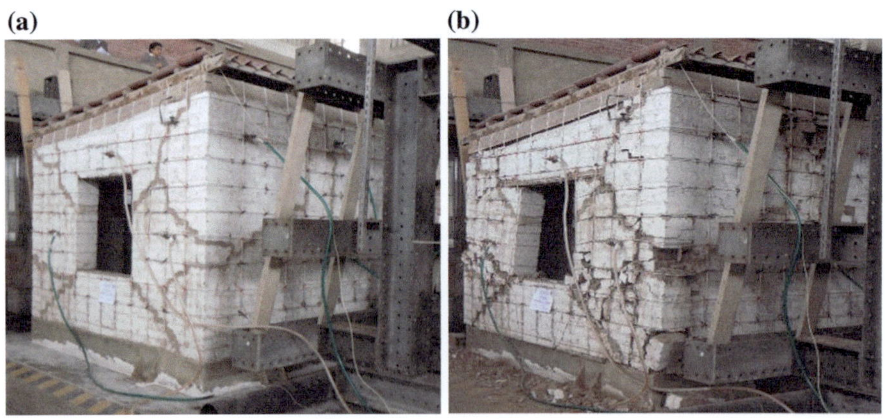

Fig. 8 Adobe model with synthetic rope as strengthening solution, placed with manual knots [43]: **a** before the test, **b** after the test

Fig. 9 Full-scale adobe wall strengthening [29]: **a** repair of cracks, **b** external mesh placement, **c** damage on the wall

1% was approximately 45 kPa (i.e. 70% of the maximum shear strength of the wall). The maximum imposed drift was 1.6%, with a corresponding displacement of 45 mm. After the repair and strengthening, the stiffness of the wall improved, becoming equivalent to that of the original wall (Fig. 10). The shear strength capacity of the wall increased by 23.4% after the retrofit, and the maximum deformation tripled. The fragility of the wall decreased after the peak force was reached, thereby increasing its ductility and energy dissipation capacity. In consecutive cycles, a lower degradation of strength was observed in the retrofitted wall. The inexpensive repair and strengthening solutions used on the wall thus proved to be very effective [28, 29, 57].

With the aim of exploring alternative strengthening techniques with minimal cost, using materials for different uses than the ones primarily assumed, an external strengthening solution with straps from used car tires was developed in a project that involved researchers from the Victoria University of Wellington and the Pontifical Catholic University of Peru [58, 59]. The tire straps were prepared by

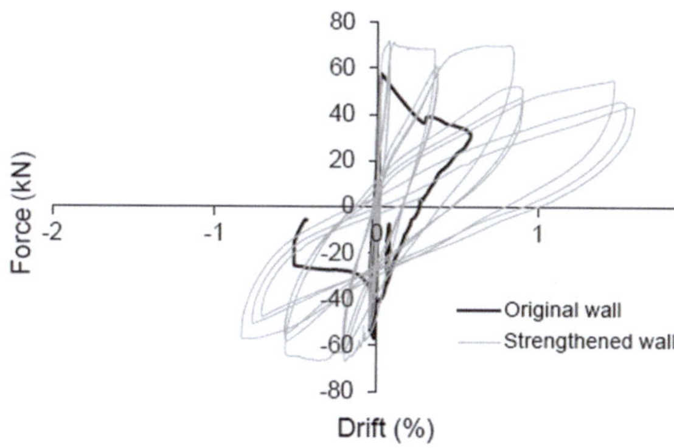

Fig. 10 Comparison between the force-drift curves of the original wall (without reinforcement) and strengthened wall [29]

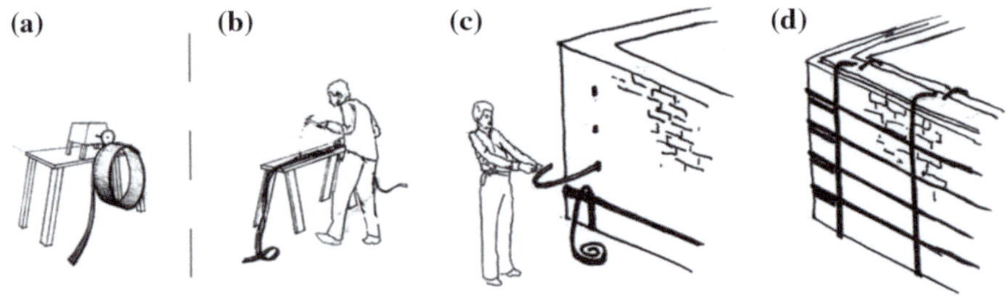

Fig. 11 Preparation and application of tire straps in adobe walls (*Credit* Matthew French)

Fig. 12 Shaking table test on an adobe model strengthened with tire straps [58]

spiral cutting a tire into a continuous strap (Fig. 11). The tire straps were then placed horizontally and vertically on the adobe walls and tied through holes in the walls. An adobe model using this strengthening solution was tested in the shaking table (Fig. 12). The results obtained showed a significant increase in the deformation capacity of the model, without collapse, thus proving the effectiveness of this strengthening solution.

3.4 *Standards and Codes*

There are several standards and codes available for earthen construction. However, few countries have standards or codes officially recognized. Furthermore, many of the existing documents are incomplete and not all of them address the seismic design issue. In general, there are very different typologies of standards and codes. Some are entirely dedicated to only one construction technique. For example, the standard NTE E.080, from Peru, is directed to the seismic design of low-cost adobe construction [60], while in Zimbabwe there is a specific code for rammed earth structures [61].

Only five documents were found that address the seismic design of earthen buildings. These documents, briefly described in the following subsections, are from Peru, Chile, New Zealand, Morocco, and India.

3.4.1 NTE E.080, Peru

The technical standard from Peru [60], applicable only to adobe structures, includes a recommendation for the composition of adobe blocks, indicating a range of percentages of clay, lime, and sand. It estimates the total seismic horizontal force, which depends on the type of foundation soil, use of the construction, seismic coefficient, and total weight of the construction. It also gives indications on how to build foundations, walls, horizontal and vertical shore systems, slabs, ceilings, and seismic reinforcements. The following reinforcement solutions are referred, with specific rules for their application: internal cane mesh, wire mesh, reinforced concrete beams and columns, and geogrid.

3.4.2 NTM 002, Chile

The NTM 002 Chilean standard [62] provides indications regarding structural projects for the alteration, restoration, rehabilitation, renovation, repair, or structural consolidation of earthen constructions. This document is applicable to adobe, rammed earth, daub (*quincha*), and also stone masonry with clay mortar. The rehabilitation recommendations clearly state that, along with the seismic reinforcement, adequate repair of cracks, reconnection of corners, and restitution of verticality of the structure should be conducted. The following strengthening systems compatible with earthen construction are indicated: adobe buttresses, steel connectors for timber, timber structures, steel/synthetic cables or strips, polymer meshes, and steel meshes.

3.4.3 NZS 4299:1998, New Zealand

New Zealand has the most complete set of norms for earthen structures [63–65]. However, in terms of seismic reinforcement solutions, only three systems are described: steel wire mesh, synthetic geogrid with a minimum specific tensile strength, and steel reinforcement bars. The application of reinforcement solutions using mesh and geogrid is described as a horizontal reinforcement in intersections or joints between walls. The steel reinforcement bars system is the reinforcement solution most addressed and thoroughly explained with comprehensive schemes [65].

3.4.4 RPCTerre 2011, Morocco

In 2011, the standard from Morocco [66] introduced important notions of the seismic behaviour of earthen constructions with detailed calculations of the seismic action. This standard is applicable to structures made with adobe, rammed earth, cob, and rubble stone masonry with earth mortar. Regarding reinforcement techniques, vertical and horizontal solutions are presented using different materials such as timber, reinforced concrete, and synthetic meshes.

3.4.5 IS 13827:1993, India

The Indian standard [67] expresses recommendations about the height of earthen buildings and other geometric properties depending on the seismic zone. Regarding the seismic strengthening, the placement of timber ring beams with proper connections at the corners and junctions of walls is recommended, along with the use of diagonal struts at corners. A mesh of bamboo or canes is also referred for vertical reinforcement of walls for the highest seismic zone defined. Strong detailing is given on the construction of earthen buildings with wood or cane structures, indicating diagonal bracing frame systems to include before applying the plaster.

4 Numerical Analysis

Adobe masonry consists of a composite material made of adobe units and mortar joints. As the seismic performance and structural behaviour of adobe construction highly depends on the type of materials used and their properties, a lack of knowledge on the material characteristics and behaviour often compromises an adequate numerical analysis [68]. Adobe masonry has a mostly brittle behaviour, which means that an elastic analysis is only able to provide information related to the areas where cracking initially occurs and not to the cracking development process. In addition, depiction of the non-linear behaviour of adobe masonry is especially complex, as its characterization is difficult to carry out and its properties and behaviour are highly variable [57].

The numerical modelling of masonry may use three main approaches [69]: macro-modelling, simplified micro-modelling, and detailed micro-modelling. Macro-modelling uses an isotropic, homogeneous, and continuous material to represent adobe units, mortar, and unit-mortar interface. In simplified micro-modelling, the units are represented by continuum elements, and the behaviour of the mortar joints and unit-mortar interface is lumped in discontinuous elements. In detailed micro-modelling, units and mortar joints are represented by continuum elements, and the unit-mortar interface is represented by discontinuous elements. As expected, macro-modelling is less time-consuming than the other two types mentioned. However, micro-modelling is a more adequate means to understand and translate the full behaviour of adobe masonry.

The constitutive models to be used are the main concern in numerical analysis [68, 70]. In the study of the behaviour of unreinforced brick masonry, Page [71] assumed elastic bricks and concentrated the non-linear behaviour on joints, with shear and tensile bond failures modes. Lotfi and Shing [72] and Lourenço and Rots [73], focusing on unreinforced masonry structures, in general, proposed other interfaces using plasticity and fracture mechanics. In Tarque et al. [68], three numerical models of the in-plane behaviour of adobe walls were developed, the first one considering nonlinearity at the mortar joints and the other two smeared non-linearity. The results were calibrated with experimental tests and it was possible to observe that the tensile strength controls the global behaviour of the adobe structure and that there is stress distribution when the maximum tensile strength is reached in one element. In Banadaki et al. [70], the interfaces were modelled by a contact with sliding and separation capability in tension. The results of the numerical analysis were compared with those obtained in experimental tests, and a good correspondence between the crack patterns was obtained.

In general, there is still a lack of experimental work on adobe construction that may allow the adequate characterization of the material and the representation of the behaviour of the walls, with and without reinforcement. More research is needed to gather sufficient information to support the development and calibration of reliable models for numerical analysis.

5 Concluding Remarks

Adobe is a traditional construction material that is still used in many parts of the world. Although this material has strong benefits and advantages, adobe masonry has a fragile behaviour and very low tensile strength. Moreover, the construction and rehabilitation of adobe structures is generally made by non-specialized staff, based on empirical knowledge passed through generations. For these reasons, these structures may have important structural problems that need to be prevented, especially in areas with relevant seismicity.

In this chapter, different strengthening techniques, developed from the mid-seventies to the present time, were presented. These techniques focus on improving the structural resistance, stability, and behaviour of adobe constructions. Given that adobe construction is massively used by no-income and low-income families and communities, there are several inexpensive retrofit solutions among the techniques presented and explained. Before electing the type of retrofit to perform, however, it is fundamental to understand the existing problems and issues. Once the adequate type of retrofit solution is selected, improvement of the structural behaviour of an adobe construction is possible and can be easily performed. In fact, simple solutions proved to be extremely effective in increasing the strength, stability, and deformation capacity of structures, and thus the adequate implementation of these solutions can save lives and prevent important social, cultural, and economic losses.

Acknowledgements This research was financially supported by Project POCI-01-0145-FEDER-007457—CONSTRUCT—Institute of R&D in Structures and Construction, funded by FEDER funds through COMPETE2020—Programa Operacional Competitividade e Internacionalização (POCI) and by national funds through Fundação para a Ciência e a Tecnologia (FCT). This work was also carried out within the framework of the project POCI-01-0145-FEDER-016737 (PTDC/ECM-EST/2777/2014), financed by FEDER funds through Programa Operacional Factores de Competitividade (COMPETE) and by national funds through Fundação para a Ciência e a Tecnologia (FCT). The funding provided is kindly acknowledged.

The authors would also like to acknowledge the contribution of:

University of Aveiro;

Laboratory of Earthquake and Structural Engineering (LESE), Faculty of Engineering of the University of Porto;

Pontifical Catholic University of Peru.

References

1. Gandreau D, Delboy L. World heritage inventory of earthen architecture—2012 WHEAP, World Heritage Earthen Architecture Programme. Grenoble: CRAterre ENSAG; 2012.
2. Houben H, Guillaud H. Earth construction: a comprehensive guide. London: ITDG Publishing; 1994.
3. Minke G. Building with earth: design and technology of a sustainable architecture. Basel: Birkhäuser-Publishers for Architecture; 2006.
4. Correia M, Dipasquale L, Mecca S. Terra Europae: earthen architecture in the European Union. Pisa: Edizioni ETS; 2011.
5. Rigassi V. Compressed earth blocks: manual of production, vol. 1. Vilefontaine: CRATerre-EAG; 1995.
6. Tolles E, Krawinkler H. Seismic studies on small-scale models of adobe houses. Report No. 91. Stanford, California; 1990.
7. Delgado M, Guerrero I. The selection of soils for unstabilised earth buildings: a normative review. J Constr Build Mater. 2007;21:237–51.
8. UNESCO. World heritage list. Paris: United Nations Educational Scientific and Cultural Organization (UNESCO); 2016. http://whc.unesco.org/en/list/.
9. Meli R. Experiencias en México sobre reducción de vulnerabilidad sísmica de construcciones de adobe. SismoAdobe2005: international seminar on architecture, construction and conservation of earthen buildings in seismic areas, Lima, Peru; 2005.
10. Fratini F, Pecchioni E, Rovero L, Tonietti U. The earth in the architecture of the historical centre of Lamezia Terme (Italy): characterization for restoration. Appl Clay Sci. 2011;53 (3):509–16.
11. Silveira D, Varum H, Costa A, Martins T, Pereira H, Almeida J. Mechanical properties of adobe bricks in ancient constructions. Constr Build Mater. 2012;28(1):36–44.
12. Silveira D, Varum H, Costa A. Influence of the testing procedures in the mechanical characterization of adobe bricks. Constr Build Mater. 2013;40:719–28.
13. Ottazzi G, Vargas J. Investigación comparativa sobre la resistencia del adobe. Proyecto Cooperativo Universidad Nacional Autónoma de México (UNAM) - Pontificia Universidad Católica del Perú (PUCP). IV Congreso Nacional de Ingeniería Civil, Chiclayo, Peru; 1982.

14. Vargas J, Bariola J, Blondet M. Seismic strength of adobe masonry, summary of a research project sponsored by the US Agency for International Development (AID). Peru: Lima; 1984.
15. Vargas J, Bariola J, Blondet M, Mehta PK. Seismic strength of adobe masonry. Mater Struct. 1986;19(112):253–8.
16. Vargas J, Blondet M. Repairing structural cracks in historic earthen building located in seismic areas. Final report. Agreement PUCP-Getty Conservation Institute; 2007.
17. Webster F, Tolles E. Earthquake damage to historic and older adobe buildings during the 1994 Northridge, California Earthquake. In: 12th world conference on earthquake engineering, Auckland, New Zealand; 2000.
18. Memari A, Kauffman A. Review of existing seismic retrofit methodologies for adobe dwellings and introduction of a new concept. In: SismoAdobe2005: international seminar on architecture, construction and conservation of earthen buildings in seismic areas, Lima, Peru; 2005.
19. JSCE, Earthquake Engineering Committee. The January 13, 2001 off the coast of El Salvador earthquake. Investigation of damage to civil engineering structures, buildings and dwellings. Tokyo: Japan Society of Civil Engineers (JSCE); 2001.
20. Mahdi T. Behavior of adobe buildings in the 2003 Bam earthquake. In: SismoAdobe2005: international seminar on architecture, construction and conservation of earthen buildings in seismic areas, Lima, Peru; 2005.
21. Blondet M. Behavior of earthen buildings during the Pisco earthquake of August 15, 2007. Oakland: Earthquake Engineering Research Institute (EERI); 2008.
22. Elnashai A, Gencturk B, Kwon O-S, Al-Qadi I, Hashash Y, Roesler J, Kim S, Jeong S-H, Dukes J, Valdivia A. The Maule (Chile) earthquake of February 27, 2010: consequence assessment and case studies. MAE Center report no. 10-04. Urbana: Mid-America Earthquake Center; 2010.
23. Gautam D, Rodrigues H, Bhetwal KK, Neupane P, Sanada Y. Common structural and construction deficiencies of Nepalese buildings. Innov Infrastruct Solut. 2016;1(1):1.
24. Blondet M, Vargas J. Investigación sobre vivienda rural, Convenio con el Ministerio de Vivienda y Construcción, Lima, Peru; 1978.
25. Vargas J, Ottazzi G. Investigaciones en adobe, Publicación DI-81-01. Lima: Pontifical Catholic University of Peru; 1981.
26. Silveira D, Varum H, Costa A, Carvalho J. Mechanical properties and behavior of traditional adobe wall panels of the Aveiro district. J Mater Civ Eng. 2015;27(9):04014253.
27. Blondet M, Torrealva D, Villa García G, Ginocchio F, Madueño I. Using industrial materials for the construction of safe adobe houses in seismic areas. In: EarthBuild2005: international earth building conference, Sydney, Australia; 2005.
28. Oliveira C, Varum H, Figueiredo A, Silveira D, Costa A. Experimental tests for seismic assessment and strengthening of adobe structures. In: 14ECEE: 14th European conference on earthquake engineering, Ohrid, Republic of Macedonia; 2010.
29. Figueiredo A, Varum H, Costa A, Silveira D, Oliveira C. Seismic retrofitting solution of an adobe masonry wall. Mater Struct. 2013;46(1–2):203–19.
30. Blondet M, Vargas J, Velasquez J, Tarque N. Experimental study of synthetic mesh reinforcement of historical adobe buildings. In: Structural analysis of historical constructions, New Delhi, India; 2006. p. 715–22.
31. Yamín L, Rodríguez A, Fonseca L, Reyes J, Phillips C. Comportamiento sismico y alternativas de rehabilitación de edificaciones en adobe y tapia pisada con base en modelos a escala reducida ensayados en mesa vibratoria. Revista de Ingeniería. 2003;18:175–90.
32. Dowling D, Samali B, Li J. An improved means of reinforcing adobe walls—external vertical reinforcement. In: SismoAdobe2005: international seminar on architecture, construction and conservation of earthen buildings in seismic areas, Lima, Peru; 2005.
33. Zavala C, Igarashi L. Propuesta de reforzamiento para muros de adobe. In: SismoAdobe2005: international seminar on architecture, construction and conservation of earthen buildings in seismic areas, Lima, Peru; 2005.

34. Wu F, Li G, Li H-N, Jia J-Q. Strength and stress–strain characteristics of traditional adobe block and masonry. Mater Struct. 2013;46(9):1449–57.
35. Agüero J, Cerón J, Gonzáles JC, Méndez MT. 2015. Análisis estructural de dos muros de adobe con diferente sistema de aparejo. 15° SIACOT: Seminario Iberoamericano de Arquitectura y Construcción con Tierra 2015, Cuenca, Ecuador.
36. Vargas J, Blondet M, Ginocchio F, Villa García G. 35 Años de investigaciones en sismo adobe: la tierra armada. Lima: Pontifical Catholic University of Peru; 2005.
37. Blondet M, Vargas J, Tarque N, Iwaki C. Construcción sismorresistente en tierra: la gran experiencia contemporánea de la Pontificia Universidad Católica del Perú. Informes de la Construcción. 2011;63(523):41–50.
38. Corazao M, Blondet M. Estudio experimental del comportamiento estructural de las construcciones de adobe frente a solicitaciones sismicas. Lima: Banco Peruano de los Constructores; 1973.
39. Blondet M, Torrealva D, Villa García G, Ginocchio F, Tarque N, Velásquez J. Refuerzo de construcciones de adobe con elementos producidos industrialmente: ensayos de simulación sísmica. Proyecto DAI 3088 - Informe Final. Lima: Pontifical Catholic University of Peru; 2006.
40. Torrealva D, Acero J. Reinforcing adobe buildings with exterior compatible mesh. The final solution against the seismic vulnerability? In: SismoAdobe2005: international seminar on architecture, construction and conservation of earthen buildings in seismic areas, Lima, Peru; 2005.
41. Torrealva D, Vargas J, Blondet M. Earthquake resistant design criteria and testing of adobe buildings at Pontificia Universidad Católica del Perú. In: Proceedings of the Getty Seismic Adobe Project 2006 Colloquium. Los Angeles: The Getty Conservation Institute; 2009. p. 3–10.
42. Blondet M, Vargas J, Sosa C, Soto J. Using mud injection and an external rope mesh to reinforce historical earthen buildings located in seismic areas. In: SAHC2014: 9th international conference on structural analysis of historical constructions, Mexico City, Mexico; 2014.
43. Blondet M, Vargas J, Tarque N, Soto J, Sosa C, Sarmiento J. Refuerzo sísmico de mallas de sogas sintéticas para construcciones de adobe. 15º SIACOT: Seminario Iberoamericano de Arquitectura y Construcción con Tierra, Cuenca, Ecuador; 2015.
44. Tolles E, Kimbro E, Webster F, Ginell W. Seismic stabilization of historic adobe structures. Final report of the Getty Seismic Adobe Project. Los Angeles: The Getty Conservation Institute; 2000.
45. Tolles E. Getty Seismic Adobe Project research and testing program. In: Proceedings of the Getty Seismic Adobe Project 2006 colloquium. Los Angeles: The Getty Conservation Institute; 2009. p. 34–41.
46. Cancino C, Macdonald S, Lardinois S, D'Ayala D, Fonseca C, Torrealva D, Vicente E. The Seismic Retrofitting Project: methodology for seismic retrofitting of historic earthen sites after the 2007 earthquake. In: Terra 2012: 11th international conference on the study and conservation of earthen architecture heritage, Lima, Peru; 2012.
47. Vera R, Miranda S. Reparación de muros de adobe con el uso de mallas sintéticas. In: SismoAdobe2005: international seminar on architecture, construction and conservation of earthen buildings in seismic Areas, Lima, Peru; 2005.
48. Costa A, Varum H, Pereira H, Rodrigues H, Vicente R, Arêde A, Costa A. Avaliação experimental do comportamento fora do plano de paredes de alvenaria de adobe. In: 5th seminar of earth architecture in Portugal, Aveiro, Portugal; 2007.
49. Varum H, Costa A, Silveira D, Pereira H, Almeida J, Martins T. Structural behaviour assessment and material characterization of traditional adobe constructions. In: AdobeUSA 2007: 4th international adobe conference of the Adobe Association of the Southwest, El Rito, New Mexico; 2007. p. 138–45.
50. Varum H, Figueiredo A, Silveira D, Martins T, Costa A. Outputs from the research developed at the University of Aveiro regarding the mechanical characterization of existing adobe constructions in Portugal. Informes de la Construcción. 2011;63(523):127–42.
51. Yamín L, Phillips C, Reyes J, Ruiz D. Estudios de vulnerabilidad sísmica, rehabilitación y refuerzo de casas en adobe y tapia pisada. Apuntes. 2007;20(2):286–303.

52. Dowling D, Samali B. Low-cost and low-tech reinforcement systems for improved earthquake resistance of mud brick buildings. In: Proceedings of the Getty Seismic Adobe Project 2006 colloquium. Los Angeles: The Getty Conservation Institute; 2009. p. 23–33.
53. United Nations: Sendai framework for disaster risk reduction 2015–2030. United Nations International Strategy for Disaster Reduction; 2015. p. 1–25.
54. MVCS. Fichas para la reparación de Viviendas de Adobe. Lima: Ministerio de Vivienda, Construcción y Saneamiento del Perú (MVCS); 2014.
55. Zegarra L, Quiun D, San Bartolomé A, Giesecke A. Reforzamiento de viviendas de adobe existentes. Primera parte: Ensayos sísmicos de muros U. In: XI national congress of civil engineering, Lima, Peru; 1997.
56. Zegarra L, Quiun D, San Bartolomé A, Giesecke A. Reforzamiento de viviendas de adobe existentes. Segunda parte: Ensayos sísmicos de módulos. In: XI national congress of civil engineering, Lima, Peru; 1997.
57. Varum H, Tarque N, Silveira D, Camata G, Lobo B, Blondet M, Figueiredo A, Rafi M, Oliveira C, Costa A. Structural behaviour and retrofitting of adobe masonry buildings. In: Costa A, Guedes JM, Varum H. editors. Structural rehabilitation of old buildings—building pathology and rehabilitation, vol. 2. Berlin: Springer; 2014. p. 37–75.
58. Charleson A. Seismic strengthening of earthen houses using straps cut from used car tires: a construction guide. Oakland: Earthquake Engineering Research Institute (EERI); 2011.
59. Charleson A, Blondet M. Seismic reinforcement for adobe houses with straps from used car tires. Earthquaque Spectra. 2012;28(2):511–30.
60. SENCICO: Norma Técnica Edificación NTE E 0.80 Adobe. Reglamento Nacional de Construcciones. Lima: Servicio Nacional de Capacitación para la Industria de la Construcción; 2000.
61. SAZ. Standards Association Zimbabwe Standard 724:2001: standard code of practice for rammed earth structures. Harare: Standards Association of Zimbabwe; 2001.
62. MINVU. Proyecto de intervención estructural de construcciones de tierra - NTM 002. Santiago: Ministerio de Vivienda y Urbanismo; 2013.
63. SNZ. NZS 4297:1998 Engineering design of earth buildings. Wellington: Standards New Zealand (SNZ); 1998.
64. SNZ. NZS 4298:1998 materials and workmanship for earth buildings. Wellington: Standards New Zealand (SNZ); 1998.
65. SNZ. NZS 4299:1998 earth buildings not requiring specific design. Wellington: Standards New Zealand (SNZ); 1998.
66. MET, MHUPV: Reglement Parasismique des Constructions en Terre - RPCTerre 2011. Kingdom of Morocco: Ministère de l'équipement et du transport and Ministère de l'habitat, de l'urbanisme, et de la politique de la ville; 2011.
67. BIS. IS 13827:1993—Indian Standard: improving earthquake resistance of earthen buildings —guidelines. New Delhi: Bureau of Indian Standards; 1993.
68. Tarque N, Camata G, Espacone E, Varum H, Blondet M. Numerical modelling of in-plane behaviour of adobe walls. In: Sismica 2010: 8th national conference on seismology and earthquake engineering, Aveiro, Portugal; 2010.
69. Lourenço P. Analysis of masonry structures with interface elements: theory and applications. Delft: Delft University of Technology; 1994.
70. Banadaki AD, Ghannad MA, Bakhshi A, Lotfalipour M. A Numerical study on seismic behavior of a typical rural house of Iran. In: 14WCEE: 14th world conference on earthquake engineering, Beijing, China; 2008.
71. Page AW. Finite element model for masonry. J Struct Eng. 1978;104(8):1267–85.
72. Lotfi HR, Shing PB. Interface model applied to fracture of masonry structures. J Struct Eng. 1994;120(1):63–80.
73. Lourenço PB, Rots JG. Multisurface interface model for analysis of masonry structures. J Eng Mech. 1997;123(7):660–8.

Repair and Strengthening of Traditional Timber Roof and Floor Structures

Jorge M. Branco, Thierry Descamps and Eleftheria Tsakanika

1 Introduction

Nowadays, the increased sensitivity towards the conservation of cultural heritage leads to the adoption of restoration techniques which guarantee as much as possible the preservation of the building authenticity and integrity, the conservation of the materials, the construction technology, the original structural system, the minimal interventions, their reversibility and compatibility with the existing parts of the buildings [1–4]. Current knowledge assumes the need to preserve and to protect as much as possible of the authentic material (e.g. minimization of replacements of timbers) using either simple techniques or more precise and sophisticated ones. Moreover the original and authentic structural systems, must be protected and preserved too, even if they will not be visible after the restoration works, as a cultural value with important advantages for the overall behaviour of the building especially in seismic areas.

In many cases, all over the world, traditional buildings involve timber structures at least as timber floors and roof systems. Damage in these structures can have different sources:

J.M. Branco (✉)
ISISE, Civil Engineering Department, School of Engineering,
University of Minho, 4800-058 Guimarães, Portugal
e-mail: jbranco@civil.uminho.pt

T. Descamps
URBAINE, Department of Mechanical and Civil Engineering,
University of Mons, Mons, Belgium
e-mail: thierry.descamps@umons.ac.be

E. Tsakanika
National Technical University of Athens, Athens, Greece
e-mail: etsakanika@arch.ntua.gr

A. Costa et al. (eds.), *Strengthening and Retrofitting of Existing Structures*,
Building Pathology and Rehabilitation 9, https://doi.org/10.1007/978-981-10-5858-5_5

- natural defects of wood;
- biological degradation;
- fire;
- environmental and atmospheric agents in particular, transient moisture content;
- original design, execution, maintenance or intervention errors;
- excessive loading.

Before any intervention, the first step is the assessment of the existing timber structure according to the materials, the elements, the joints (strength and stiffness) and its overall behaviour. Proper assessment of the material (risk of decay or insect attack) with appropriate techniques is obviously of major importance and therefore the study of recent state-of-the-art concerning the diagnostic procedures, is highly recommended [5, 6]. Techniques are in profusion. They provide huge support for the structural understanding of the whole load bearing system, however their use and choice must be discussed among all scientists involved in the assessment process to ensure the right agreement between the aim and the diagnostic or intervention techniques that will be used. This assessment may lead to the replacement of a portion or the whole member. On the other hand, in the case that the member or joint is kept in service and reinforcement is needed, an accurate assessment of the state of conservation of the structural elements (timber or metal ones) is crucial too.

Many retrofitting techniques have been developed and several of them have been reported in relevant manuals, books and scientific papers. However, while some applications were largely used and taught, others have been slowly left aside, with the risk of forgetting them and losing an important and valuable part of our heritage concerning the construction technology. For instance, the retrofitting of historical timber elements with prestressed systems has been performed several times in the past, but still there are many uncertainties and difficulties that discourage new applications and limit their development. For new constructions or for restoration projects, the post-tensioning with steel elements has been proposed since the beginning of the 19th century [7] as a specific way of applying the prestresses. Some inspiring cases, realized on historical timber trusses by Italian conservators, have already shown the potential of this strengthening method, as the roof structure of the theatre of Sarteano [8], or the one of Savona [9]. Regarding existing timber single elements several studies on post-tensioning methods have looked for the optimal layout to improve the load-bearing capacity of these timber elements. Different configurations are still under analysis and the application are very broad from the strengthening of simple short-span beams [10, 11], to the retrofitting of wooden bridges [12].

Old timber floors as timber roofs suffer mainly from decay problems at their support areas (timber parts of the beams embedded in the external walls) and from excessive in-plane bending deflections since they are usually one-span beams designed to bear moderate loads, compared to current uses or a new use with heavy load requirements. The structural refurbishment of traditional/historic timber floors can be achieved, in order to increase the bending stiffness and strength of the main elements, by using steel or FRP plates or bars or by using and activating as structural members other elements, such as a concrete slab, timber planks or timber wood-based panels (plywood). The structural behaviour of the resulting timber composite structure is governed by the strength and stiffness of the mechanical fasteners that connect the existing timber beams to the new elements. Another important aspect to be keenly considered is the timber floor diaphragm effect, which may affect the structural performance of a traditional masonry building subjected to lateral seismic loads: the common configuration of existing timber floors or ceilings with a crossly arranged single layer of wooden planks or the use of plywood panels consist a common solution for an in plane shear strengthening, in order to ensure a redistribution of lateral seismic loads and an efficient connection of the load bearing walls improving the seismic performance of the whole building (box-behaviour) [13–16].

It must be highlighted that this paper will focus mainly on reinforcement methods of traditional roof and floor structures and not on intervention techniques used for replacing timber elements that suffer from biological attack.

2 Timber Roofs

Three major pathologies may affect timber roofs: first decay problems usually at the support areas (see timber floors), damage or a lack of strength or stiffness of joints, damage or a lack of strength or stiffness of a single element or of the whole timber structure.

2.1 *Timber Joints*

It is important to be mentioned the huge amount of different types of joints that exist, a testimony of the diversity and richness of the timber cultural heritage and in parallel, the difficulty of studying, repairing and reinforcing them. Timber joints depend on the structural system that support, on the limited lengths of the lumber, the possibilities and requirements for transportation and erection, the local traditions and the state of technology and craftsmanship of each period and area.

In the past, the actions taken by carpenters to strengthen joints were based on experience and precise observations of failure modes encountered in real structures and a good understanding of their behaviour and their weakest points. In some cases this led to an improvement in the designing of the joints and one can say that many carpentry joints are an evolution or a "reinforcement" of older primary joints. For example, a notched joint with a tenon can be considered as a "reinforcement" or an improvement of a tenon joint because the notch increases the load bearing capacity of the joint (Fig. 1).

In the past, joints were not designed to transfer loads through shear, with metal fasteners such as nails, screws or bolts. Their ability to carry the loads was achieved

Fig. 1 Notched joints with tenons

through friction and mainly the direct contact of special cuttings and notches formed usually at the end of the connected members. The few metal fasteners that were used in parallel with carpentry joints, were ensuring the good fitting and contact of the members at the area of the connection (see Fig. 8). These types of joints are called carpentry joints. The last years, various new reinforcement techniques such as the use of screws (including self-tapping-screws), metal elements (plates, strips, stirrups), glued composites (glass or carbon fibres, weft knitted textiles) and glued-in rods and bars, or even full injection with fluid adhesives among others have been proposed.

Feio et al. [17] have tested full-scale notched and skewed tenon joints under compression in order to assess the local failure in compression at an angle to the grain and the slipping of the joint. Failure modes observed in the tested joints are damages due to compression in the brace which are localized at the tenon end or distributed along the full contact length. An out-of-plane bulging of wood under the contact length was observed. In some cases, damages in compression associated with shear failure were observed too (Fig. 2).

When observed on-site, this type of failure mode mainly highlights a poor design of the joint (too small contact areas) or unexpected compression forces in one element. No reinforcement can repair a damage in compression perpendicular or at an angle to the grain and usually the replacement of a part or the whole elements is required.

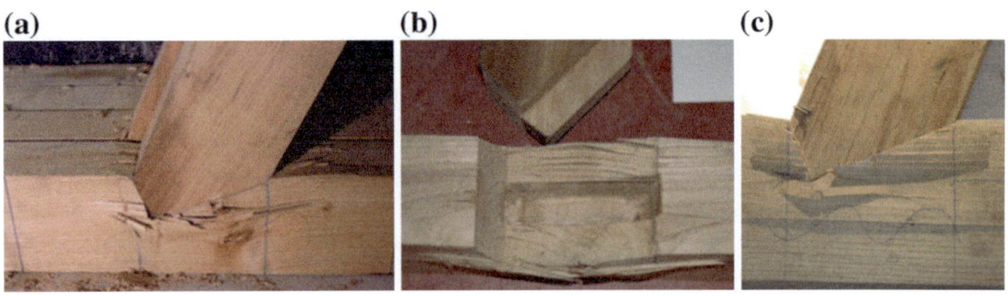

Fig. 2 Typical experimental failure patterns of notched and skewed tenon joints under compression: **a** joint collapsed in compression, with uniform distribution of damage, **b** joint collapsed in compression, with out-of-plane bulging, **c** combined failure in compression and shear parallel to the grain at the tenon [17]

Whatever the joint is, in order to ensure the transmission of the loads from one member to the other and the required strength and stiffness, it is important to keep all the surfaces of the joints in close contact. In the case of reverse loads, uplift, poor construction (Fig. 3c), or high shrinkage of the wood elements, joints may develop gaps between connected elements, reducing the contact areas. One traditional reinforcement technique consists in placing a wooden wedge or a piece of timber to ensure perfect contact between the tenon and the mortise or the timber elements (Fig. 3). This wooden wedge should be made of hardwood (for strength and stiffness) and its moisture content (MC) should be as close as possible to that of the reinforced wooden elements in order to avoid any shrinkage of the wedge.

Pinned tenon joints also have a very low bearing capacity in tension since only the wooden pin acts. If it is possible (e.g. replacement of the member), a traditional technique can be used which consists in fashioning the new joint with a dovetail tenon increasing the strength in tension. If the element remains in place and in service, a binding metal strip may be used as reinforcement in tension. In order to avoid cracks, the strip can be screwed or bolted under the supporting beam and the spacing, the distances from the end and the edges of the connected timbers must be respected (Fig. 4).

Fig. 3 **a** Tenon and mortise joint: additional wooden wedge used to ensure a tight contact between the tenon and the mortise, **b** additional wooden piece to ensure the contact of the timber elements, **c** proposal for the reinforcement of a poorly constructed heel joint. Addition of a new wedge and screws

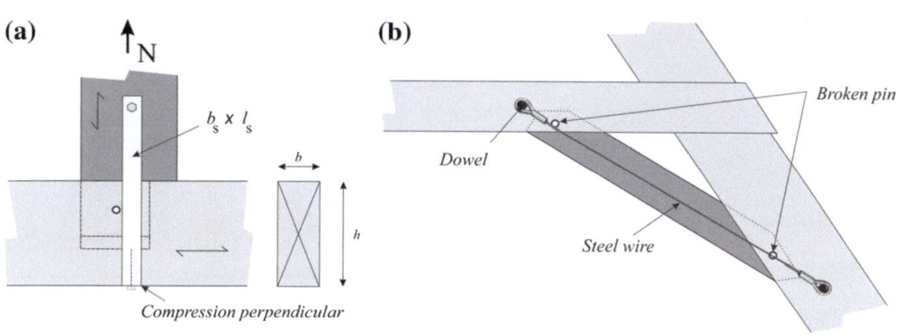

Fig. 4 Reinforcement of a tenon joint in tension adding metal elements

The strengthening of existing notched joints, the most common connection in traditional roofing systems, mainly aims to avoid shear failure in the front portion of the notch. Most of the time, an end beam repair is required too because of decay and a wooden prosthesis must be used to replace the degraded material (Fig. 5).

The prosthesis can be made using timber elements (solid or glued-laminated timber) [15, 18] (Figs. 5, 6 and 7), steel elements or resin mechanically jointed to sound wood with steel fasteners or different types of glued-in rods (see timber floors).

As mentioned before, in past times, binding strips, stirrups, nails and bolts were used to ensure the contact of the connected timber members and to avoid their dismantling under reverse loads in severe wind or seismic events (Fig. 8). The intervention concerning the metal elements that were used in the original construction of the joints, or, added later, usually included the substitution of the old type metal connectors (gypsy nails, binding strips, etc.) by new ones (screws, bolts, new type strips etc.) and either the treatment of the original metal or the use of a new one.

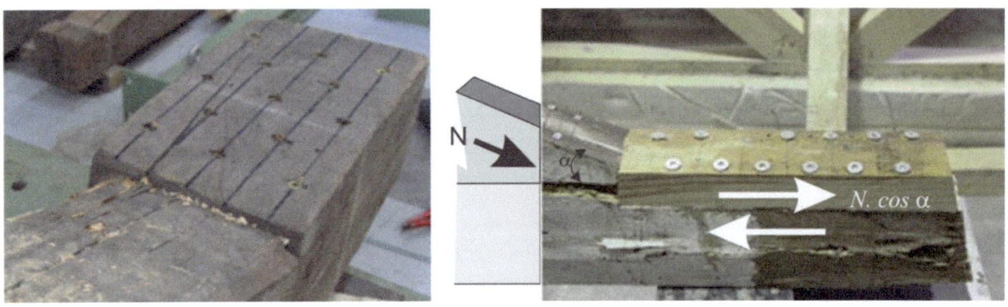

Fig. 5 Notched joints (connection of the rafter to the tie-beam) reinforced to shear stresses in the frontal part of the notch with a screwed prosthesis

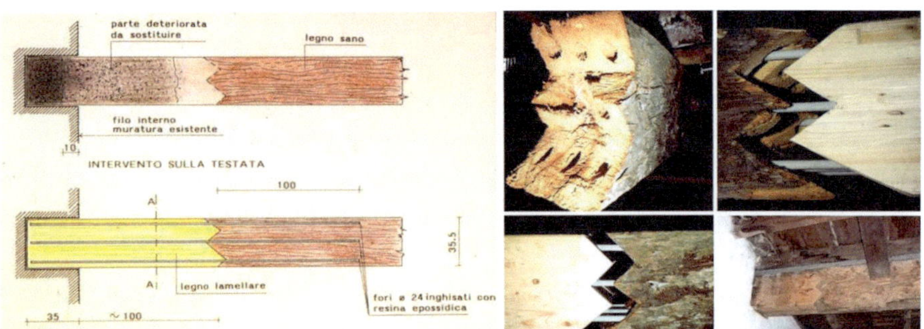

Fig. 6 Intervention on the historic roof of Valentino Castle (17th century), Torino. Italy. Prosthesis at the end of the tie-beams using glued laminated timber connected to the sound wood by fiber-glass rods [19]

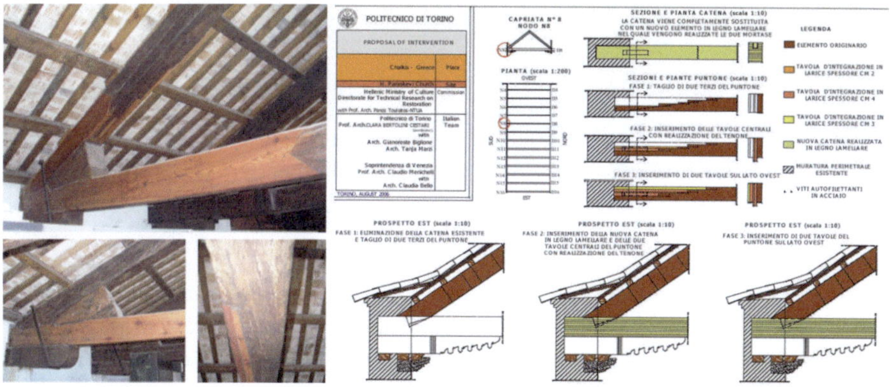

Fig. 7 A prosthesis proposal built up from timber boards (thickness 2 or 4 cm), of the same species as the sound wood, connected to each other and to the sound timber with a mechanical system (self-threading stainless steel screws) and a bi-components resin [15, 18]

Fig. 8 Examples of original metal elements used in notched joints

The strengthening techniques used nowadays try to reproduce the old techniques even when using new metal plates, strips and fasteners like screws and self-tapping screws (Fig. 9). These kinds of interventions can affect the stiffness of the joint, and they should be checked too and designed accordingly.

Strengthened joints with metal devices were tested by Branco et al. under monotonic and cyclic loading [20]. The purpose was to uncover any advantages and drawbacks in the behaviour of the joint and of the strengthening as well as to look at different types of strengthening. The four types of strengthened joints tested are modern implementations of traditional techniques. All the tests conducted have concluded that all the strengthening techniques are efficient and the metal devices

Fig. 9 Contemporary strengthening interventions on notched joints reproducing old techniques

carry a part of the loads too, improving the load-carrying capacity of the joint. However, the improvement in terms of strength and stiffness may varies substantially for different technics. The least efficient regarding both maximum force and stiffness is the solution with the external tension ties (Fig. 10).

In dovetail-lap joints loaded in tension, the splitting of timber is a common failure mode. The traditional reinforcement of those joints consists in adding fasteners (bolts, nails, screws, etc.) restoring the shear mechanism provided by the pin (Fig. 11). The design of this strengthening technique is based on the calculation of the shear resistance of the new fasteners. This intervention affects the stiffness of the joint (displacement of the centre of rotation). Binding strips or steel wire may also be used.

The easiest way to reinforce a scarf joint in tension or realize a prosthesis connecting the new member with the sound old part (Fig. 12) can be achieved by adding metal fasteners (screws or bolts). On the contrary, the compression forces must be transferred by the contact areas of the timber elements. The addition of metal fasteners must secure the contact of these timbers. For the same reason, in

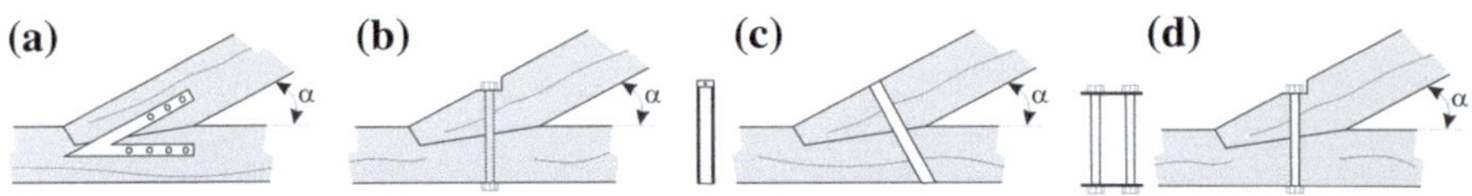

Fig. 10 Traditional strengthening techniques of notched joints: **a** metal stirrups, **b** internal bolt, **c** binding strip, **d** external tension ties

Fig. 11 Traditional reinforcement of dovetail-lap joints under tension loads by adding wooden dowels

previous historic periods, wooden pegs were used. In case of high loads, lateral metal, timber or wood-based plates can be added to improve the load-bearing capacity of the joint and to increase the stiffness. Both types of reinforcements are used in restoration works (Fig. 12). In both cases the distances between the fasteners and between the fasteners and the ends or edges of the timbers must be followed according to relevant rules usually found in modern standards. Otherwise failures may occur.

Under bending, the rule of thumb that the weak point is the risk of premature splitting of wood is encountered here too (joints cut with right angles are less suitable due to the concentration of stresses at the corner of the cuttings). Self-tapping screws can also be used to strengthen splittings at the area of a scarf joint (Fig. 13a). From this point of view, scarf joints are better than halved-scarf joints. Under tension only, reinforcement screws can be driven only in the overlapping area (Fig. 13d). This reinforcement can be checked using Johansen's equations assuming that the tensile load is completely carried by the screws.

In the case of the Trait-de-Jupiter joint it is common to add metal connectors passing through the joint depth to reinforce the joint (Fig. 14d). Another solution with glued in rods is presented in Fig. 14.

It is important to be mentioned that the reinforcement method depends on the loading condition of the joint and the type of stresses that need to be transferred to the connected members (tension, compression, shear, or/and bending).

Fig. 12 Scarf joints reinforced with metal fasteners and plates. Prosthesis with scarf joint and bolts realized at the heel joint of a roof

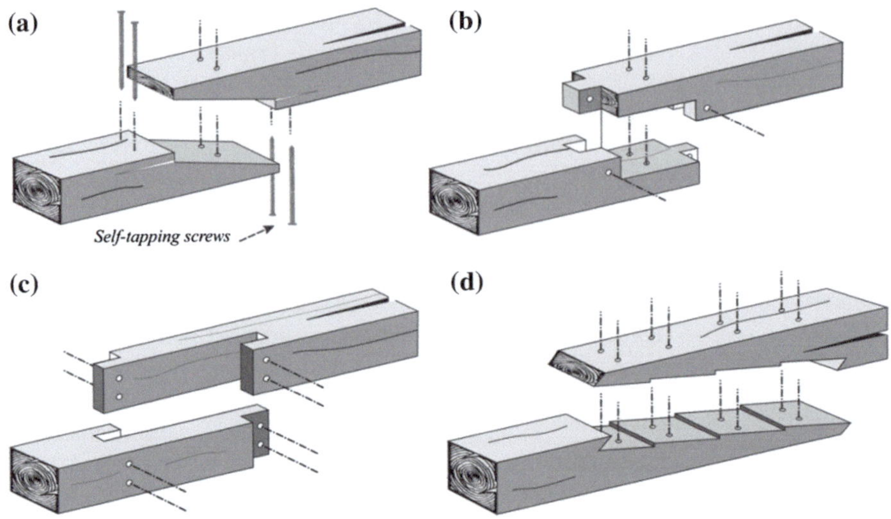

Fig. 13 **a** Scarf joint reinforcement perpendicular to the grain with self-tapping screws, **b** reinforcement of bending strength (weak axis) with a cog (half cogged scarf joint), **c** face-halved scarf joint, **d** multiple scarf joint with under-squinted ends

Fig. 14 Scarf joint reinforced with glued in rods: steel rods are glued in both timber members and connected with a long nut. *Credits* Pascal Lemlyn. Restauration du Moulin de l'abbaye de la Paix Dieu, Institut du patrimoine Wallon, Belgique

2.2 *Timber Roof Members*

When large deflections are observed on timber roof members, interventions mostly consist in increasing the stiffness of the element (and consecutively its strength) to control the deflection and prevent the bending failure that may occur at the end. Pre-stressing emerges as a global reinforcement technique. One may notice that these techniques described below could of course be applied to timber floor elements In the bibliography produced since the second half of the 19th century, it is possible to appreciate how the empirical work of many engineers has created such a broad selection of layouts and structural solutions (Figs. 15 and 16).

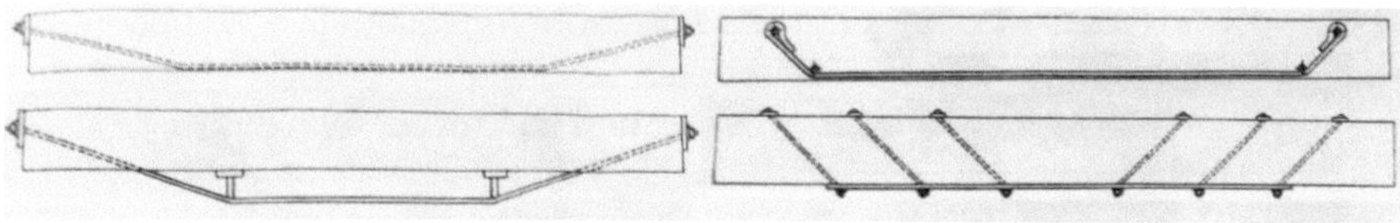

Fig. 15 Typologies for strengthening existing timber beams using pre-stressing techniques [7]

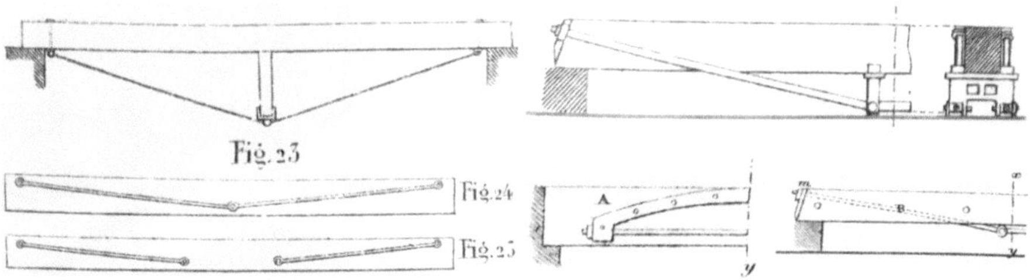

Fig. 16 Reinforced beams with metal tendons, by A.R. Emy [7]

Technical manuals refer to this kind of reinforcements with mild opinions, mainly pointing out the difficulties of installing the outer tendons at the head of the beam (due to the fact that they are embedded into the walls [7]. Of course, improvements of the technics to be used have been done since the early beginning giving confidence to the use of this kind of reinforcement. The major issue that still has to be clarified is how the shrinkage, the deformation of timber under loads perpendicular or at an angle to the grain and the rheological behaviour of the material may affect the loss of pre-stressing (influence of the environmental and material initial conditions) [21]. An advantage of these reinforcements is their reversibility.

In order to find a systematic approach to post-tensioning restoration for wooden structures (floors or roofs) it is necessary to wait until the second half of the 20th century. All the main actors belong to the Italian school, which is known as a very active centre for restoration theories and application developments. Some examples were reported in manuals, written in the second half of the 20th century, which are still a reference for present professionals (Fig. 17).

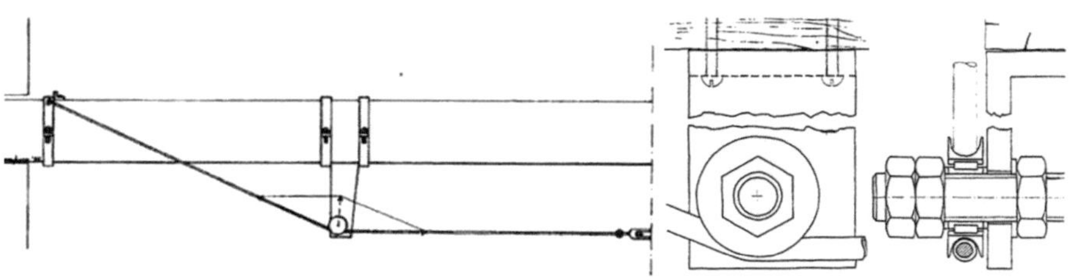

Fig. 17 Technical details for the strengthening of a beam through post-tensioning [8]

Similarly, real case of studies of restoration offer to practitioners an opportunity to apply some improved reinforcements and develop the empirical knowledge of the 19th and 20th century which is sometimes far to be fully understood and definitely need further studies (Figs. 18, 19 and 20).

The most common type of roof in Byzantine and post Byzantine buildings is a spatial "*post and beam*" system, which functions in a completely different way from the well-known types of king post trusses, which seem to be more common at Italy and other European countries than they were at countries around Eastern

Fig. 18 Reinforcement of the rafter of a truss in the theater of Sarteano by Tampone, 1977 [8]

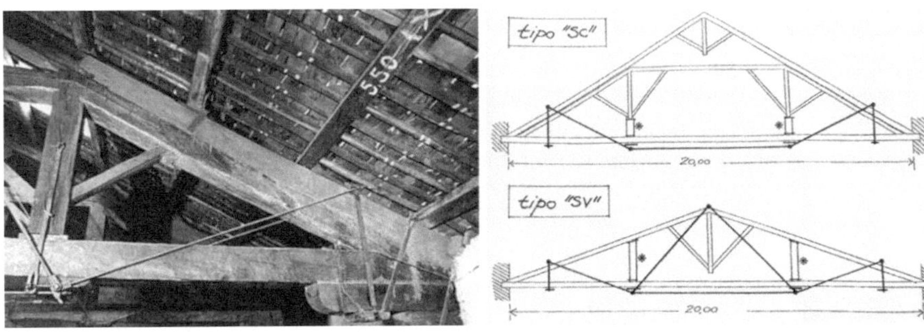

Fig. 19 Detail and design schemes of the project of Savona Theater. Courtesy of the designer, Ing. L. Paolini

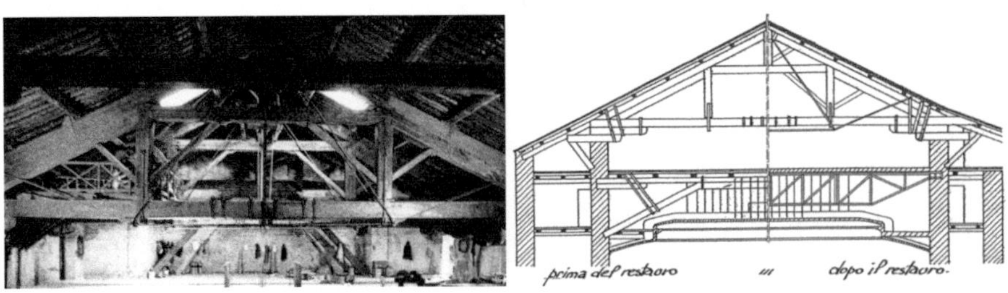

Fig. 20 Full view of the truss, and realized project of Savona Theater. Courtesy of the designer, Ing. L. Paolini

Fig. 21 Post and beam type of roofs of Post-Byzantine mansions in Greece [14, 16]. Excessive deformation of the longitudinal horizontal beams that support the rafters. Reinforcement by adding just new timber struts

Mediterranean (Byzantine and Ottoman empire). The loads are transferred from the rafters through a three-dimensional (spatial) system of beams and posts (vertical or inclined) on the horizontal timbers, which rest not only on the outer walls, but mainly on the internal ones (Fig. 21). In the "post and beam" system, the connections of the vertical and inclined posts are capable of transferring compression forces but not tension. A typical failure of these roofs is the deformation of the longitudinal horizontal beams that support the rafters (Fig. 21), because of the absence of adequate posts. The improvement of the original load-bearing system can be accompliced by the addition of new struts (more dense supporting) (Fig. 21), an easy to apply and reversible intervention maintaining the original beam at its position [13, 14, 16].

3 Timber Floors

Traditional timber floors are composed mainly by timber boards nailed on the beams (Fig. 22), in few cases laterally restrained by secondary elements (see Fig. 24). Ceilings may exist too (Fig. 22). To allow the interruption of the main beams at singular points, like stairs and fireplaces, secondary beams were introduced, forming the so-called stair shaft (Fig. 22).

The main beams are simply supported or fully supported on the walls, most of the time made of masonry in old buildings. As a practical rule, it has been seen [22] that the beams are supported in the entire wall thickness or at 2/3 of that dimension, but this is not always the case. Natural slate may be placed as a support under the

Fig. 22 Simple (vernacular/traditional architecture) and more elaborate timber floors (historic buildings/monuments). Stair shaft in timber floor [22]

timber beam. Its rough surface is suitable to prevent moisture. To guarantee an adequate connection between the floor slab and the walls, especially in seismic areas, metal devices are used. Those metal devices can present various forms but the main purpose is to use flat steel bars nailed or screwed to the timber beams of the slab, having in the opposite edge a special geometry to improve the connection with the masonry wall (Fig. 23). On the contrary, in many countries eastern than Greece, floor beams are connected on a system of timber ties (lacings), embedded in the masonry walls, for the seismic improvement of the building [13, 14, 16] (Fig. 24).

To prevent any biological degradation, the beams ends are normally painted with oil, leads solutions or tar.

In some cases, beams ends are placed over wall plates, in particular over light-framed timber partition walls, with the aim to ensure the uniform distribution of the load over them. The lateral restraining of the main beams is usually made by strutting that can be herring-bone or solid (see Fig. 21).

3.1 Damages and Strengthening Techniques

Despite the presence of structural damage in timber floors, it is unusual to observe a failure on these structures. This can be explained by the system effect given by the

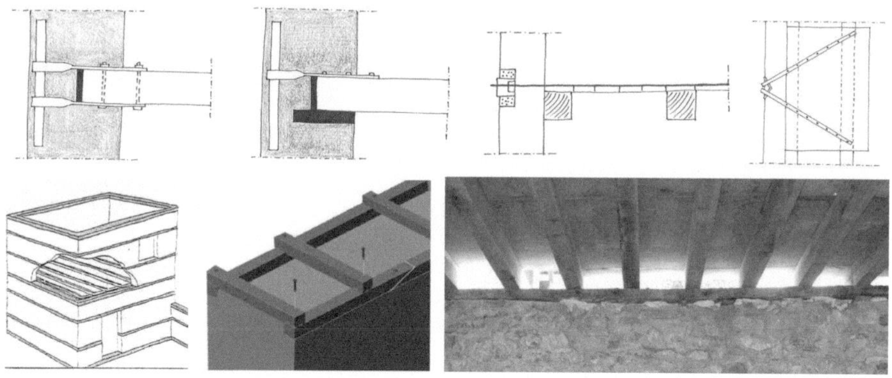

Fig. 23 Examples of metal devices used to ensure the connection between the main beams and the masonry wall [23]. Timber ties that connect the floor beams to the masonry walls [13, 14, 16]

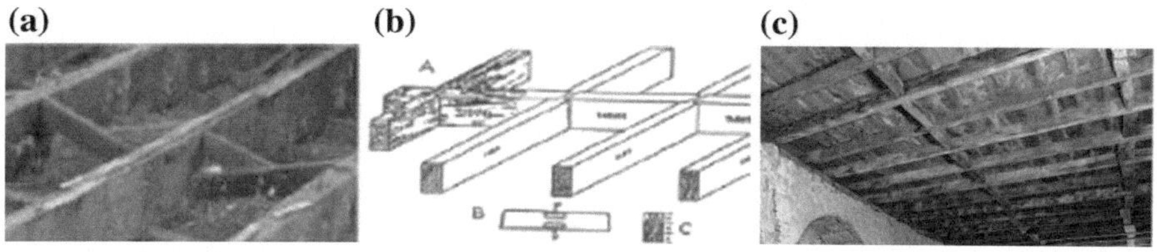

Fig. 24 Examples of floor strutting. **a** Herring-bone and **b**, **c** solid

floor boards and in some cases, mainly in important buildings by the high safety level normally applied in the past in the design of those structures. In domestique architecture (traditional buildings, houses) the dimensions of timber are usually small). Very often, as in roofs, intervention on timber floors involves end-beams repairs. End-beams, embedded in masonry walls, are the more exposed zones to biological agents. If a high level of humidity is present in masonry e.g. due to infiltration, when disconnections or damages occur in roofing elements (tiles etc.) and the adsorbed moisture cannot be evaporated because of lacking in ventilation, the suitable conditions for biotic attacks are established and therefore degradation of end-beams can be expected (Fig. 25).

This fact is well known among structural restorers and engineers familiar with timber. The first considerations about the effects of moisture content onto the end-beams were due to Vitruvio [24], whereas the suggestion of preparing aerated supports for hosting beam heads were firstly attributable to Alberti [25].

Nowadays, several examples of techniques and methods can be found in literature for the repair and reinforcement of timber end-beams [8, 15, 26]. Since the early seventies, many companies have developed materials and techniques to repair decayed timber elements (design of a prothesis) (Figs. 6, 7 and 26). All of them aim to restore the load bearing support that has been lost because of the material decay. Fire damage, may cause a reduction in the member cross-section too, resulting in

Fig. 25 Examples of decayed old timber end beams that can be found both in floor and roof systems

inadequate strength and stiffness. The elements that substitute decayed timber can be recognizable, visible or not and the elements that connect them to the sound wood can be either external or internal too.

It is possible to identify two distinct groups of techniques according to the material to be used:

Timber, wood based materials and steel elements (steel sections, plates, rods etc.). The intervention consists in the substitution of the decayed part by a new element which can be made of wood (solid or glued-laminated) (Figs. 6, 7 and 26a) or in steel (Fig. 26b). Any new timber should be of the same species as the original one and have the same moisture content. The cuttings of the timber prosthesis and its on-site application may vary depending on the used method and on many other parameters that have to be taken into account (aesthetics, presence of decorative elements, access, fire protection etc.).

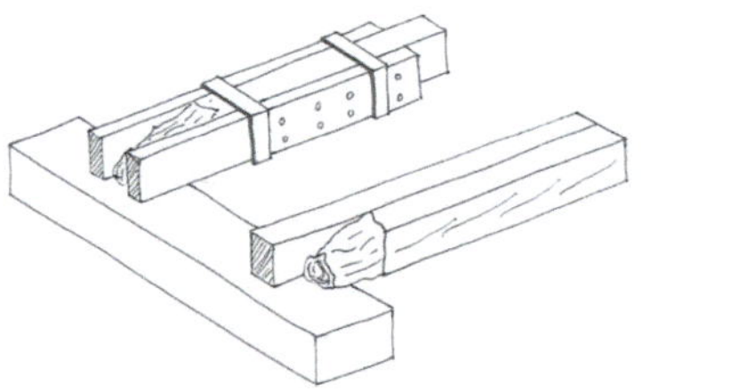

(a) New timber elements connected to the sound part with steel straps, plates and steel fasteners

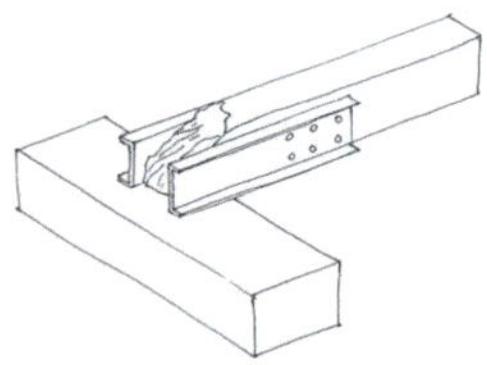

(b) New steel elements replacing the decayed timber

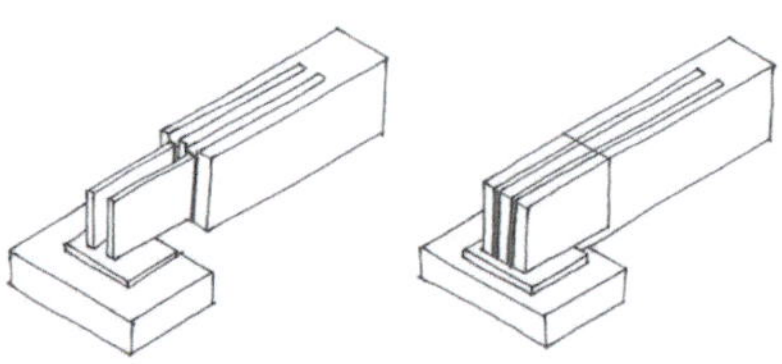

(c) Glued-in plates connecting the prosthesis to the sound part

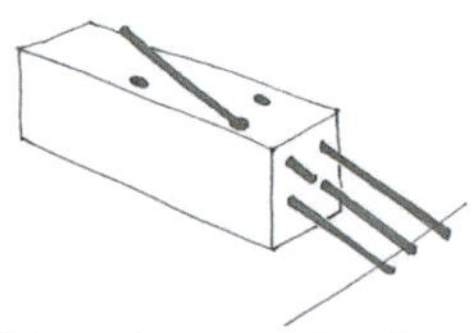

(d) Glued-in rods connecting the prosthesis made either of wood or resin

Fig. 26 Examples of end-beams repair techniques

Resin and so-called fibre reinforced polymers (*FRP*). The decayed part of the beam is replaced by a resin prosthesis connected to the remaining part of the beam element using glued-in plates (Fig. 26c), horizontal or inclined rods of different types (Fig. 26d).

When the aesthetic value is of low importance and no fire resistance is required, the resin part can be visible or the connection of the prosthesis can be external. However, usually the prosthesis is connected to the sound part of the element with glued-in plates or glued-in rods. Plates and rods can be in FRP (glass, carbon or aramid) but the steel ones (stainless usually) are the more appreciated due to their reduced cost. The use of prosthesis became well accepted mainly due to its low intrusion level, simplicity and the good aesthetic value that can be provided from some of the used techniques. The use of timber prosthesis compared to resin or steel elements is considered closer to the principles concerning the conservation of historic timber structures.

The design rules for this kind of repairs are based on simple and empiric design rules and technical data provided in the cases that resins are used by resin manufacturers (European Technical Approval). Those rules are probably simple and conservative. For example, the one used for the Beta-System considered a strain distribution in repaired beam similar to that one of sound wood, and moreover imposed that the moment of resistance of the sound timber is equal to that one of rods used for the intervention. In such a way the total area of rods could be calculated. The length of anchorage is simply calculated from the resin-to-rods allowable bond stress and from the shear strength of sound timber (so, rigid bond between resin and wood is assumed as related to the design safety value of shear strength for wood). Apart from these empirical approaches, based on the allowable stresses, the adoption since 1994 of Eurocode 5 (EN 1995-1-1:2004) [27] changed the way to handle the design approach of this kind of interventions. The common approach is substantially similar to that described in the former standard 1995-2:1997 [28], and not appearing in the current version of 1995-2:2008 [29]. European countries have obviated to this normative deficiency through National Application Documents (e.g. the Italian CNR-DT 206/2007 [30]).

When the load-carrying capacity of the whole floor has to be increased or if the vertical deflection of the beams is too high, the introduction of additional elements is a common technique (reinforcement) (Figs. 27 and 28).

Steel plates, timber or wood based boards nailed or screwed to the main beams is an effective way to repair or strengthen timber beams. In most of the cases, those new elements are made of wood, wood-based products or steel. In these cases, additional elements are introduced between the main beams (placed parallel to them) with or without any connection to them or placed perpendicular to the main beams with the aim to reduce their span. In this last case, the new elements present significant cross section as result of the considerable load that they have to support (Fig. 27).

The reinforcement techniques commonly adopted in the practice consist in coupling the existing beams using concrete or wooden slabs placed usually over them: different configurations are possible depending on the slab material and

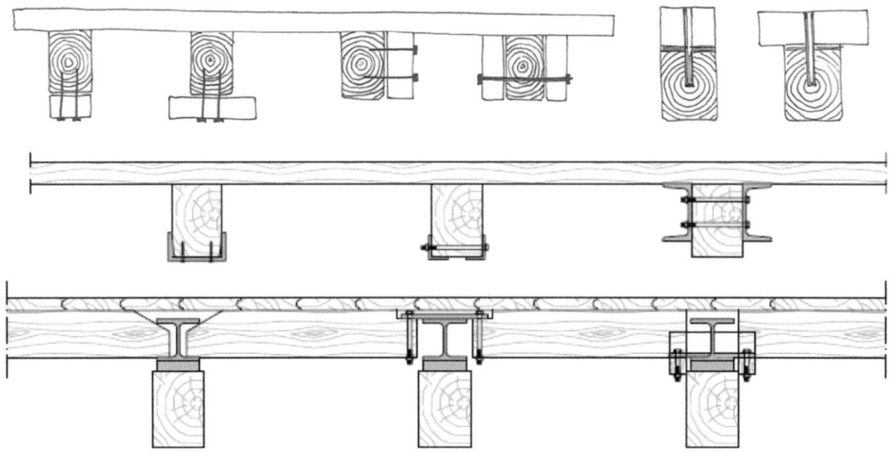

Fig. 27 Examples of reinforcement techniques of timber floor beams

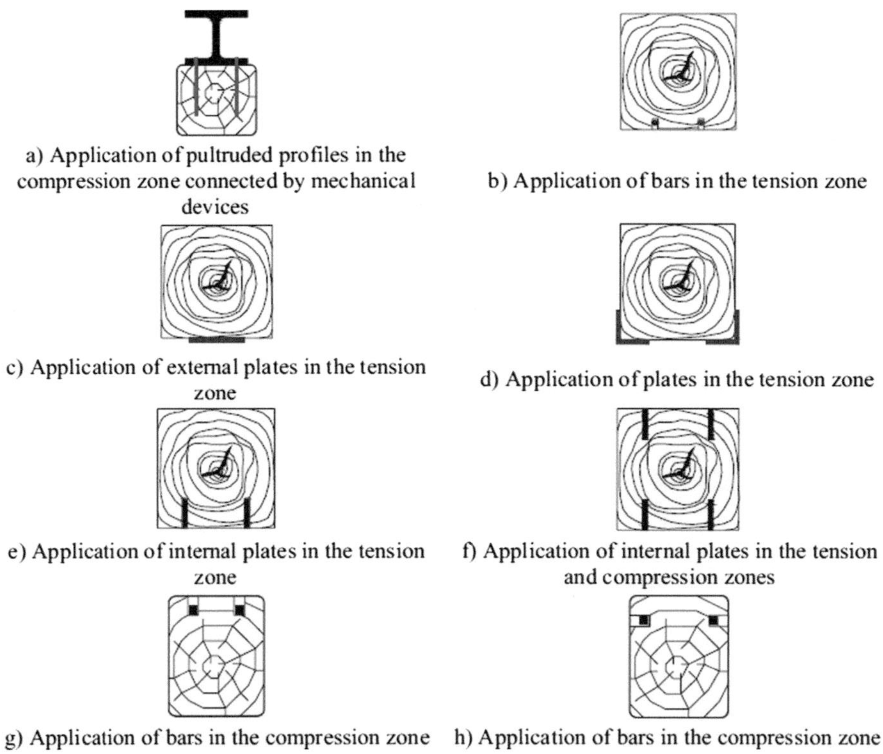

Fig. 28 Examples techniques for flexural strengthening using FRP bars and plates bonded to the external surface of the beams possibly with the addition of mechanical connectors or inside special slots cut into the beams (CNR-DT 201/2005) [31]

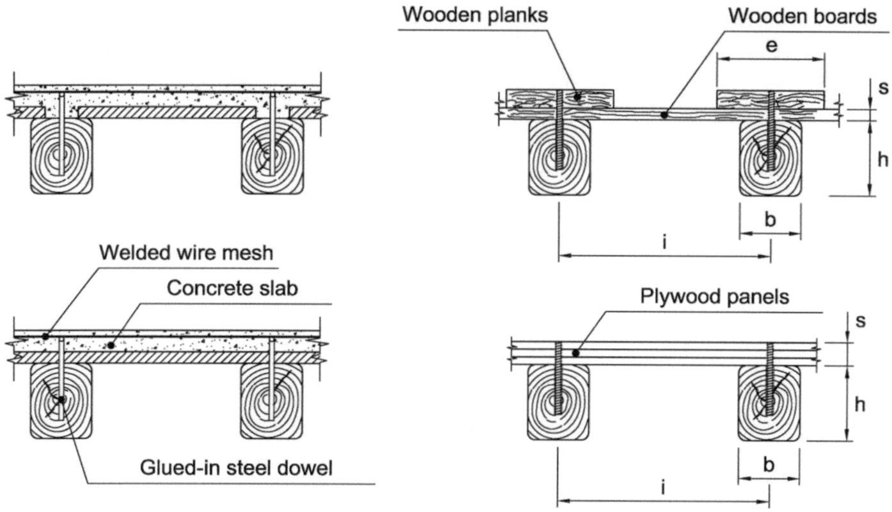

Fig. 29 Examples of reinforcement techniques of timber floors [32]

connection system (Fig. 29). In this way, it is possible to rely on a composite T section beam.

One of the most widely used and effective techniques for strengthening floors is based on the connection of a new 40–50 mm height concrete topping on the existing timber joists. Different types of metal fasteners, notched shear keys or slotted-in perforated plates generally assure the effective collaboration between the two different materials [33–40]. The new composite section ensures a significant floor stiffness upgrade, while the concrete topping connected with the vertical walls is able to give an effective diaphragmatic action to the floor improving the lateral load resistance of masonry buildings in seismic areas. The use of concrete also allows load distribution to take place, provides acoustic and fire insulation and increases the natural frequency of the floor. On the other hand, a concrete slab must be thin since it adds undesirable additional weight on the floor, and consequently increase the seismic and the foundation loads. This technique, while simple and very efficient, is now often considered not sufficiently reversible: particularly in Italy it is frequently not approved by the Cultural Heritage Offices to be used in buildings of historical value.

In timber-to-timber composite sections, the use of traditional materials and dry assembly methods are in agreement with the restoration issues of compatibility, reversibility or recoverability of the intervention. Moreover the additional loads are quite small. For all composite sections the mechanical characteristics of the connection are the main factor which influences the structural response. Design of composite sections requires the consideration of partial composite action, due to the impossibility of achieving an extremely rigid shear connection between web and flange (deformable shear connection between web and flange). Analysis can follow the Eurocode 5 [27] approximate 'gamma method', where an effective flexural stiffness $(EI)_{ef}$ for the composite section is calculated as a function of the stiffness of the shear connection, taking into account the slip between the flange and the joist. Also the 'shear analogy method', where the composite beam is divided into two virtual components coupled with stiff bars can be used to determine internal forces [41–43].

The type of prosthesis or reinforcement method may vary depending on many parameters that have to be taken into account (cultural values, aesthetics, presence of decorative elements, access, fire protection, on-site application, cost etc.).

3.2 In Plane Structural Behaviour

It is important to assess the floor or the roof (usually at the ceiling level), diaphragm's in-plane stiffness, as it can affect the structural performance of a traditional masonry building subjected to lateral loads.

The common configuration of existing timber floors with a crosswise single layer of wooden planks might need an in-plane shear strengthening and mainly a connection mechanism to the vertical load bearing elements (masonry or timber-framed walls) especially to the ones that are parallel to the floor or roof beams in order to ensure an efficient distribution of the lateral seismic load through all bearing walls [16, 18, 32].

In Italy, Spain and Portugal, the connection to the wall is achieved using metal devices (Figs. 23 and 32b). In Greece and in most of the countries eastern than Greece, for thousands of years a horizontal continuous system of timber ties/lacings, a timber grid, running around all masonry walls, is tying the building as a belt in several levels (Figs. 23 and 30). On these timber ties the floors and the roofs were nailed improving the connection and the collaboration of the walls (box behaviour). The improvement of the diaphragmatic action of the timber roofs and floors is considered a successful reinforcement of the buildings against severe earthquake events [13–16].

Fig. 30 Roofs nailed on timber ties that connect the vertical load bearing walls of the buildings (masonry or timber-framed) improving their seismic resistance [15, 16]

As mentioned before, there are several techniques for strengthening existing timber floors. Their effectiveness in terms of in-plane stiffness too differs depending on the strengthening technique used [44–47]. The most effective ones, in terms of in-plane stiffness, are those which reinforce the compression side of the floor cross-section. The basic idea is to improve the in-plane stiffness of the floor system by implementing a more efficient T-section as for vertical loads. Among the various possibilities, the addition of a concrete layer over the timber structure is common practice [45, 48]. Another possibility is to use timber or wood-based materials (e.g. plywood) (Fig. 31), instead of a concrete slab to enhance the in-plane stiffness [16, 32, 46, 47].

In fact, an old technique for strengthening timber floor systems is the addition of a second layer of wooden boards perpendicular to the existing ones (Fig. 32). This second layer is used to recover part of the existing deformation (as a false floor) to increase the bending stiffness and contribute to the in-plane stiffness of the floor.

An alternative to the traditional technique of adding a second layer of floorboards to strengthen the floor placed transversally to the existing ones consists to place either floorboards placed diagonally or plywood panels, over the existing planks and under the new ones, since the last layer has to be made by boards, At the level of the roof, the plywood panels can be added between the horizontal tie-beams and the planks of the ceiling (Fig. 31) [16]. CLT (cross laminated timber) panels can be used over the floor beams too and screwed onto them. It must be noticed that, thanks to the enhanced mechanical performances of CLT, namely good in- and out-plane load bearing capacities and two way action capability, it is possible to

Fig. 31 Use of plywood panels over the existing floor boards and under the second layer of boards for the pavement. Use of plywood panels between the horizontal tie-beams of the roof and the boards of the ceiling for the improvement of the diaphragmatic action of the floor and the roof

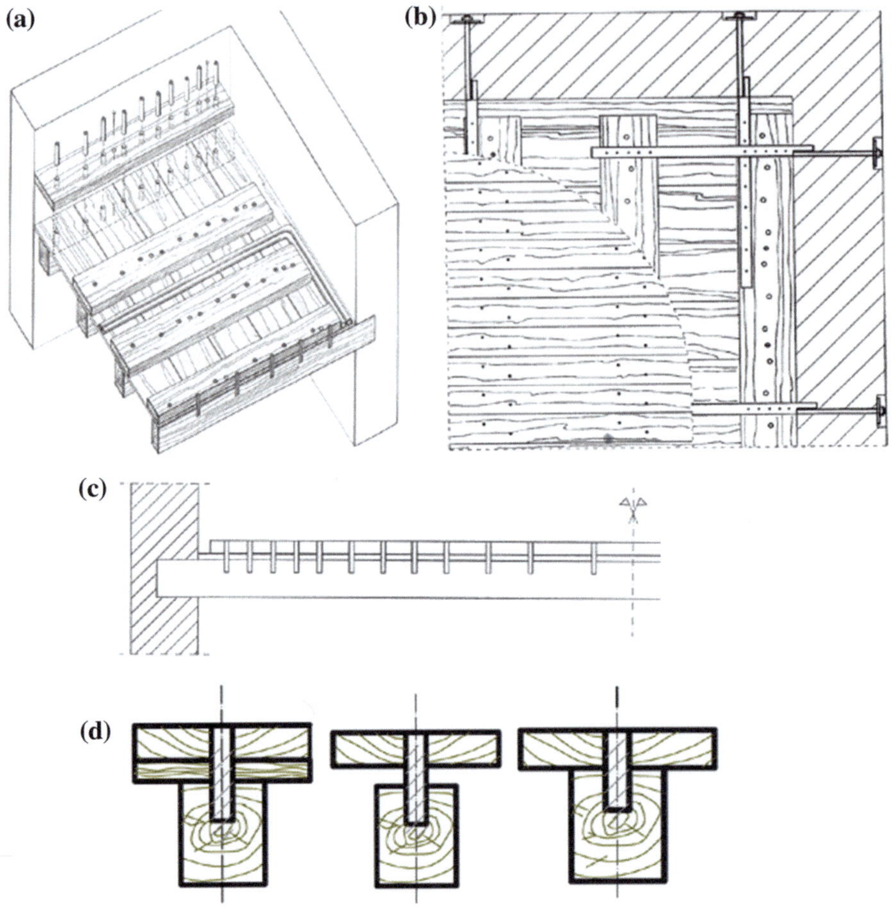

Fig. 32 Stiffening intervention with dry hardwood pins to connect planks and/or boardings to original timber floors (**a**, **b**) and detail of connection to walls (**b**), distribution of dowels along a main beam (**c**), and different working of connections depending on contribution of boarding (**d**), combined flexural and shear (*left*), bending (*centre*), pure shear (*right*)

replace an entire timber floor system composed of beams and boards with CLT panels. However, the techniques more promising and more compatible to the principles of conservation of traditional/historic buildings is the ones that keep the existing beams or in general the original structural system. Five full-scale timber floors were tested in order to analyse the in-plane behaviour of these structural systems [49]. The main objective was an assessment of the effectiveness of in-plane strengthening using cross-laminated timber (CLT). The use of CLT panels is revealed to be an effective way to increase the in-plane stiffness of timber floors, through which the behaviour of the composite structure can be significantly changed, depending on the connection applied, or modified as required.

4 Conclusions

When working on old timber structures, the fact that the structure has survived for decades or centuries without failure it may be sufficient proof of its load bearing capacity if the use is not changed and it has to be taken into account if any or what kind of intervention is needed. On the other hand, this maybe not be sufficient proof for the future when new use and new imposed loads are introduced. If the decay of timber elements is too large, then local replacement of the decayed part is clearly the only solution. If repairs are necessary, specific reliable on-site assessment techniques are required to determine the appropriate level of intervention needed. This point remains very important to evaluate the replacement, repair and strengthening solutions along with the cultural significance of each case, the know-how and the associated project costs. Evaluation of the durability of the intervention works carried out with new innovative techniques is necessary too. Reinforcement may help to achieve several aims, for example, increasing the load bearing capacity (strength and stiffness) when is needed, or increasing the ductility of timber members or timber structures.

For joints, reinforcements help to reduce gaps in order the mechanisms of transmitting the loads to work properly, overcome timber weaknesses by increasing the shear strength and the tensile strength perpendicular to the grain, also helping to reduce the propagation of cracks. It is important to be pointed out that the interventions in joints (repairs or reinforcements) must not change their stiffness and consequently the overall original behaviour of the original structural system.

For timber beams, reinforcements help to restore the load bearing capacity that has been lost because of the material decay at the support area, to increase the moment of resistance (and so limit the deflection too) or to increase the in-plane stiffness for lateral loads.

The study of reinforcement techniques is not yet included in European standards as Eurocode 5 but only, for specific aspects, in the National Annexes of some countries. Investigations on that promising topic have helped to figure out how to overcome timber weaknesses and have resulted in the proposal of design models and reinforcement methods. Some of the most important and applicable outcomes will probably be integrated in the revision of Eurocode 5 to help engineers to restore timber floors or timber roofs. But since existing structures and mainly historic structures are not covered by the new standards suitable for mainly new built structures, it is urgent and of great importance relevant European Standards to be developed too. These Standards have to provide the necessary tools for structural engineers, members of a multi-disciplinary team that have to work together in a restoration project, to evaluate the existing condition of a historic timber structure and moreover to select the proper interventions using innovative and/or simple techniques that will save the authenticity of our architectural heritage, including the authenticity of the "invisible" in many cases load bearing system.

References

1. Venice Charter. Second international congress of architects and technicians of historical monuments, Venice, May 25–31, 1964.
2. Krakow Charter. International conference on conservation Krakow, Krakow; 2000.
3. ISCARSAH ICOMOS Committie. Principles for the analysis, conservation and structural restoration of architectural heritage 2003. IIWC ICOMOS Committie: Principles for the preservation of Historic Timber Strucures; 1999.
4. UNI Ente Nazionale Italiano di Unificazione. UNI 11138: cultural heritage—wooden artefacts—criteria for the preliminary evaluation, the design and the execution of works, Milano, Italy; 2004.
5. Kasal B, Tannert T. In situ assessment of structural timber. RILEM State-of-the-Art Reports, vol. 7, 129 p; 2011.
6. Cruz H, Yeomans D, Tsakanika E, Macchioni N, Jorissen A, Touza M, Mannucci M, Lourenço PB. Guidelines for the on-site assessment of historic timber structures. Int J Archit Herit Conserv Anal Restor. 2015;9(3):277–89.
7. Gasparini DA, da Porto F. Prestressing of 19th century wood and iron truss bridges in the U.S. In: Huerta S, editor. Proceedings of the first international congress on construction history, Madrid; 2003. p. 978–86.
8. Tampone G. Il restauro delle strutture di legno. Milano: Hoepli; 1996. p. 321.
9. Paolini L, Messina C, Marradi AM. Il restauro della copertura del teatro Chiabrera a Savona, in "Legno e restauro", a cura di G. Tampone, Firenze: Messaggerie toscane; 1989.
10. Song J-K, Kim S-Y, Yang K-H, Han S-A. Flexural behavior of post-tensioned glued-laminated timber with wire rope. Seoul, s.n.; 2007.
11. Gesualdo FA, Rezende RB, Cunha TA. Numerical and experimental evaluation of a double inverted trussed. Constr Build Mater. 2014;50:736–43.
12. Al-hayek HWH. Flexural behaviour of post-tensioned timber beams. Winnipeg: University of Manitoba; 2014.
13. Tsakanika E. Methodology concerning the restoration of historical buildings. Case studies: the Turkish Mansion and the Hagi Mehmet Aga Mosque in Rhodes. In: Proceedings of international conference of the IIWC "Conservation of Historic Wooden Structures", Florence, vol. 2; 2005. p. 194–203.
14. Tsakanika E. Byzantine and post-byzantine historical timber roofs in Greece. Typical failures, misunderstanding of their structural behaviour, restoration proposals. In: Proceedings of XVI international conference and symposium of the IIWC "From Material to Structure", Florence; 2007. http://www.icomos.org/iiwc/2007.htm.
15. Bertolini C, Touliatos P, Miltiadou N, Delinikolas N, Crivellaro A, Marzi T, Tsakanika E, Pignatelli O, Biglione GG. The timber roof of Hagia Paraskevi Basilica in Halkida, Greece: multi-disciplinary methodological approach for the understanding of the structural behaviour. Analysis and diagnosis. In: Proceedings of XVI international conference and symposium of the IIWC "From Material to Structure", Florence; 2007. http://www.icomos.org/iiwc/2007.htm.
16. Tsakanika-Theohari E, Mouzakis H. A post-Byzantine mansion in Athens. The restoration project of the timber structural elements. In: Proceedings of 11th world conference on timber engineering, (WCTE 2010), vol. 2. Riva del Garda, Trentino, Italy; 2010. p. 1380–9.
17. Feio AO, Lourenço PB, Machado JS. Testing and modeling of a traditional timber mortise and tenon joint. Mater Struct. 2014;47:213–25.
18. Menichelli C, Adami A, Balletti C, Bertolini Cestari C, Bettiol G, Biglione G et al. Le strutture lignee dell'arsenale di Venezia. Sudi et restauri. In: Proceedings of XXV Convegno Internazionale Scienza et Beni Culturali – Conservare e restaure il legno, Bressanone, Italy (in Italian); 2009.
19. Bertolini Cestari C, Invenizzi S, Marzi T, Spano A. Numerical survey, analysis and assessment of past interventions on historical timber structures: the roof of Valentino Castle.

In: Proceedings of international conference on structural health assessment of timber structures, SHATIS'15, Wroclaw (Polond), vol. 2; 2015.
20. Branco JM, Cruz PJS, Piazza M. Experimental analysis of laterally loaded nailed timber-to-concrete connections. Constr Build Mater. 2009;23(1):400–10.
21. Sanabra M, Capellà J. The four ages of early prestressed concrete structures. PCI J. 2014;59 (4):93–121.
22. Segurado JES. Trabalhos de carpintaria civil. Livraria Bertrand, Lisboa: Biblioteca de Instrução Profissional; 1942.
23. Teixeira JJL. Descrição do sistema construtivo da casa burguesa do Porto entre os séculos XVII e XIX. Contributo para uma história da construção arquitectónica em Portugal, Provas de aptidão pedagógica e capacidade científica. FAUP, Porto; 2004.
24. Barbaro D. I dieci libri dell'architettura di M. Vitruvio (anastatic printing from the original of 1567). Milano: Il Polifilo; 1997.
25. Portoghesi P, Orlandi G, Alberti LB. De re aedificatoria. Milano: Il Polifilo; 1966.
26. Arriaga F, Peraza F, Esteban M, Bobadila I, Garcia F. Intervención en estructuras de madera. AITIM: Espanha; 2002. p. 512.
27. EN 1995-1-1:2004. Eurocode 5. Design of timber structures. Part 1-1: general—common rules and rules for buildings. European Committee for Standardization, Brussels.
28. ENV 1995-2:1997. Eurocode 5: Design of timber structures. Part 2: bridges. European Committee for Standardization, Brussels.
29. EN 1995-2:2008. Eurocode 5: Design of timber structures. Part 2: bridges. European Committee for Standardization, Brussels.
30. CNR-DT 206/2007. Istruzioni per la Progettazione, l'Esecuzione ed il Controllo di Strutture di Legno. Consiglio Nazionale delle Ricerche, Roma.
31. CNR-DT 201/2005. Guidelines for the design and construction of externally bonded FRP systems for strengthening existing structures. Italian National Research Council, Rome; 2007.
32. Branco JM, Tomasi R. Analysis and strengthening of timber floors and roofs. In: Costa A, Miranda Guedes J, Varum H, editors. Structural rehabilitation of old buildings. Berlin: Springer; 2013. P. 235–58. ISBN: 978-3-642-39685-4. doi:10.1007/978-3-642-39686-1.
33. Turrini G, Piazza M. Una tecnica di recupero dei solai in legno. Recuperare. 1983;5:396–407.
34. Ronca P, Gelfi P, Giuriani E. Behavior of a wood–concrete composite beam under cyclic and long term loads. In: Proceedings of the international conference on structural repair and maintenance of historic buildings, Seville, Spain; 1991. p. 263–75.
35. Natterer J, Hamm J, Favre P. Composite wood–concrete floors for multi-story buildings. In: Proceedings of the 4th international wood engineering conference, New Orleans, Omnipress, Madison, Wisconsin, USA, vol. 3; 1996. p. 3431–5.
36. Gelfi P, Giuriani E, Marini A. Stud shear connection design for composite concrete slab and wood beams. J Struct Eng. 2002;128(12):1544–50.
37. Giuriani E. L'organizzazione degli impalcati per gli edifici storici. L'edilizia. 2004;134:30–43.
38. Gutkowski R, Brown K, Shigidi A, Natterer J. Laboratory tests of composite wood–concrete beams. Constr Build Mater. 2008;22(6):1059–66.
39. Bathon L, Graf M. A continuous wood–concrete-composite system. In: Proceedings of world conference of timber engineering, Whistler, BC; 2000.
40. Clouston P, Bathon L, Schreyer A. Shear and bending performance of a novel wood-concrete composite system. J Struct Eng. 2005;131(9):1404–12.
41. Kreuzinger H. Platten, Scheiben und Schalen: Ein Berechnungsmodell für gängige Statikprogramme. bauen mit holz. 1999;1:34–9.
42. Deutsches Institut für Normung. DIN 1052: Entwurf, Berechnung und Bemessung Deutsches Institut für Normung: DIN 1052. Berechnung und Bemessung: Entwurf; 2004.
43. Kuhlmann U, Michelfelder B. Optimised design of grooves in timber-concrete composite slabs. In: Proceedings of the 10th world conference on timber engineering, Portland, Oregan, USA; 2006.

44. Modena C, Valluzzi MR, Garbin E, da Porto F. A strengthening technique for timber floors using traditional materials. In: 4th structural analysis of historical constructions, Proc. Intern. Symp., Padova, Italy. Balkema, Rotterdam; 2004. p. 911–21.
45. Piazza M, Tomasi R, Baldessari C, Acler E. Behaviour of refurbished timber floors characterized by different in-plane stiffness. In: Structural analysis of historic construction: preserving safety and significance. VI international conference on structural analysis of historic construction, vol. 2. London, Bath: Taylor & Francis; 2008. p. 843–50.
46. Valluzzi M, Garbin E, Benetta MD, Modena C. Experimental assessment and modelling of in-plane behavior of timber floors. SAHC 2008. Structural analysis of historical constructions. Bath, UK; 2008.
47. Angeli A, Piazza M, Riggio M, Tomasi R. Refurbishment of traditional timber floors by means of wood-wood composite structures assembled with inclined screw connectors. In: 11th world conference on timber engineering—WCTE 2010. Riva del Guarda, Trentino, Italy; 2010.
48. Tomasi R, Baldessari C, Piazza M. The refurbishment of existing timber floors: characterization of the in-plane behaviour. In: Prohitech Conference, 21–24 June, Rome; 2009.
49. Branco JM, Kekeliak M, Lourenço PB. In-plane stiffness of timber floors strengthened with CLT. Eur J Wood Wood Prod. 2015;73(3):313–23. doi:10.1007/s00107-015-0892-2.

Strengthening of RC Buildings with Steel Elements

J.M. Castro, M. Araújo, M. D'Aniello and R. Landolfo

1 Introduction

In Europe, especially in the Southern part and in Mediterranean basin, existing reinforced concrete (RC) buildings are often affected by a number of structural inadequacies and deteriorated conditions. Therefore, structural engineering consultants and practitioners are oftentimes faced with a rather challenging situation of having to assess and strengthen an existing building for a number of different reasons, such as the need for extending the dimensions of the building, changing of its use, repairing visible damage and deformations or improving seismic resistance. Nevertheless, despite these various reasons for strengthening, most codes, such as the Eurocodes, simply address the structural assessment and strengthening problems from an earthquake-resistant perspective, as discussed in detail by Romão et al. [1], and little guidance is provided to deal with more general cases, such as low-rise gravity buildings located in areas of low or very low seismicity. As a result, practitioners are usually forced to rely on engineering judgment [2], commonly resorting to codes that have been developed to address the design of new buildings, which, for instance, do not cover structural condition inspection issues,

J.M. Castro (✉) · M. Araújo
CONSTRUCT-LESE, Civil Engineering Department,
Faculty of Engineering, University of Porto, 4200-465 Porto, Portugal
e-mail: miguel.castro@fe.up.pt

M. Araújo
e-mail: maraujo@fe.up.pt

M. D'Aniello · R. Landolfo
Department of Structures for Engineering and Architecture,
University of Naples "Federico II", 80134 Naples, Italy
e-mail: mdaniel@unina.it

R. Landolfo
e-mail: landolfo@unina.it

A. Costa et al. (eds.), *Strengthening and Retrofitting of Existing Structures*,
Building Pathology and Rehabilitation 9, https://doi.org/10.1007/978-981-10-5858-5_6

or to a number of handbook guidelines and existing bibliography. Moreover, even the most recent versions of codes dealing with the seismic assessment and strengthening of buildings, such as Part 3 of Eurocode 8 (EC8-3) [3], reveals a number of inconsistencies in the assessment procedures [4–8] and provides scarce information with respect to the design of the strengthening techniques. For instance, in the case of RC buildings, EC8-3 solely provides guidance for the strengthening of RC members using jacketing approaches [1] and no further reference is provided regarding strengthening interventions at the storey or global levels. Although this might be related with the fact that EC8-3 simply defines its compliance and acceptance criteria at the member level, the European code does not consider the eventual unfeasibility, disruptiveness and cost of implementing such local strengthening techniques. The preparation of a future generation of codes, particularly of EC8-3, that specifically address different strengthening solutions suitable to deal with different local and global performance requirements seems to be therefore necessary. This chapter aims, on the one hand, to address this issue by not only discussing various strengthening needs commonly associated to existing RC buildings, but also by contextualizing their design from a code perspective. Hence, most of the observations regarding code-based performance requirements are based on the prescriptions of the current set of Eurocodes.

Concerning the strengthening interventions and seismic retrofitting solutions, several techniques and design strategies have been deeply investigated and adopted in the past, either to improve or to retrofit RC buildings, such as bonding and/or jacketing with composite materials, using special devices, modifying the structural scheme by adding RC shear walls, adding steel elements and sub-systems, etc. Among the wide range of upgrading options, the use of steel elements offers a diversity of strengthening applications and might be considered as a rational alternative to other strengthening techniques due to their ease of design and detailing and cleaner, less intrusive and disruptive implementation. A number of studies and research initiatives have been conducted in the past with the aim of providing specific information in terms of the performance of RC buildings subjected to different strengthening techniques involving the use of steel elements [9–13]. Among the various existing guidelines, FEMA 547 [14] is one of the very few documents that provides thorough guidance to practitioners in terms of the most common structural deficiencies found in RC buildings with different typologies (e.g., moment resisting frames, shear walls or precast buildings). The FEMA guideline proposes strengthening techniques for each structural deficiency and makes considerations on construction, cost, disruption and proprietary concerns. Still, despite the pertinence of presenting general strengthening techniques per specific building weakness, it is important to highlight that the effectiveness of each retrofitting solution is case dependent and varies with the level of seismic hazard, as discussed by Asteris et al. [15]. A recent study based on the assessment of a typical existing RC building constructed in the 50s in Lisbon, carried out by Furtado et al. [16], showed that strengthening RC buildings using steel elements is a very efficient solution. Indeed, the rational design of steel systems allows minimizing structural demands, by implementing conventional steel bracing systems, and also improving the cost-efficiency, through the use of steel bracing systems with energy dissipation devices. The latter solution was demonstrated by Furtado et al. [16] to be as efficient as concrete jacketing of RC columns or as adding RC shear walls, although costing four times less and around 10% of the property value of the building. Thus, those Authors [16] proved that steel elements are a rational solution from a strengthening standpoint and are highly worth to be considered by structural engineering consultants and practitioners. Hence, this chapter also aims at presenting a state-of-the-art review on different strengthening techniques using steel-based solutions, from traditional local and global interventions using bolted or bonded steel plates and jackets or concentric steel braces to more innovative dissipative systems based on eccentric braces, buckling-restrained braces or metal shear panels.

2 Assessment and Strengthening Needs in Existing RC Buildings

2.1 *Code-Based Assessment Approaches and Common Practices*

Seismic strengthening of existing RC buildings is one of the most common and challenging strengthening problems that designers might have to face. Past earthquake events have dramatically tested the seismic performance of the existing RC building stock of many different countries, revealing not only the high seismic vulnerability of pre-code buildings designed to simply withstand gravity loads [17], but also, and critically, the lack of guidance to designers in the repair and strengthening process [2]. Codes that specifically address the seismic assessment and strengthening of existing buildings are relatively recent and unknown to many practitioners, introducing new performance-based concepts that might imply the use of advanced and complex nonlinear methods of analysis. Searer et al. [18, 19] have alarmingly demonstrated a clear distrust and lack of confidence of American practitioners in the American seismic assessment and strengthening code [20]. Similarly, a survey recently conducted among Portuguese practitioners by Araújo and Castro [21] has shown that although half of the respondents have previously dealt with a seismic safety assessment problem, only 8% of those respondents used EC8-3, 37% used Part 1 of Eurocode 8 (EC8-1) [22] and the vast majority still applied the former national code (RSA [23]) that defines seismic actions in combination with Eurocode 2 [24] or other design and detailing national codes. Among the reasons for not applying EC8-3, 35% of the participants responded to be unaware of its existence, while 36% consider it too complex to be applied in practice. For what concerns the use of nonlinear static and dynamic procedures, it was seen that half of the respondents are not familiar with those type of procedures and only 12% have actually applied them in practice.

Therefore, it is clear that, when dealing with the assessment and strengthening of RC buildings, structural engineering consultants and practitioners commonly resort to: (i) linear elastic analysis procedures; (ii) codes that prescribe general design and detailing rules; (iii) and, seismic codes that aim at the design of new buildings. Such practices are conceptually different from the approaches followed by assessment and strengthening codes [1] and should be implemented with great caution as they: (i) do not recognize the level of knowledge of the structural condition of the existing building; (ii) rely on safety factors that only account for the uncertainty in the properties of materials produced under known quality specifications, which may not reflect those of existing buildings; (iii) imply the pre-selection of a certain ductility class (e.g., seismic design to EC8-1) which is associated to a behaviour factor, q, that has been derived for new buildings with known and controlled dissipative mechanisms. Further validation studies using nonlinear analysis should be carried out on the strengthened building to assess its overall ductility and compliance with the requirements of the selected ductility class; (iv) are based on the q-factor approach and linear methods of analysis that may not provide the exact picture of the most probable failure mechanism of the existing building; (v) and, finally, when Eurocode 2 or other design and detailing national codes are adopted in combination with national standards that define seismic actions, do not consider specific seismic detailing that assures the achievement of acceptable levels of ductility at the member or joint level. The latter point is in fact critical when strengthening techniques at columns are employed, which might aim at avoiding the formation of possible soft-storey mechanisms by relying on beams to control the performance of the existing building.

A brief overview on the differences between the conceptual principals that set the basis of the various assessment approaches followed by Portuguese practitioners is presented in Table 1. It may be readily drawn from Table 1 that assessing existing buildings to EC8-1 might be too demanding for the structure, as more stringent performance requirements are prescribed by the European code, particularly related with the serviceability limit states. In fact, the aim of EC8-3 is not to attempt upgrading an existing building to comply with design code provisions, since the strengthening costs would be prohibitive, but rather to ensure acceptable levels of damage [21].

Furthermore, whilst EC8-3 establishes acceptance criteria at a member level that consist of controlling deformation demands in ductile members and strengths in brittle members, EC8-1 establishes global criteria, e.g. control of second-order effects, that might imply significant lateral strength and stiffness requirements for the structure [27]. These global criteria could easily result in over-strengthened and more expensive solutions. In turn, the Portuguese RSA [23], which may be representative of other seismic codes in use since the 80s, does not refer to any serviceability verifications and defines a seismic action that is associated to a return period twice that of EC8-1.

Besides seismic strengthening, other issues, such as the change in usage of the existing building, the need for extending the building dimensions or the development of visible damage, may imply the implementation of further strengthening

Table 1 Comparison between different seismic assessment approaches

	National code (RSA) with Eurocode 2	Part 1 of Eurocode 8	Part 3 of Eurocode 8
Knowledge of the structure	• Not applicable	• Not applicable	• Definition of three knowledge levels and confidence factors [5, 7, 21, 25, 26]
Performance requirements	• Ultimate limit state: safety of people and/or of the structure. Internal failure or excessive deformation of the structure or members	• No-(local-) collapse limit state [26]. Prevention of collapse of any structural member and retention of structural integrity and residual load capacity in the event aftermath • Damage limitation state. Mitigation of property loss by limiting non-structural damage. Structural elements should retain their full strength and stiffness and need no repair [26]	• Near collapse limit state. Heavily damaged structure with low residual strength and stiffness • Significant damage limit state. Significantly damaged structure with some residual strength and stiffness • Damage limitation limit state. Lightly damaged structure with structural elements prevented from significant yielding
Global criteria	• Analysis including geometric imperfections and second-order effects	• Structural regularity • Ductility class • Control of second-order effects and inter-storey drifts at the damage limitation state • Weak beam-strong column condition	• Not applicable
Local criteria	• Force-based design and detailing of members	• Force-based design and detailing of ductile and brittle members	• Control of deformations (chord rotations) in ductile members [4, 21, 26] • Control of strengths in brittle members

measures. However, as mentioned above, Eurocodes do not provide general assessment and strengthening specifications, leading practitioners to rely on their engineering judgment. A rational approach would be to apply the concepts introduced by EC8-3 in the collection of the information necessary for structural assessment and to follow, for instance, the general rules provided by each specific Eurocode (e.g., EC2) by taking into account the mean values of material properties divided by the EC8-3 confidence factor and specific partial factors, as well. Issues related with the sustainability and life-time of the strengthened building should be equally addressed [28, 29], as for example the eventual capacity deterioration due to environmental effects or fatigue and their consequent costs.

2.2 Structural Deficiencies and Strengthening Needs

Older non-ductile gravity-load designed RC buildings oftentimes conform to irregular architectural layouts, and are known to be potentially vulnerable to earthquakes [30]. The lack of ductility and lateral stiffness and strength, the absence of resistance hierarchy criteria, the poor detailing (e.g. lack of stirrups and insufficient amount of reinforcements) are some of the most common structural deficiencies found in this type of buildings. The seismic strengthening of older RC buildings located in seismic prone-regions is a critical issue and it should be taken as a first priority by public authorities and stakeholders.

As noted by Fardis [26], the aim of strengthening is to assure that all relevant elements fulfil the general verification inequality $E_d \leq R_d$, where E_d refers to demands and R_d to capacities, which could be achieved either by reducing demands or by increasing capacities. Different strengthening measures can be implemented to verify this inequality depending on the deficiencies exhibited by the structure. An overview of the most common structural deficiencies and subsequent strengthening measures is presented in Table 2. These measures may be grouped into: (i) local strengthening measures; (ii) removal or reduction of irregularities and structural discontinuities; (iii) global structural stiffening and reinforcement measures; (iv) and, finally, mass reduction measures. The strengthening systems should be equally designed to comply with the general code prescriptions presented in Table 2.

Although EC8-3 does not explicitly define lateral stiffness and strength requirements, the inter-storey drift limits proposed by EC8-1 to control non-structural damage at the serviceability limit state and the criterion prescribed by EC8-1 to assess second-order effects might be adopted. Also, in those cases where the capacity of columns is increased by relying on beams to govern the seismic response of the existing building (e.g., avoid a soft-storey mechanism), the weak beam-strong column condition (clause 4.4.2.3 of EC8-1) should be assured. Obviously, the reliability of such condition will depend on the quality of the information collected from the members framing into that same joint. Moreover, it should be importantly emphasized that the seismic assessment of existing buildings should only be conducted employing linear analysis if the EC8-3 applicability criterion is verified using a limit value for ρ_i (the ratio between the demand D_i obtained from the analysis and the corresponding capacity C_i for the i-th 'ductile' primary element of the structure) in the range of 2–3, and if the EC8-1 weak beam-strong column condition is assured. Otherwise, nonlinear analysis or q-factor analysis using a q factor value equal to 1.5 should be employed, as suggested by EC8-3. These considerations are made following a recent study conducted by Araújo and Castro [21], wherein it was shown that linear analysis can result in local deformation demands with errors greater than 60% if vertical regularity in the distribution of demands is not assured. Finally, it should be noted once again that the structural deficiencies and strengthening needs presented in Table 2 are oriented

Table 2 Common deficiencies, strengthening needs and code-based compliance criteria

Structural deficiencies	Strengthening needs	Code-based compliance criteria
• Lack of member or joint ductility under seismic conditions • Lack of local and/or global strength and stiffness • Insufficient in-plane wall shear strength or flexural capacity • Insufficient reinforcement	• Selective local strengthening at a member level • Add new dissipative elements to provide stable failure mechanisms or increase lateral stiffness and strength • Reduce seismic demand by removing upper stories or adding seismic isolation or supplemental damping • Remove components creating short columns	• Verify local deformation and strength capacities by fulfilling the general safety inequality ($E_d \leq R_d$) at all limit states • Skip ductility verifications if sufficient reinforcement and detailing is provided according to EC8-1 • Control and inclusion of second-order effects in structural analysis (e.g. EC8-1 and EC2) • Control of inter-storey drifts at the serviceability limit state according to EC8-1 • Verify the weak beam-strong column condition as proposed in EC8-1 • Consider the EC8-1 structural regularity criteria for good practice • Verify the EC8-3 linear analysis applicability criterion [4, 21]
• Structural irregularity in plan and elevation • Presence of a weak storey • Change in failure mechanism (e.g., development of soft-storey) as a result of falling down of non-structural components (e.g., coupled infill walls)	• Add strength and stiffness at storey levels to achieve vertical regularity in demands • Ensure diaphragm effect and sufficient connection between elements to guarantee the load path • Use of expansion joints to transform a single irregular building into various regular ones • Add new balancing elements to reduce torsional effects • Uncouple infill walls or use them as structural walls	

towards seismic strengthening. More general cases would most probably require local or global increase in stiffness and strength.

3 Traditional Structural Strengthening Techniques Using Steel Elements

3.1 Increase of Member Capacities Through Bonding or Bolting Steel Plates and Steel Jacketing

The enhancement of member capacities is oftentimes a cost-effective option when only a reduced number of members in a structure are deficient and when columns within weak storeys or at a certain side of the building are to stiffened and strengthened to correct vertical irregularities or torsional imbalance in plan [26]. In general, the strengthening of members should be conducted with the aim to correct its inadequate shear, flexural and axial compression capacities, as well as its irregular plastic hinge confinement and lap splice [31].

A common strengthening technique that aims at achieving such objectives consists on the use of bonded external steel plates applied to the soffit of the existing member. This strengthening measure is usually adopted due to its ease of application, limited disruption and low price of the materials involved in the process. However, despite the fact that external plating of RC elements may be seen as a rational approach to enhance RC members' ductile flexural capacity, a number of previous studies [32–34] have shown that such strengthening technique is highly prone to premature plate peeling failure before the development of the members' full flexural capacities, if not properly designed, as a result of diagonal shear cracks and vertical flexural cracks at the plate curtailment location. To address this issue, Aykac et al. [32] have recently conducted an extensive experimental test campaign on a number of RC beams strengthened with soffit steel plates anchored with the help of bolts or side plates (collars) with different heights. The bottom plate was epoxy-bonded to the beam in every specimen. For illustrative purposes, Fig. 1a presents the layout of some specimens. Design aspects were discussed by Aykac et al. [32] and further recommendations were provided, namely: (i) external plates terminating at the shear spans are more susceptible to shear peeling; (ii) thin plates are less liable to peel off and do not need additional anchorage; (iii) thick plates anchored with collars were seen to be the most effective strengthening solution,

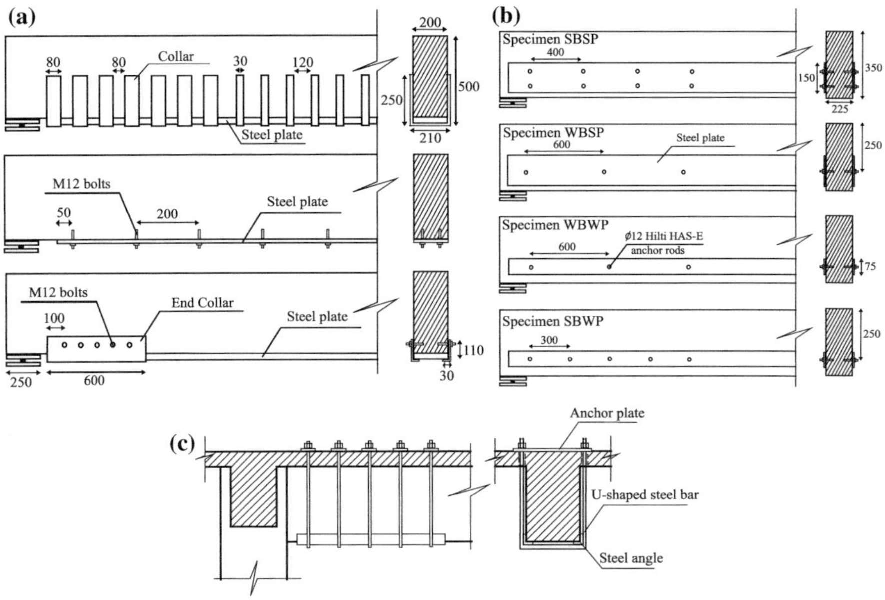

Fig. 1 Strengthening of RC beams by means of: **a** bonded steel plates (adapted from Aykac et al. [32]), **b** bolted side-plates (adapted from Su et al. [37]), **c** and *U-shaped bars* welded to steel angles (adapted from Monitorul Oficial al României [39]). Dimensions in mm

which may be further enhanced by including discrete collars along the length of the member, as represented in Fig. 1a. The side legs of the end collars should be extended at least to two-thirds of the height of the beam. Likewise, Zhang et al. [35] investigated the feasibility of strengthening two-way RC slabs by externally bonding steel plates. It was concluded that not only such strengthening technique can greatly increase both cracking and ultimate strengths, but also plate peeling failure does not occur as opposed to plated beams. To facilitate the implementation of such technique on site when the plates are too large to be handled, orthogonally placed steel strips may be used to achieve the same amount of external reinforcement in the plated region. The suitability of designing steel plate bonding to slabs based on yield line analysis was equally verified by Zhang et al. [35].

Another commonly adopted and attractive solution that aims to increase the strength and shear capacity of existing RC beams consists on the use of bolted side-plates. Contrarily to bonding external plates at the soffit of the beam, anchoring bolted side-plates, as represented in Fig. 1b, might be a logical approach as it does not require the removal of infilled walls and allows propping up the existing RC member. This technique works by relying on the slip between the RC beam and the steel plates, which mobilizes anchor bolts and transfers forces to the steel plates. Significant differential deformation between the steel plates and the RC beam can still occur, particularly when plastic hinging initiates [36], and thus the effect of this partial interaction in the composite behaviour of the beam should be accurately accounted for as it could lead to unsafe design. The mobilization of this partial interaction between the steel plates and the RC beam depends on the strengths of the bolts and plates and controls the most probable failure mechanism. According to Su et al. [37], three failure mechanisms may develop: (i) a concrete-controlled mechanism; (ii) a bolt-controlled mechanism; (iii) and, a plate-controlled mechanism. The first mechanism occurs when strong bolts and strong plates (see specimen SBSP in Fig. 2b) are used. Members develop their full strength capacity by exploiting the coupled behaviour between the concrete and the steel plates and brittle failure is expected to occur due to concrete crushing. The second mechanism occurs when weak bolts (see specimens WBSP and WBWP in Fig. 2b) are employed. The members reach their full strength capacity similarly to concrete-controlled members and fail when the ultimate deformation of the bolts is reached. Finally, the third mechanism is observed when strong bolts and weak plates (see specimen SBWP in Fig. 2b) are used. Steel plates are expected to yield before the achievement of the member's full strength and to govern the post-peak softening behaviour. Su et al. [37] concluded that, although a significant strength enhancement is achieved when the SBSP arrangement is adopted, the SBWP arrangement is recommended due to its ductile failure mode. The use of bolted side-plates in the strengthening of existing RC coupling beams was equally assessed and validated by Su and Zhu [38]. Finally, it should be referred that the Romanian P 100-3/2008 code [39] is one of the very few normative documents that provide guidance and detailing in terms of the strengthening of existing members. In addition to the detailing of bolted side-plates with height equal to the clear web of the RC beam bonded with epoxy resin (special attention should be paid to this

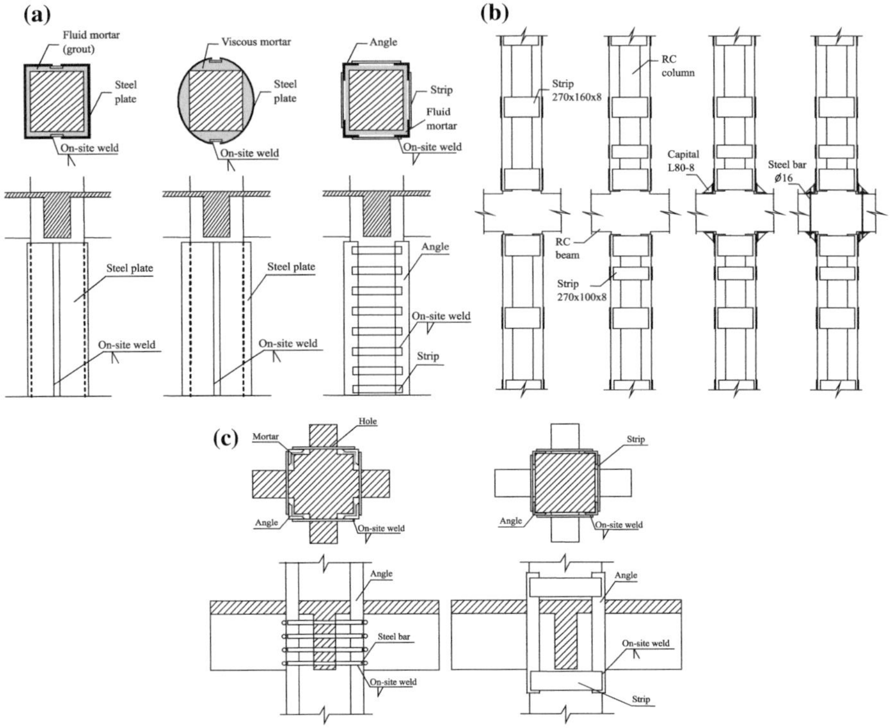

Fig. 2 Strengthening of RC columns and joints by means of steel jacketing: **a** thin-walled jackets and cages strips welded to four corner angles (adapted from Monitorul Oficial al României [39]), **b** details tested at the Technical University of Valencia (adapted from Refs. [42–45]), **c** joint jacketing (adapted from Monitorul Oficial al României [39]). Dimensions in mm

strong plate solution according to the results obtained by Su et al. [37]) or bonded U-shaped steel plates that embrace the beam's soffit with pressure-injected cement or epoxy mortar, the Romanian code also proposes a simple strengthening technique that consists of embracing the beam's soffit with U-shaped steel bars anchored to the slab and welded to steel angles, as represented in Fig. 1c. The anchor plates should have a thickness greater than 5 mm and should be spaced by at least 200 mm.

For what concerns the strengthening of existing RC columns using steel elements, steel jacketing is the most rational and widely adopted solution. Although being more expensive than concrete jacketing, this technique is simple, familiar to the industry and readily available [26]. As briefly defined in EC8-3, this technique consists of caging existing rectangular RC columns with four corner angles to which continuous steel plates or discrete steel strips are welded (see Fig. 2a). The corner angles may be epoxy-bonded to the concrete or simply made to adhere to the concrete without gaps along the height of the column. The 10–20 mm gap between the plate and the surface of the column should be grouted with non-shrink mortar

[26] (see Fig. 2a) and the steel strips can be pre-heated in the range between 200 and 400 °C prior to welding, to enhance the confining effect. Alternatively, thin-walled steel jackets may be used. This solution is particularly relevant for the case of circular columns and consists of semi-circular halves fitting closely around the column and field-welded along two vertical seems. Clearly, this strengthening technique aims at adding up additional confinement to that already provided by transverse hoops, increasing the member's shear capacity and correcting deficient lap-splices [40]. Moreover, as thoroughly discussed by Montuori and Piluso [41], the confinement effect may equally enhance the member's deformation capacity and laterally as well as restrain steel rebars from buckling. The corner steel angles may also provide an effective increase of longitudinal reinforcement. An extensive experimental campaign was carried out at the Technical University of Valencia [42–45] to investigate the various parameters that influence the effectiveness of steel jacketing. The following conclusions were drawn: (i) the type of mortar adopted (cement or epoxy) does not affect the behaviour of the column; (ii) unloading the column prior to strengthening improves its behaviour, although not significantly; (iii) the addition of strips of smaller size at the column ends (Fig. 2b) considerably improves its axial strength and ductility [42]; (iv) the inclusion of capitals (Fig. 2b) in axially loaded members may enhance their axial capacity, although it may also cause local failure at the joint. Capitals connected by means of steel bars through the joint (Fig. 2b) can be used to ensure a convenient transfer of forces between the angles located on both sides of the joint [43, 44]. This solution enhances both flexural and deformation capacities; (v) and, finally, if the influence of the beam-column joint is eliminated, e.g., using capitals connected with steel bars, the capacity of the member could be accurately estimated from Eurocode 4 [46]. In turn, as noted by Fardis [26], if a 25–50 mm gap between the end of the jacket and the member end exists, the yield and the moment resistance of the strengthened member should be taken equal to that of the original member. EC8-3 provides guidance for quantifying the increase in the member's shear capacity and for detailing regions where deficient lap-splices are present. Moreover, the benefits evidenced by Garzón-Roca et al. [45] in continuing the steel angles along the joint may also enhance the shear capacity of the joint itself. Some detailing is provided by the Romania P 100-3/2008 code, as represented in Fig. 2c.

Finally, it should be referred that, according to Nagasprasad et al. [47], the use of an end strip at the column base with height equal to 1.5 times the height of intermediate strips allows the achievement of a desirable moment capacity in strengthened columns. A design procedure has also been proposed by those Authors [47].

New strengthening techniques have been recently investigated to increase the capacity of existing RC columns and joints. By recognizing the negative stress-lagging effect between the original column and the new jacket due to pre-existing load, Wang and Su [48] presented a novel pre-cambered steel plate strengthening approach that aims at alleviating such negative effects. In turn, Said and Nehdi [49] proposed a strengthening technique for RC joints that consists of using local steel braces. This technique was shown to significantly delay brittle joint shear failure and moved plastic hinging away from the face of the column, thus enhancing the anchorage of beam bottom reinforcement.

3.2 Reduction of Demands Through Lateral Strengthening and Stiffening with Concentric Steel Bracing Systems

As opposed to local strengthening, global interventions can be considered a cost-effective solution when disruption of occupancy and damage to partitions, architectural finishes and other non-structural components are of concern [26]. Previous earthquakes have shown that business interruption, non-structural components and contents losses oftentimes overcome structural losses. Current assessment codes, such as EC8-3, do not explicitly address this issue, simply limiting the level of damage to structural components subjected to frequent earthquakes.

A commonly adopted technique that aims at increasing the lateral stiffness and strength of a specific storey or overall structure consists of adding steel bracing systems to the existing RC building. One advantage of this strengthening technique is the reduced impact on foundations [14, 26], especially if compared to other design solutions such as the use of RC shear walls. The braces are typically designed to provide earthquake resistance [26], although Sousa and Castro [10] showed that allowing the existing building to undergo inelastic deformations may be beneficial from an economical and structural stand point. In practice, structural engineers tend to adopt different approaches and structural analysis procedures when faced with the design of the bracing system [14]. According to Fardis [26], if linear analysis is adopted, two independent gravity- and seismic-alone analyses should be conducted for the existing and retrofitted buildings, respectively, and the results should be superimposed afterwards. Instead, if nonlinear analysis is used, both actions should be taken as acting simultaneously. Typically, steel bracing systems consist of concentric X-diagonal, V or inverted-V braces placed within (Fig. 3a) or outside the plane of the existing RC building (Fig. 3i). Whilst the first solution facilitates the composite behaviour of the final system, the latter could be beneficial for architectural and disruption reasons. Youssef et al. [50] demonstrated the benefits of concentric steel bracing systems in increasing the seismic performance of existing RC frames and provided further insights on design methodologies.

A major concern that has been the focus of extensive research in the past is related with the transfer of forces between the bracing system and the existing RC frame. Massumi and Tasnimi [51] experimentally tested five types of connections between the gusset plate located at the end of the brace and the RC joint, concluding that: (i) braces simply bolted at the end of columns (see specimen BF12 of Fig. 3b) exhibit significant cyclic strength deterioration and do not allow the new system to achieve its maximum potential strength capacity; (ii) bolted connections

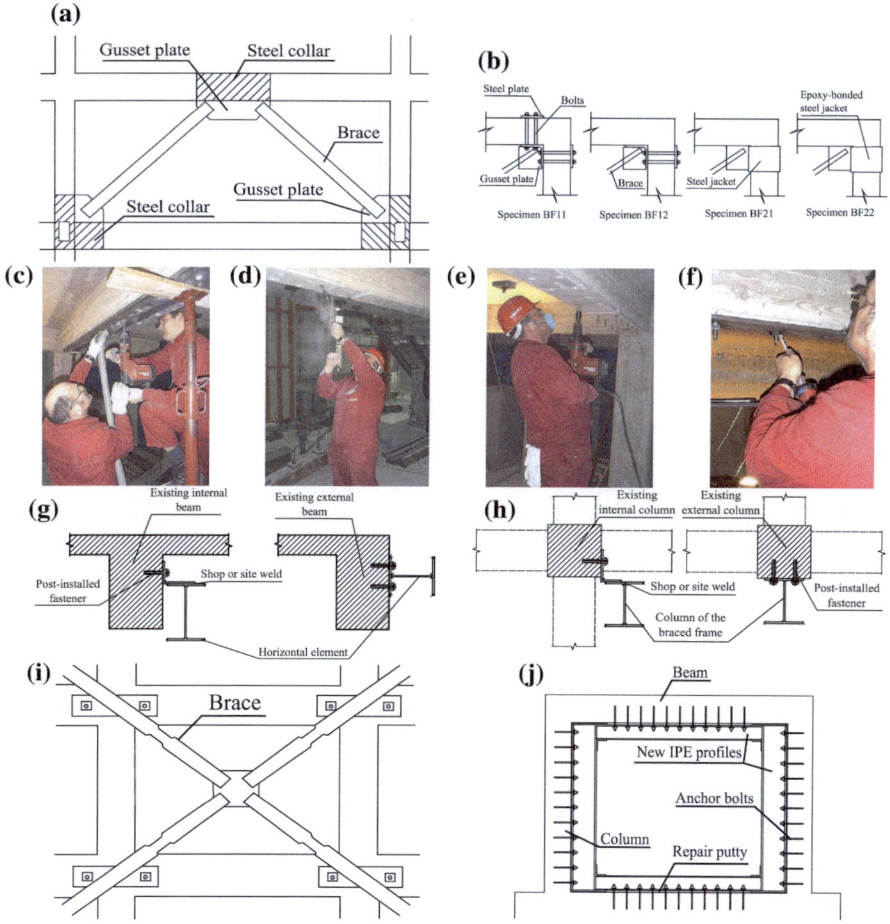

Fig. 3 Strengthening of existing RC frames: **a** schematic representation of the steel bracing system adapted from Monitorul Oficial al României [39], **b** types of local connections to existing elements (adapted from Massumi and Tasnimi [51]), **c–f** implementation of new steel elements through epoxy-grouted post-installed fasteners (images taken by Varum [11]), **g** connection between the new horizontal steel member and the existing RC beam (adapted from Refs. [14, 39]), **h** connection between the new steel column and the existing RC column (adapted from Refs. [14, 39]), **i** attachment of the bracing system to the façade of the building (adapted from Monitorul Oficial al României [39]), **j** infill steel frame tested by Ozcelik and Binici [52]

to both column and beam ends (see specimen BF11 of Fig. 3b) were seen to provide good results, similarly to those found by Youssef et al. [50]; (iii) connections based on caging the column end without any bonding adhesive (see specimen BF21 of Fig. 3b) were seen to be prone to slippage, thus not being able to transfer conveniently the forces between the braces and the existing structure, and should be avoided; (iv) and, higher strength capacities and energy dissipation were obtained when epoxy-bonded jacketing solutions (see specimen BF22 of Fig. 3b) were adopted.

A comprehensive discussion on the design of this type of connections is reported in [53]. Nevertheless, despite the less intrusive character of these local connections, they might be insufficient when high lateral (e.g., inertia) forces are expected to develop. A common alternative solution consists of installing the diagonal braces within a steel frame made of steel profiles (i.e. beams and columns) firmly attached to the surrounding concrete members [14, 26]. This connection is usually done by means of epoxy-grouted post-installed fasteners, as shown in Fig. 3c–f, consisting of three phases: (i) drilling of partial-depth holes using the annuli of the steel members as template (Fig. 3c); (ii) removal of the steel member and finishing and cleaning of the holes (Fig. 3d); (iii) and fixing of the new member (Fig. 3f) after grouting the holes with epoxy and inserting the fasteners (Fig. 3e). Other techniques to attach the new steel elements to the existing RC frame have been studied by Ishimura et al. [54], wherein existing procedures to design this type of strengthening intervention have been equally assessed and validated. It should be referred that the installation of new braced frames within the existing RC frame, as adopted by Varum [11] (Fig. 3c–j), might be preferable when a single storey is to be strengthened, since it does not allow continuity along the height of the building [14]. Otherwise, the new braced frame should be placed alongside the existing RC frame, as illustrated in Fig. 3g, h, or externally at the façades of the buildings, as represented in Fig. 3i, for which some detailing in the connections may be found in the Romanian P 100-3/2008 code. Finally, experimental testing conducted by Ozcelik and Binici [52] has demonstrated that strengthening deficient pre-code existing RC buildings with infill steel frames alone is still a technically sound solution. The experimental setup adopted by those Authors is schematically represented in Fig. 3j. Based on the strengthening of an existing five-storey RC building located in Istanbul, the Authors have demonstrated the adequacy of this technique.

4 Innovative Structural Strengthening Techniques Using Steel Dissipative Systems: Outcomes from the ILVA-IDEM Project

The topic of strengthening existing RC buildings is rather pertinent and has been the focus of significant research in the last decades. Traditional concentric steel bracing systems were seen to provide a satisfactory enhancement of the seismic performance of deficient buildings by relying on steel braces to dissipate the energy input by the ground motion. However, conventional steel braces exhibit a complex behaviour triggered by local and global instability phenomena and fracture. Therefore, to overcome this issue from both structural and economic point of view, innovative strengthening techniques that incorporate energy dissipation devices, such as friction, viscoelastic or metallic dampers, have been developed [9]. These innovative solutions have the advantage of reducing the level of demands in the original structure by concentrating the energy dissipation on devices that can be easily replaced following the seismic event. To better demonstrate the features of some of these innovative solutions, the outcomes from the experimental campaigned carried out by the University of Naples "Federico II" group within the ILVA-IDEM project [55–60] are herein briefly presented. The ILVA-IDEM project represented a unique research opportunity to conduct full-scale testing of various strengthening techniques implemented in a real two-storey RC building destined for demolition [55, 56]. Three innovative strengthening techniques using metal elements were adopted: (i) eccentric braces; (ii) buckling-restrained braces; (iii) and, shear panels in steel or pure aluminium.

4.1 Eccentric Braces

Strengthening systems using eccentric braces (EBs) typically consist of an inverted Y-shaped assembly wherein a dissipative link is attached to the existing RC slab or beam [57, 58], as shown in Fig. 4a, b). A key aspect in the design of eccentric braces is related to the definition of the cross-section and length of the link, which determines the shear plastic strength of the storey and the stiffness of the bracing system, respectively [57]. Three eccentric bracing solutions were investigated in the ILVA-IDEM project by designing the link according to different criteria. In the first case, the braces were designed according to EC8-1, but neglecting capacity design criteria. The shear strength capacity was estimated based on the first-yielding definition provided by Popov and Engelhardt [61] and a link length that satisfied the conditions of being the shortest, achieving the maximum inelastic link shear rotation with the first plastic hinge forming in the existing building. The EC8-1 inter-storey drift limit at the serviceability limit state was also considered in the design [9]. An end-plate bolted connection between the link and the slab was adopted (Fig. 4c), whereas the braces were connected to the frame by means of gusset plates bolted to the beam-to-column joint (Fig. 4d, e). In the second case, the link end-plate connection was additionally designed using capacity design principles, leading to an increasing in its thickness. Finally, in the third case, the link was designed to increase the system ductility by forcing the plastic deformations to be confined within the links. A steel built-up section without web stiffener was designed so that the shear strength of connections would be two times greater than the average yield strength of the links. This solution resulted on the need of increasing the steel grade of the bolts. The following conclusions were drawn [57, 58]: (i) the three tested eccentric bracing systems were seen to provide promising results, substantially increasing the lateral stiffness and strength of the existing RC building; (ii) although the link designed under capacity design principles lead to the highest increase in the lateral strength of the system, it failed suddenly due to shear failure of the bolts connecting the link to the braces. In turn, the most dissipative

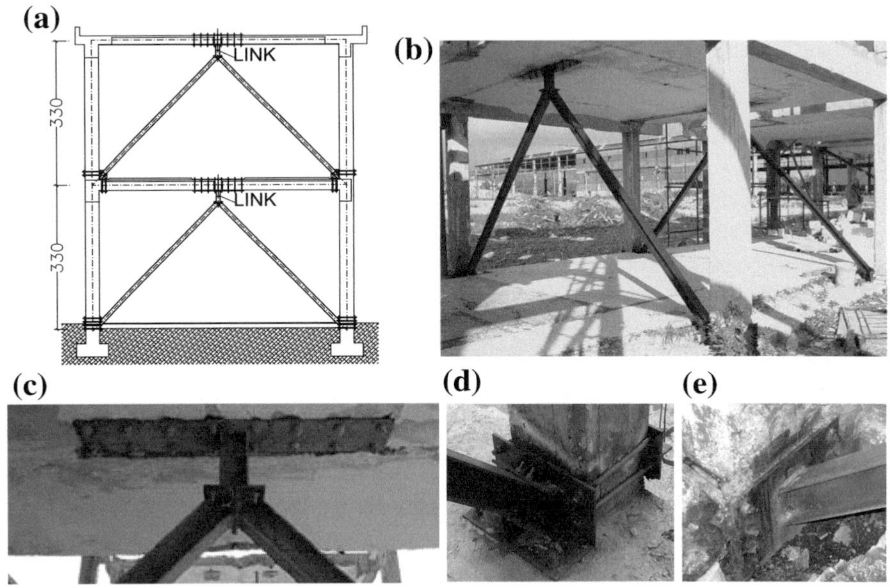

Fig. 4 Strengthening of the existing RC building studied in the ILVA-IDEM project with eccentric braces: **a** the adopted solution, **b** the strengthened building, **c** link-to-slab connection, **d** brace-to-RC beam-column joint, **e** brace-to-RC column base connection

built-up link provided the best performance enhancement both in terms of increasing the system's lateral stiffness and strength and the energy dissipation capacity; (iii) and, finally, the link's shear over-strength was found to play a critical role on the overall strength of the system and to be significantly larger than previously expected. Recently, Della Corte et al. [62] have shown that the link shear over-strength mostly depends on the combined effect between tensile axial forces, developing due to axial deformation restraints and nonlinear geometrical effects, and the link geometrical properties, namely the ratios between the flange and web areas and between the link length and the cross-section depth.

4.2 Buckling-Restrained Braces

Buckling-Restrained Braces (BRBs) have been the subject of extensive research and consist of a core axial load-bearing steel element, subdivided into a central yielding part and two non-yielding segments at the core ends, involved by a casing buckling-restraining element that aims at preventing global brace buckling and restraining high-mode core buckling [59]. This advantageous ability of preventing overall buckling phenomena allows designing BRBs for yielding at low inter-storey drift levels, anticipating dissipative action and providing substantial and stable energy dissipation capacity to existing RC buildings. Two types of BRBs were largely investigated in the scientific literature, whose main difference consists in the adopted casing, namely: (i) unbounded braces, in which the restraining tube is filled with either concrete or mortar and an unbounding layer is added to the contact surface between the core and the filling concrete with the aim of allowing the core to freely slide inside the casing element and expand in compression [63]; (ii) and all-steel braces, which typically consist of two or more restraining tubes placed in direct contact with the yielding steel core. In both types of device, an air gap between the brace and the restraining tube is present to allow relative deformations between both members [59]. All-steel BRBs have the advantage of being lighter than unbounded braces and can be designed to be detachable, facilitating inspection and maintenance. Further details on the seismic design of this type of elements can be found in [63–65].

Figure 5a, b show the overall arrangement of BRB system implemented in the ILVA-IDEM project RC building. Bolted connections were adopted to connect the braces to the RC beam-column joints (see Fig. 5c). Two BRB solutions have been investigated in the test campaign: (i) a first one, represented in Fig. 5d, that consists of two buckling-restraining rectangular steel tubes in contact with an internal core element and a couple of internal plates welded to the tube walls to complete restraining action; (ii) and a second one, shown in Fig. 5e, composed by a core tapered in a more gradual manner, thus reducing the buckling length of the end portion of the core, to provide extra flexural stiffness for out-of-plane buckling strength and to allow full axial yielding of the core element. The second BRB was conceived to be fully detachable by joining the two buckling-restraining tubes together using bolted stiffened elements (see Fig. 5e, f). The following conclusions were drawn [57–59]: (i) the available theoretical models were seen to provide adequate design and performance control of BRBs; (ii) BRBs with a core tapered more sharply (see Fig. 5d) were seen to be more susceptible to local buckling at the brace ends, producing strong flexural deformation and failure of the closing plates. Although ensuring the expected increase in the lateral stiffness and strength of the existing building, this type of BRBs was characterised by a rather limited ductility capacity; (iii) detachable BRBs with a core tapered in a more gradual manner were seen to provide, not only the expected lateral stiffening and strengthening of the existing frame, but also a substantial increase in its deformation capacity, which was around two times that of the bare RC structure. However, local buckling of one end plate was still observed, even satisfying capacity design criteria, which resulted from strong local geometrical imperfections and a flexible end-restraint; (iv) at last, the brace-to-RC joint connections were seen to influence the loading/unloading stiffness of the system. Still, it was not possible to derive a general conclusion since these effects depend on the technique adopted to drill the holes required to fill in the mortar on site. Further design recommendations to BRB systems have been provided within the Steelretro project [13].

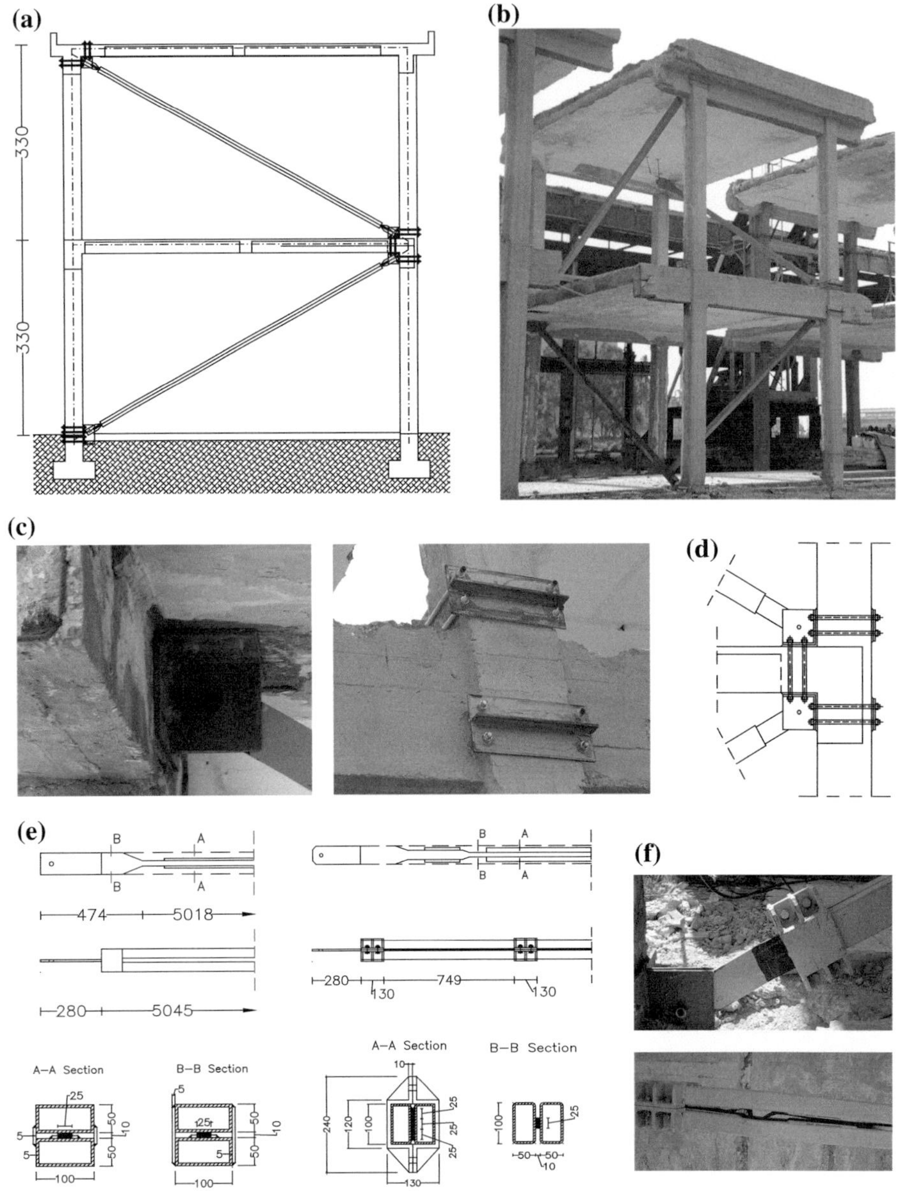

Fig. 5 Strengthening of the existing RC building studied in the ILVA-IDEM project with buckling restrained braces: **a** adopted solution, **b** the strengthened building, **c** brace-to-RC beam-column joint, **d** first un-detachable all-steel BRB tested, **e**, **f** and second detachable all-steel BRB tested. Dimensions in mm

4.3 Metal Shear Panels

Innovative strengthening systems using metal shear panels (MSPs) offer a rational and innovative strengthening solution as they combine significant strength, stiffness and ductility enhancements with limited weight, low erection costs and space constraints. However, contrary to steel bracing systems, MSPs still require deeper attention and further research investigations [60]. MSPs typically consist of a series of steel plates located around a certain service area or within perimetral frames with the aim of creating a stiffening core capable of resisting horizontal demands. They are essentially sub-divided into two main categories: (i) compact shear panels; (ii) slender shear panels. The former type aims at dissipating the energy input by ground motions through the development of a pure shear-resisting mechanism [60]. In such cases, shear buckling phenomena should be avoided by using low yield strength steel or pure aluminium materials, properly endowed with flexural stiffeners. The slender shear panels are made of thin steel plates connected to the members of a surrounding steel frame by means of welded or bolted connections that aim at improving the strength and stiffness of the original structure. This type of MSPs are expected to buckle under elastic regime and to activate a tension field mechanism characterized by diagonal (i.e., parallel to the direction of the principal stresses) tensile bands at the web that transfer the lateral shear forces. This mechanism produces a hysteretic behaviour characterized by a pronounced pinching effect [57, 60]. A design procedure that relies on the participation of the original RC building in addition to that of MSPs for resisting external horizontal actions has been proposed by De Matteis [60]. Extensive numerical analysis on the behaviour of slender MSPs has also been conducted by Formisano et al. [66]. If the strengthening system is to be designed as a dual system, particular attention has to be paid to the assessment of the capacity of the existing RC frame to accommodate the forces resulting from tension field mechanism developing in the MSP.

According to the Steelretro project report [13], load transfer beams can be a favourable solution to enable transferring the forces to parts of the structure with sufficient capacity. This was the solution adopted in the ILVA-IDEM project, as depicted in Fig. 6a, wherein two shear panels with a DX56D steel (Fig. 6b) and an EN AW1050 A aluminium alloy (Fig. 6c) were adopted [60]. The performance of the strengthened RC building was seen to be significantly improved by both MSPs [57, 60], although the aluminium MSPs provided a more satisfactory dissipative behaviour. Both cases relied on a dual system, in which a combined plastic mechanism characterized by hinging at the RC beam-column joints and plastic deformation of the tensioned diagonals of the MSPs (Fig. 6b, c) was developed.

Mazzolani [57] carried out a comparison between the efficiency of the three innovative strengthening techniques herein briefly presented (i.e., EBs, BRBs and MSPs) and the response of the original bare RC building. Overall, the interpretation of the results presented by [57] allowed concluding that each single strengthening technique leads to a substantial increase in the lateral stiffness and strength of the existing building. The increase was more pronounced when EBs were adopted,

Fig. 6 Strengthening of the existing RC building studied in the ILVA-IDEM project with shear panels: **a** image of the strengthened building, **b** and **c** plastic mechanisms developed in the aluminium and steel panels, respectively (image taken from Mazzolani [57])

although these systems, as mentioned before, were more prone to brittle behaviour due to shear failure of the bolts connecting the link and the braces. BRBs and MSPs, in turn, provided a similar enhancement in the lateral stiffness and strength of the building, although, still, the highest increase in the deformation capacity of the original building was achieved using detachable BRBs with a core tapered in a more gradual manner. Interestingly, a multi-criteria decision making analysis recently carried out by Formisano and Mazzolani [67], that accounted for various optimization conditions (i.e., intervention costs, vulnerability reduction, intervention feasibility, disturbance to occupiers, functional-to-aesthetic compatibility, reversibility and protection from damage), has shown that, for the specific case of the RC building investigated in the ILVA-IDEM project, MSPs seem to be the optimum strengthening solution, followed by BRBs and EBs.

5 Final Considerations

The strengthening of existing RC buildings is an issue of paramount importance to which most structural engineering consultants and practitioners are oftentimes exposed to, particularly when major topics such as Urban Rehabilitation or Disaster Risk Reduction are opened for debate by public authorities, scientific community and stakeholders. However, despite the relevance that strengthening of the existing building stock has in the economic, social and cultural environment of a region, very few guidelines and codes addressing this specific subject are available in Europe, and those that have been issued simply focus on the seismic assessment and strengthening of existing buildings. This lack of clarity and guidance from current Eurocodes lead practitioners to commonly rely on their engineering judgment and to resort to assessment procedures that are conceptually suitable to design new structures, but not for the assessment of existing ones. Therefore, this chapter aimed at providing a brief and concise discussion on the assessment procedures most commonly adopted by practitioners and at clarifying the adequacy of such approaches. Nevertheless, despite the brief recommendations provided herein, the use of approaches that are not specifically dedicated to the assessment of existing buildings should be employed with caution. The revision of current codes and the proposal of a future generation of documents that more clearly address this issue are significantly needed. Such activities have been recently set as a first priority by many scientific committees, such as the ECCS-TC13 seismic design committee.

This chapter has also discussed the use of steel elements as a rational, cost-effective and efficient strengthening solution. A number of strengthening techniques, from traditional to more innovative approaches, have been presented, alongside with a brief discussion on some relevant issues that influence their design and performance. Notwithstanding the information provided in this chapter, the Authors highly encourage the Reader to seek for further information on the specificities of each strengthening technique. In spite of the numerous numerical and experimental studies already carried out in the past focusing on the assessment of each strengthening technique, guidelines and codes that systematize this information and provide guidance on the design and detailing of those techniques still do not exist in Europe. Nevertheless, due to the nature and creativeness of the engineering community, it is expected that new strengthening and retrofitting techniques will be proposed in the coming years and decades that will surely represent further advances in finding the most cost-effective solutions to adopt in the intervention of existing RC buildings.

Acknowledgements Part of this work reports to research financially supported by Project POCI-01-0145-FEDER-007457—CONSTRUCT—Institute of R&D in Structures and Construction, funded by FEDER funds through COMPETE2020—Programa Operacional Competitividade e Internacionalização (POCI) and by national funds through Fundação para a Ciência e a Tecnologia (FCT).

References

1. Romão X, Penna A. Code-based procedures for seismic safety assessment and retrofit. In: Costa A, Arêde A, Varum V, editors. Strengthening and retrofitting of existing structures. Building pathology and rehabilitation. Berlin: Springer; 2017.
2. Rodriguez M, Park R. Repair and strengthening of reinforced concrete buildings for seismic resistance. Earthq Spectra. 1991;3(7):439–59.
3. CEN. ENV 1998-3 Eurocode 8: design of structures for earthquake resistance—part 3: assessment and retrofitting of buildings. Brussels: European Committee for Standardization; 2005.
4. Araújo M, Castro JM. On the quantification of local deformation demands and adequacy of linear analysis procedures for the seismic assessment of existing steel buildings to EC8-3. Bull Earthq Eng. 2016;14(6):1613–1642.

5. Romão X, Delgado R, Guedes J, Costa A. A comparative application of different EC8-3 procedures for the seismic safety assessment of existing structures. Bull Earthq Eng. 2010;8:91–118.
6. Romão X, Gonçalves R, Costa A, Delgado R. Evaluation of the EC8-3 confidence factors for the characterization of concrete strength in existing structures. Mater Struct. 2012;45(11): 1737–58.
7. Franchin P, Pinto PE, Rajeev P. Confidence factor? J Earthq Eng. 2010;14(7):989–1007.
8. Pinto PE. The Eurocode 8—part 3: the new European code for the seismic assessment of existing structures. Asian J Civ Eng. 2005;6(5):447–56.
9. D'Aniello M. Steel dissipative bracing systems for seismic retrofitting of existing structures: theory and testing. Ph.D. dissertation, Università degli Studi di Napoli Federico II, Naples; 2007.
10. Sousa R, Castro JM. Seismic design of RC frames retrofitted with concentric steel braces. In: Mazzolani F, Herrera R, editors. STESSA 2012: behaviour of steel structures in seismic areas. London: Taylor and Francis; 2012.
11. Varum H. Seismic assessment, strengthening and repair of existing buildings. Ph.D. dissertation, Universidade de Aveiro, Aveiro; 2003.
12. Varum H, Dias-Teixeira F, Marques P, Pinto AV, Bhatti AQ. Performance evaluation of retrofitting strategies for non-seismically designed RC buildings using steel braces. Bull Earthq Eng. 2013;11(4):1129–56.
13. RFCS. Steel solutions for seismic retrofit and upgrade of existing constructions (Steelretro). Brussels: Research Found for Coal and Steel; 2013.
14. FEMA. FEMA 547 Techniques for the seismic rehabilitation of existing buildings. Washington: Federal Emergency Management Agency; 2006.
15. Asteris PG, Chronopoulos MP, Chrysostomou CZ, Varum H, Plevris V, Kyriakides N, Silva V. Seismic vulnerability assessment of historical masonry structural systems. Eng Struct. 2014;62–63:118–34.
16. Furtado A, Rodrigues H, Varum H, Costa A. Evaluation of different strengthening techniques' efficiency for a soft storey building. Eur J Environ Civ Eng. 2015;. doi:10.1080/19648189.2015.1119064.
17. Silva V, Crowley H, Varum H, Pinho R, Sousa L. Investigation of the characteristics of Portuguese regular moment-frame RC buildings and development of a vulnerability model. Bull Earthq Eng. 2015;13(5):1455–90.
18. Searer GR, Paret TF, Freeman SA. ASCE-31 and ASCE-31: what good are they? In: Proceedings of the ASCE Structures Congress; 2008.
19. Paret TF, Searer GR, Freeman SA. ASCE 31 and 41: Apocalypse now. In: Proceedings of the ASCE Structures Congress; 2011.
20. ASCE. Seismic evaluation and retrofit of existing buildings (ASCE/SEI 41-13). Reston: American Society of Civil Engineers; 2014.
21. Araújo M, Castro JM. A critical review of European and American provisions for the seismic assessment of existing steel moment-resisting frame buildings. J Earthq Eng. 2017;. doi:10.1080/13632469.2016.1277568.
22. CEN. ENV 1998-1 Eurocode 8: design of structures for earthquake resistance—part 1: general rules, seismic action and rules for buildings. Brussels: European Committee for Standardization; 2004.
23. RSA. Regulamento de Segurança e Acções. Decreto-Lei nº 235/83. Lisboa: Imprensa Nacional – Casa da Moeda; 1983 (in Portuguese).
24. CEN. ENV 1992-1-1 Eurocode 2: design of concrete structures. Part 1-1: general rules and rules for buildings. Brussels: European Committee for Standardization; 2004.
25. Romão X, Delgado R, Costa A. Practical aspects of demand and capacity evaluation of RC members in the context of EC8-3. Earthq Eng Struct Dyn. 2010;39(5):473–99.
26. Fardis MN. Seismic design, assessment and retrofitting of concrete buildings based on EN-Eurocode 8. In: Ansal A, editor. Geotechnical, geological and earthquake engineering. Berlin: Springer; 2010.

27. Peres R, Castro JM. Comparison of European and American approaches for consideration of P-Δ effects in seismic design. In: Proceedings of the 14th European Conference on Earthquake Engineering; 2010.
28. Landolfo R, Cascini L, Portioli F. Sustainability of steel structures: towards an integrated approach to life-time engineering design. Front Archit Civ Eng China. 2011;5(3):304–14.
29. Kappos A, Dimitrakopoulos E. Feasibility of pre-earthquake strengthening of buildings based on cost-benefit and life-cycle cost analysis with the aid of fragility curves. Nat Hazards. 2008;45(1):33–54.
30. Liel A, Deierlein G. Cost-benefit evaluation of seismic risk mitigation alternatives for older concrete frame buildings. Earthq Spectra. 2013;29(4):1391–411.
31. FIB. Seismic assessment of retrofit of reinforced buildings. Lausanne: Fédération Internationale du Bétion; 2003.
32. Aykac S, Kalkan I, Aykac B, Karahan S, Kayar S. Strengthening and repair of reinforced concrete beams using external steel plates. J Struct Eng. 2013;139(6):929–39.
33. Oehlers DJ. Reinforced concrete beams with plates glued to their soffits. J Struct Eng. 1992;188(8):2023–38.
34. Leung CKY. Delamination failure in concrete beams retrofitted with a bonded plate. J Mater Civ Eng. 2001;13(2):106–13.
35. Zhang J, Teng J, Wong Y, Lu Z. Behaviour of two-way RC slabs externally bonded with steel plate. J Struct Eng. 2001;127(4):390–7.
36. Siu WH, Su RKL. Effects of plastic hinges on partial interaction behaviour of bolted side-plated beams. J Constr Steel Res. 2010;66(5):622–33.
37. Su RKL, Siu WH, Smith ST. Effects of bolt-plate arrangements on steel plate strengthened reinforced concrete beams. Eng Struct. 2010;32(6):1769–78.
38. Su RKL, Zhu Y. Experimental and numerical studies of external steel plate strengthened reinforced concrete coupling beams. Eng Struct. 2005;27(10):1547–50.
39. Monitorul Oficial al României. Cod de proiectare seismică—Partea A III-A—Prevederi pentru evaluarea seismică a clădirilor existente. Indicativ P 100-3/2008. Bucureşti: Ministerul Dezvoltării Regionale şi Locuinţei; 2009 (in Romanian).
40. Aboutaha RS, Engelhardt MD, Jirsa JO, Kreger ME. Retrofit of concrete columns with inadequate lap splices by the use of rectangular steel jackets. Earth Spectra. 1996;12(4): 693–714.
41. Montuori R, Piluso V. Reinforced concrete columns strengthened with angles and battens subjected to eccentric load. Eng Struct. 2009;31(2):539–50.
42. Giménez E, Adam JM, Ivorra S, Calderón PA. Full-scale testing of axially loaded RC columns strengthened by steel angled and strops. Adv Struct Eng. 2009;12(2):169–81.
43. Giménez E, Adam JM, Ivorra S, Calderón PA. Influence of strips configuration o the behaviour of axially loaded RC columns strengthened by steel angles and strips. Mater Des. 2009;30(10):4103–11.
44. Adam JM, Giménez E, Calderón PA, Ivorra S. Experimental study of beam-column joints in axially loaded RC columns strengthened by steel angles and strips. Steel Compos Struct. 2008;8(4):329–42.
45. Garzón-Roca J, Ruiz-Pinilla J, Adam JM, Caderón PA. An experimental study on steel-caged RC columns subjected to axial force and bending moment. Eng Struct. 2011;33(2):580–90.
46. CEN. ENV 1994-1-1 Eurocode 4: design of composite steel and concrete structures. Part 1-1: general rules and rules for buildings. Brussels: European Committee for Standardization; 2004.
47. Nagaprasad P, Sahoo DR, Rai DC. Seismic strengthening of RC columns using external steel cage. Earthq Eng Struct Dyn. 2009;38(14):1563–86.
48. Wang L, Su RKL. Theoretical and experimental study of plate-strengthened concrete columns under eccentric compression loading. J Struct Eng. 2013;139(3):350–9.
49. Said A, Nehdi M. Rehabilitation of RC frame joints using local steel bracing. Struct Infrastruct Eng. 2008;4(6):431–47.

50. Youssef MA, Ghaffarzadeh H, Nehdi M. Seismic performance of RC frames with concentric internal steel bracing. Eng Struct. 2007;29(7):1561–8.
51. Massumi A, Tasnimi AA. Strengthening of low ductile reinforced concrete frames using steel x-bracings with different details. In: Proceedings of the 14th World conference on earthquake engineering; 2008.
52. Ozcelik R, Binici B. A case study of strengthening of deficient RC building with internal steel frame. In: Papadrakakis M, Fragiadakis M, Plevris, editors. 3rd ECCOMAS thematic conference on computational methods in structural dynamics and earthquake engineering; 2011.
53. Maheri MR, Hadjipour A. Experimental investigation and design of steel brace connection to RC frame. Eng Struct. 2003;25(13):1707–14.
54. Ishimura M, Sadasue K, Miyauchi Y, Yokoyama T, Fujii T, Minamo K. Seismic performance evaluation for retrofitting steel brace of existing RC buildings with low-strength concrete. In: Proceedings of the 15th World conference on earthquake engineering; 2012.
55. Della Corte G, Fiorino L, Mazzolani FM. Lateral-loading tests on a real RC building including masonry infill panels with and without FRP strengthening. J Mater Civ Eng. 2008;20(6):419–31.
56. D'Aniello M, Della Corte G, Mazzolani FM. Experimental tests of a real building seismically retrofitted by special buckling-restrained braces. In: Proceedings of the American Institute of Physics (from Seismic Engineering Conference Commemorating the 1908 Messina and Reggio Calabria Earthquake 2008), Part: Part 1-2, vol. 1020; 2008, p. 1513–1520.
57. Mazzolani FM. Innovative metal systems for seismic upgrading of RC structures. J Constr Steel Res. 2008;64(7–8):882–95.
58. Mazzolani FM, Della Corte G, D'Aniello M. Experimental analysis of steel dissipative bracing systems for seismic upgrading. J Civ Eng Manag. 2009;15(1):7–19.
59. Della Corte G, D'Aniello M, Landolfo R. Field testing of all-steel buckling-restrained braces applied to a damaged reinforced concrete building. J Struct Eng. 2014;141(1):D4014004. doi:10.1061/(ASCE)ST.1943-541X.0001080.
60. De Matteis G, Formisano A, Mazzolani FM. An innovative methodology for seismic retrofitting of existing RC buildings by metal shear panels. Earthq Eng Struct Dyn. 2009;38 (1):61–78.
61. Popov EP, Engelhardt MD. Seismic eccentrically braced frames. J Constr Steel Res. 1988;10:321–54.
62. Della Corte G, D'Aniello M, Landolfo R. Analytical and numerical study of plastic overstrength of shear links. J Constr Steel Res. 2013;82:19–32.
63. Lin PC, Tsai KC, Chang CA, Hsiao YY, Wu AC. Seismic design and testing of buckling-restrained braces with a thin profile. Earthq Eng Struct Dyn. 2015;. doi:10.1002/eqe. 2660.
64. Takeuchi T, Ozaki H, Tasui R, Sutcu F. Out-of-plane stability of buckling-restrained braces including moment transfer capacity. Earthq Eng Struct Dyn. 2013;43(6):851–69.
65. NIST. Seismic design of steel buckling-restrained braced frames. A guide for practicing engineers. NIST GCR 15-917-34. Gaithersburg: National Institute of Standards and Technology, U.S. Department of Commerce; 2015.
66. Formisano A, Mazzolani FM, De Matteis G. Numerical analysis of slender steel shear panels for assessing design formulas. Int J Struct Stab Dyn. 2007;7(2):273–94.
67. Formisano A, Mazzolani FM. On the selection by MCDM methods of the optimal system for seismic retrofitting and vertical addition of existing buildings. Comput Struct. 2015;159:1–13.

Strengthening of RC Buildings with Composites

Giorgio Monti and Floriana Petrone

1 Introduction

Still today the number of existing structures designed without considering seismic action or poorly designed/constructed is significantly larger than the number of structures conceived and built to resist seismic loads. For this reason, the research in the field of earthquake engineering of the last two decades has directed its focus towards the development of effective and efficient strengthening techniques of existing buildings.

As a first consideration, there is a substantial difference between methods for designing new buildings and the development of approaches for retrofitting existing structures. If on one hand well-established procedures are available for designing new structures according to the capacity design principles, on the other hand no unified or official methods for providing the existing structures with a sough ductility level are available. In addition, any retrofitting process is based on the assessment of the current capability of the structural system to dissipate energy. This requires a detailed analysis of the structure, aimed at identifying the actual material properties and geometry as well as deficiencies/mistakes and then at determining the optimal way to fix them. Therefore, the retrofitting process should

G. Monti (✉)
Department of Structural Engineering and Geotechnics,
Sapienza Università di Roma, Via A. Gramsci, 53, 00197 Rome, Italy
e-mail: giorgio.monti@uniroma1.it

G. Monti
College of Civil Engineering, Nanjing Tech University,
30 South Puzhu Rd., Nanjing 211816, China

F. Petrone
Energy Geosciences Division, Lawrence Berkeley National Laboratory,
1 Cyclotron Rd, Berkeley, CA 94720, USA
e-mail: FlorianaPetrone@lbl.gov

A. Costa et al. (eds.), *Strengthening and Retrofitting of Existing Structures*,
Building Pathology and Rehabilitation 9, https://doi.org/10.1007/978-981-10-5858-5_7

be interpreted as an original process, specific to each structure: as such, it does not (and cannot) consist in checking the compliance of the structure with Code provisions, but in a more comprehensive performance assessment, before and after the strengthening.

In this framework, FRP composites represent one of the technologies employed to locally strengthen structural elements (beams, columns, walls and joints). The first studies in this field date back to the beginning of the 1990s and still researchers strive at finding new solutions for enhancing the safety of existing constructions, seen as valid alternatives to more usual techniques, such as, mortar injections, concrete jacketing, steel tying and plating, base isolation and integrative (dissipative or not) bracings.

2 Materials

Continuous fibre-reinforced materials with polymeric matrix (commonly known as FRP) are composite, heterogeneous, and anisotropic materials with a (prevalent linear) elastic-brittle behaviour, widely used for strengthening civil structures. Their main advantages can be summarized in: light weight, high strength and corrosion-resistance. Composites for structural strengthening are available in several geometries, from laminates used for structural members with regular surface to bi-directional fabrics easily adaptable to non-regular shapes.

This chapter gives an overview on composite materials, with an in-depth analysis of the main constituents (fibre, matrix, and adhesive), and their main physical and mechanical properties along with a documented reference to the principal design equations for flexural, shear, torsion and axial strengthening of RC members. A full understanding of pros and cons is necessary to optimize the use of FRP and mitigate their disadvantages; this is of particular relevance to ensure durability of FRP strengthening applications where traditional materials, such as concrete and masonry, are paired with high technology materials.

2.1 *Characteristics of Composites*

In general composite materials are made of two or more basic-components (phases) of different nature and "macroscopically" distinguishable. At least two of the phases have different physical and mechanical properties, so to provide FRP composites with features different from those of the single component.

FRPs are made of (i) organic polymeric matrix and (ii) reinforcing fibres, whose main characteristics are summarized in Table 1. The matrix can be considered an isotropic material, whereas the reinforcing phase, with the exception of glass fibres, an anisotropic material. As shown in the table, Young's modulus and tensile strength of carbon fibers can be significantly higher than those of typical construction materials, making FRPs more effective from a structural point of view, especially when the weight of the structure becomes a critical issue.

Table 1 Comparison between properties of fibres, resin, and steel (typical values)

	Young's modulus	Tensile strength	Strain at failure	Coef. of thermal exp.	Density
	E	σ_r	ε_r	α	ρ
	[GPa]	[MPa]	[%]	[10^{-6} °C^{-1}]	[g/cm^3]
E-glass	70–80	2000–3500	3.5–4.5	5–5.4	2.5–2.6
S-glass	85–90	3500–4800	4.5–5.5	1.6–2.9	2.46–2.49
Carbon (high modulus)	390–760	2400–3400	0.5–0.8	−1.45	1.85–1.9
Carbon (high strength)	240–280	4100–5100	1.6–1.73	−0.6 to −0.9	1.75
Aramid	62–180	3600–3800	1.9–5.5	−2	1.44–1.47
Polymeric matrix	2.7–3.6	40–82	1.4–5.2	30–54	1.10–1.25
Steel	206	250–400 (yield) 350–600 (failure)	20–30	10.4	7.8

In general, FRP composites can be synthetically described by the following properties:

- Geometry: shape and thickness.
- Fibre orientation with respect to the symmetry axes of the material.
- Fibre concentration (volume fraction).

In most cases, composites are non-homogeneous and anisotropic materials.

Fibre-reinforced composites can be conveniently divided into two categories:

- Fabrics
- Laminates

Fabrics are usually single- or multi-layer strips/sheets, very flexible in bending and few tenths of a millimetre thick. They usually come in rolls as dry materials to be later glued to the elements. Laminates are stiff in bending and few millimetres thick, made of several layers already glued. They usually come in long and narrow plates.

Structural failure of FRP composites is often due to lack of bond between matrix and fibres. Therefore, the FRP material manufacturer or suppliers should take special care in choosing the most appropriate component to ensure bond.

2.2 *Fibres and Matrices Used in Composites*

The most common fibres used in composites are glass, carbon, and aramid. Their unique mono-dimensional geometry provides FRP laminates with stiffness and strength higher than those of three-dimensional FRP. This is due to the density of defects, which is lower in mono-dimensional configurations than in three-dimensional ones.

As for the matrices, thermoset resins are the most commonly used in the production of FRP materials. They are usually available in a partially polymerized state with fluid or pasty consistency at room temperature. When mixed with a proper reagent, they polymerize to become a solid vitreous material. The reaction can be accelerated by adjusting the temperature. Thermoset resins have several advantages, including low viscosity that allows for a relative easy fibre impregnation, good adhesive properties, room temperature polymerization characteristics, good resistance to chemical agents, and absence of melting temperature. The main disadvantages are: limited range of operating temperatures, with the upper bound limit given by the glass transition temperature, brittle behaviour, and sensitivity to moisture during field applications. The most common thermosetting resins for civil engineering are the epoxy, polyester and vinylester resins. Considering that the material is mixed directly at the construction site and achieves its final structural characteristics through a chemical reaction, it should always be handled by specialized personnel.

2.3 FRP Strengthening Systems

FRP systems suitable for external strengthening of structures may be classified as follows:

- Pre-cured systems: manufactured in various shapes by pultrusion or lamination. Pre-cured systems are directly bonded to the structural member.
- Wet lay-up systems: manufactured with fibres lying in one or more directions, e.g. FRP sheets or fabrics, and impregnated with resin at the construction site.
- Prepreg (pre-impregnated) systems: manufactured with unidirectional or multidirectional fibre sheets or fabrics pre-impregnated at the manufacturing plant with partially polymerized resin. They may be bonded to the member to be strengthened with (or without) the use of additional resins.

2.4 Mechanical Properties of FRP Strengthening Systems

In FRP composites, fibres provide both capacity and stiffness, whereas the matrix ensures the distribution of the load among the fibres and protects the same fibres from corrosion/deterioration. Most FRPs are made of fibres with high strength and stiffness, and fail at strains lower than those of the matrix.

Figure 1 shows the stress-strain relationship of fibre, matrix and FRP. The resulting FRP composite has a stiffness lower than that of fibres and fails at the same strain of the fibres, $\varepsilon_{fib,\max}$.

Table 2 summarizes the mechanical properties of a pre-cured laminate compared to the average values of the corresponding fibres.

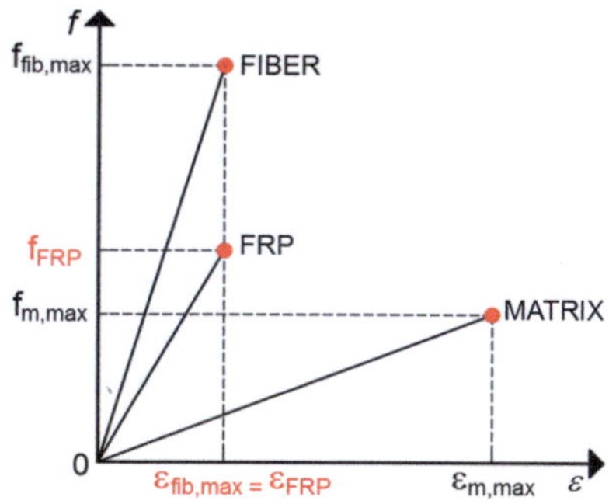

Fig. 1 Stress-strain relationship of fibres, matrix and FRP

Table 2 Comparison between mechanical properties of a pre-cured laminate and fibres

Pre-cured systems	Modulus of elasticity [GPa]		Ultimate strength [MPa]		Ultimate strain [%]	
	FRP	Fibre	FRP	Fibre	FRP	Fibre
	E_f	E_{fib}	f_f	f_{fib}	ε_{fu}	$\varepsilon_{fib,u}$
CFRP (low modulus)	160	210–30	2800	3500–4800	1.6	1.4–2.0
CFRP (high modulus)	300	350–500	1500	2500–3100	0.5	0.4–0.9

The values of the Young's modulus, E_f, and the strength, f_f, of FRP at failure are lower than those of the fibre itself, whereas the ultimate tensile strain is essentially the same, since the failure of the fibre determines FRP's failure.

3 Basis of Design for FRP Strengthening

The design of any structural strengthening through FRPs must meet the requirements of serviceability, durability and resistance to ordinary loads and exceptional actions. For example, in case of fire, the strengthening has to be designed to resist for the prescribed exposure time.

The design working life of the strengthened structure is the same as that of new structures, meaning that the design actions are those of the current Codes for new constructions.

Safety verifications are performed for Serviceability Limit State (SLS) and Ultimate Limit States (ULS), following the format of the partial safety factor method, established in EN 1990 [3], where the design properties of materials and products are derived from the characteristic values, divided by the appropriate partial safety factor.

A fundamental aspect in assessing the safety of existing structures is the treatment of all uncertainties, mainly related to (1) materials mechanical properties, (2) geometry of the structure, and (3) evaluation of possible materials deterioration. As per EN 1990 [3], the design properties X_d of the materials used in the structure are calculated as function of the number of tests performed to acquire information on them:

$$X_d = \frac{\eta}{\gamma_m} m_X (1 - k_n V_X) \tag{1}$$

where η (<1) is a conversion factor, accounting for special design conditions, γ_m is the material partial safety factor, m_X is the mean value of the property X resulting from n experimental tests, k_n is a factor that accounts for the epistemic uncertainty of each X property depending on n, and V_X is the coefficient of variation (CoV), usually available for most common materials (e.g. 0.10 for steel, to 0.20 for concrete and to 0.30 for masonry and timber). For Ultimate Limit States verifications, the partial factor γ_m of FRP takes on different values depending on the failure mechanism: in case of FRP rupture, $\gamma_m = 1.0$; whereas in case of FRP debonding γ_m ranges between 1.2 and 1.5 in consideration of the possibility that debonding can actually occur, based on tests performed by the FRP supplier and as-built conditions.

Equation (1) deserves some additional comments regarding the determination of the parameters. Concrete mechanical properties, usually affected by the highest uncertainties, may be estimated using non-destructive tests (for example by measuring the ultrasonic pulse speed in conjunction with rebound tests). The reliability of these measurements largely depends on the correlation between the indirect quantity actually measured (speed, rebound, etc.) and the mechanical value sought (strength, modulus, etc.). Additional information gained by comparison to destructive tests carried out on the same structure can be used to better calibrate such correlation, thus reducing the risk of systematic errors; however, the number of destructive tests should be kept low, both for economic reasons and to limit any damage to the structure.

A similar situation arises when determining quantity and arrangement of reinforcement. In existing structures, built in the absence of rules imposing detailed working drawings, it is very unlikely that any direct information be available on the geometry as well as on the distribution of the reinforcement. In such cases, the amount of reinforcement can be estimated on the basis of a simulated project developed according to the Code in force at the time. It is acknowledged that such estimate is highly uncertain and needs to be validated by means of in-situ investigations, which can be either direct (clear exposure of the steel reinforcements by elimination of cover concrete and any other material covering them) or indirect (for example by magnetic inductance measurements using pacometer). Since direct measurements are partially destructive and imply damage to the structure, the considerations reported above about the limited number of tests for assessing the material properties hold also for the number of test needed to characterize the reinforcement.

Once the material properties of the existing structure are assessed and the materials adopted for strengthening are selected, the design capacity of the strengthened structure is given by:

$$R_d = \frac{1}{\gamma_{Rd}} R\{X_{d,i}; a_{d,i}\} \tag{2}$$

where $R\{\cdot\}$ is the function describing the relevant mechanical model considered (e.g., flexure, shear, confinement, etc.) and γ_{Rd} is the partial factor acounting for uncertainties in the above capacity model, set equal to 1 for flexure, to 1.2 for shear and to 1.1 for confinement. The arguments of the function are sets of mechanical and geometrical properties, $X_{d,i}$ and $a_{d,i}$, respectively, representing the design value of the i-th quantity (for geometrical properties, nominal values are usually adopted).

Another aspect is related to the safety assessment in case of exceptional actions, as fire. If the strengthening is designed for a predefined fire exposure time (i.e. $E_d \neq 0$, where E_d is the design value of the indirect thermal action due to fire), the service actions of the *frequent combination*, instead of *quasi-permanent combination,* have to be considered. However, the capacity of the structural elements, appropriately reduced to account for the fire exposure time, should be computed with the partial factors relevant to the exceptional situation.

4 Reinforced Concrete Structures

4.1 Anchorage

When strengthening RC members with FRP composites, the role of bond between concrete and FRP is of great relevance due to the brittleness of the loss of adhesion, the so-called "debonding" failure mechanism. According to the capacity design criterion, such failure should not precede flexural or shear failure of the strengthened member.

In general, debonding may involve different components of the strengthened structure and may take place: (1) within laminates and sheets applied to concrete for flexural/shear strengthening, (2) within the adhesive, (3) between concrete and adhesive (4) in the concrete itself, or (5) within the FRP reinforcement with different fibre inclination angles (e.g., at the interface between two adjacent layers bonded each other). When proper installation is performed, since the adhesive strength is typically much higher than the concrete tensile strength, debonding most likely takes place within the concrete itself.

Debonding failure modes for laminates or sheets used for flexural strengthening can be classified in the following four categories, schematically represented in Fig. 2:

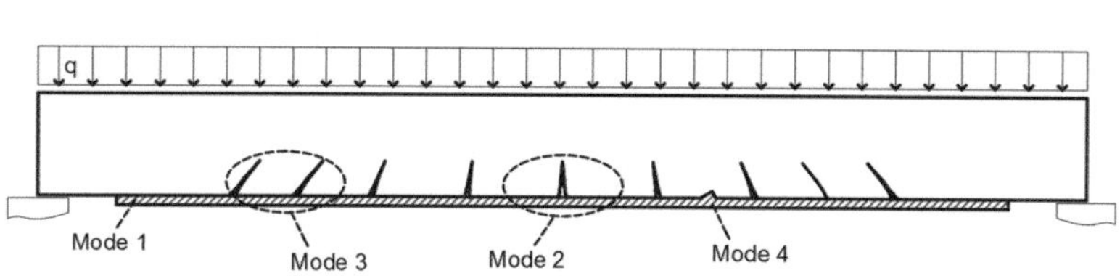

Fig. 2 FRP flexural strengthening: debonding failure modes

Mode 1: Laminate/sheet end debonding
Mode 2: Intermediate debonding, caused by flexural cracks
Mode 3: Debonding caused by diagonal shear cracks
Mode 4: Debonding caused by irregularities and roughness of concrete surface

In the following, reference is made to Modes 1 and 2 only, as they are the most frequent in ordinary design situations.

Before any flexural or shear strengthening design takes place, the evaluation of the maximum force that can be transferred from concrete to FRP and the calculation of shear and normal stresses at the concrete-FRP interface are required. The former is necessary when designing for ULS, and the latter when designing for SLS.

With reference to a typical bond test, as represented in Fig. 3, the ultimate value of the force transferred to the FRP system prior to debonding depends on the length, l_b, of the bonded area. The optimal anchorage length, l_{ed}, is defined as the length corresponding to the maximum force F that can be transferred, meaning that even if this length was increased, there would be no increase in the transferred force.

The optimal anchorage length (in mm) is given as:

$$l_{ed} = 0.10\sqrt{\frac{\pi^2 E_f t_f}{2\Gamma_{Fd}}} \geq 200 \text{ mm} \tag{3}$$

where E_f and t_f are Young's modulus and thickness of the FRP, respectively, and:

$$\Gamma_{Fd} = \frac{k_b k_G}{CF}\sqrt{f_{cm} f_{ctm}} \tag{4}$$

is the design value of the specific fracture energy, expressed as function of f_{cm} and f_{ctm}, which are the mean values of the concrete compressive and tensile strength, respectively, CF, which is an appropriate confidence factor that depends on the attained knowledge level of the existing structure, k_G, which is equal to 0.023 mm for preformed composites and to 0.037 for on-site impregnated composites, and k_b given as:

$$k_b = \sqrt{\frac{2-(b_f/b_w)}{1+(b_f/b_w)}} \geq 1 \tag{5}$$

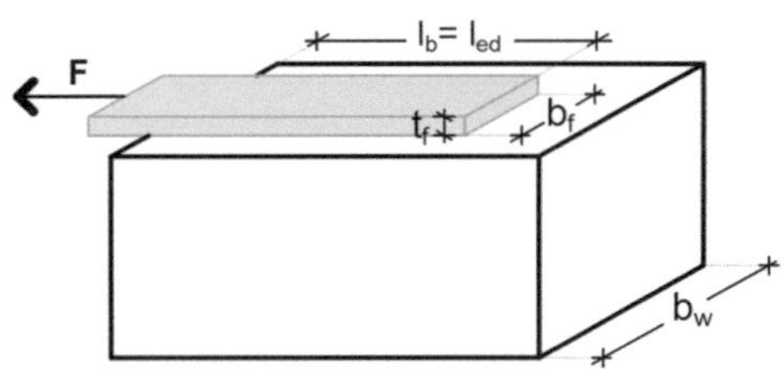

Fig. 3 Maximum force transferred between FRP and concrete

However, if the ratio between the FRP and concrete width, $b_f/b_w < 0.25$, see Fig. 2, then $k_b = 1.18$.

The design debonding strength for mode 1 is:

$$f_{fdd,1} = \frac{1}{\gamma_{fd}} \sqrt{\frac{2E_f \Gamma_{Fd}}{t_f}} \tag{6}$$

where γ_{fd} is the partial factor for debonding, ranging between 1.2 and 1.5.

The design debonding strength for mode 2 is:

$$f_{fdd,2} = 1.25 f_{fdd} \tag{7}$$

where in the computation of f_{fdd}, $k_G = 0.10$ mm should be assumed.

4.2 *Flexural Strengthening*

Flexural strengthening is necessary for structural members subjected to a bending moment that exceeds the flexural capacity. Only the case of uniaxial bending (i.e. when the moment axis coincides with a principal axis of inertia of the cross-section) is addressed here.

Flexural strengthening with FRP materials may be carried out by applying one or more laminates/ sheets to the tension side of the element.

The flexural capacity is attained when either the concrete compressive strain or the FRP tensile strain reaches its ultimate value, that is $\varepsilon_{fd} = \min(\eta_a \varepsilon_{fu}/\gamma_f, f_{ffd}/E_f)$, where the first value corresponds to concrete crushing and the second to FRP debonding, as previously defined. The flexural capacity is then expressed as:

$$M_u = \psi\, b\, x f_{cd}(d - \lambda x) + A_{s2}\sigma_{s2}(d - d_2) + A_f \sigma_f d_1 \tag{8}$$

where the neutral axis position x is found by solving:

$$0 = \psi\, b\, x f_{cd} + A_{s2}\sigma_{s2} - A_{s1} f_{yd} - A_f \sigma_f \tag{9}$$

where ψ and λ are non-dimensional coefficients representing the magnitude and the position of the compressive concrete resultant, respectively.

However, the capacity after strengthening cannot be greater than twice the initial capacity. Moreover, according to the capacity design approach, flexural strenghtening should be designed to avoid the activation of shear failure mechanisms.

Because a member strengthened with FRP is generally loaded at the time of FRP application, the existing strain state in the structure should be taken into account.

4.3 Shear and Torsion Strengthening

Shear strengthening is necessary when the shear demand is greater than the member shear capacity, evaluated considering the contributions of both concrete and steel transverse reinforcement. It may also be necessary after designing a flexural strengthening, in order to re-establish the strength hierarchy between bending and shear failure mechanisms.

Shear strengthening shall be verified at ULS only. Shear strengthening is usually realized by applying one or more layers of FRP, externally bonded to the surface of the structural member to strengthen. External FRP reinforcement can be applied in a discontinuous fashion, with gaps between following strips, or continuously, with strips next to each other.

Figure 4 shows two allowed FRP strengthening configurations: U-wrapped, and completely wrapped beams.

For U-wrap strengthening of rectangular or T-sections, delamination of the end portions of FRP reinforcement can be avoided by using laminates/sheets and/or bars installed in the direction of the member's longitudinal axis. In such case, the behaviour of U-wrap strengthening can be considered equivalent to that of a completely wrapped member, provided that the effectiveness offered by these technological solutions is demonstrated by the applicator.

The design shear strength of the strengthened element is based on the variable angle truss model and is expressed as:

$$V_{Rd} = \min\{V_{Rd,s} + V_{Rd,f}, V_{Rd,c}\} \tag{10}$$

where $V_{Rd,s}$ and $V_{Rd,f}$ are the contributions of transverse steel and FRP to shear-tension capacity, respectively, and $V_{Rd,c}$ is the contribution of concrete to shear-compression capacity. A method for evaluating the actual contribution of each component to the shear strength can be found in [4].

The FRP contribution to the overall strength is based on the selected strengthening configuration. For U-jacketing and wrapping:

$$V_{Rd,f} = \frac{1}{\gamma_{Rd}} \cdot 0.9\,d \cdot f_{fed} \cdot 2\,t_f \cdot (\cot\theta + \cot\beta) \cdot \frac{w_f}{p_f} \tag{11}$$

Fig. 4 Two configurations for shear strengthening

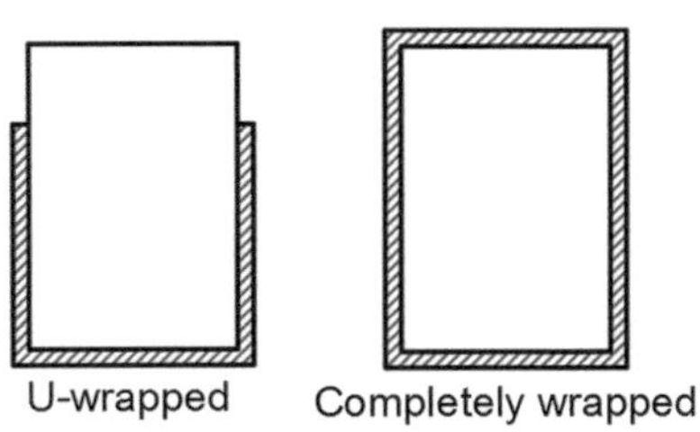

with d = cross-section effective depth, t_f = thickness of the FRP strip/sheet with angle β, θ = crack angle, p_f, w_f = FRP strip spacing and width, respectively, measured orthogonally to the fibre direction β and f_{fed} = the so-called "effective debonding strength". For the case of U-jacketing and wrapping, respectively, f_{fed} is given by:

$$f_{fed} = f_{fdd} \cdot \left[1 - \frac{1}{3} \frac{l_e \sin\beta}{\min\{0.9\,d, h_w\}}\right] \tag{12}$$

$$f_{fed} = f_{fdd} \cdot \left[1 - \frac{1}{6} \frac{l_e \sin\beta}{\min\{0.9\,d, h_w\}}\right] + \frac{1}{2}(\phi_R \cdot f_{fd} - f_{fdd}) \cdot \left[1 - \frac{l_e \sin\beta}{\min\{0.9\,d, h_w\}}\right] \tag{13}$$

where f_{fd} is the FRP design strength, h_w is the beam web depth and:

$$\phi_R = 0.2 + 1.6\,\frac{r_c}{b_w} \quad \text{with} \quad 0 \le \frac{r_c}{b_w} \le 0.5 \tag{14}$$

is a coefficient that depends on the ratio of the rounding radius r_c to the beam web width b_w.

With regard to the strengthening in torsion, it is achieved through the application of wrapping strips/sheets at an angle of 90° to the element axis. The design torsional strength of the strengthened element is given as:

$$T_{Rd} = \min\{T_{Rd,s} + T_{Rd,f}, T_{Rd,\max}\} \tag{15}$$

where $T_{Rd,s}$ and $T_{Rd,f}$ are the transverse steel and FRP contribution, respectively, and $T_{Rd,max}$ is the torque producing collapse in the compressed diagonal concrete strut. The FRP contribution to the torsional strength is given as:

$$T_{Rd,f} = \frac{1}{\gamma_{Rd}} \cdot 2 f_{fed} \cdot t_f \cdot b \cdot h \cdot \frac{w_f}{p_f} \cdot \cot\theta \tag{16}$$

where f_{fed} is given by Eq. (12) or (13).

4.4 Confinement

Appropriate confinement of RC members may improve their structural performance, by increasing the ultimate capacity and strain of structural members subjected to axial -or slightly eccentric—loads.

Ductility and capacity under combined bending and axial force, when FRP reinforcements are placed with the fibres lying along the longitudinal axis of the member, should be verified.

Confinement of RC members can be realized with FRP sheets arranged along the member perimeter as either continuous or discontinuous external wrapping.

The increase of axial capacity and ultimate strain of FRP-confined concrete depends on the applied confinement pressure, which is function of the member cross-section and FRP stiffness.

FRP-confined members (FRP is linear-elastic up to failure), unlike steel confined members (steel has an elastic-plastic behaviour), exert a lateral pressure that increases with the transversal expansion of the confined members.

In case of elements with circular cross-section of diameter D, the confined/unconfined concrete strength ratio is:

$$\frac{f_{ccd}}{f_{cd}} = 1 + 2.6\left(\frac{f_{l,eff}}{f_{cd}}\right)^{2/3} \tag{17}$$

while the ultimate concrete strain is:

$$\varepsilon_{ccu} = 0.0035 + 0.015\sqrt{\frac{f_{l,eff}}{f_{cd}}} \tag{18}$$

where both depend on the confinement pressure exerted by the FRP sheet, given as:

$$f_{l,eff} = k_{eff} \cdot f_l \quad \text{with } f_l = \frac{1}{2}\,\rho_f\,E_f\,\varepsilon_{fd,rid} \tag{19}$$

where k_{eff} is an efficiency factor (≤ 1), E_f is, again, the FRP modulus of elasticity, $\varepsilon_{fd,rid}$ is the FRP reduced design strain, defined in the following, and ρ_f is the geometric strengthening ratio, which is function of the cross-section shape (circular or rectangular), that is:

$$\begin{aligned} \rho_f &= \frac{4 t_f b_f}{D \cdot p_f} && \text{circular sections} \\ \rho_f &= \frac{2 \cdot t_f \cdot (b+d) \cdot b_f}{b \cdot d \cdot p_f} && \text{rectangular sections} \end{aligned} \tag{20}$$

being t_f and b_f the thickness and the width of the generic FRP strip, p_f the centre-to-centre distance between strips, D the diameter of the circular cross-section, and b and d the dimensions of the rectangular cross-section.

The efficiency factor is given as:

$$k_{eff} = k_H \cdot k_V \cdot k_\alpha \tag{21}$$

where k_H is the horizontal efficiency factor, equal to 1.0 for circular sections and to:

$$k_H = 1 - \frac{b'^2 + d'^2}{3 \cdot A_g} \tag{22}$$

for rectangular sections, with $b' = b - 2r_c$, $d' = d - 2r_c$ and A_g = area of the cross-section; and k_V is the vertical efficiency factor, calculated as:

$$k_V = \left(1 - \frac{p'_f}{2\,d_{\min}}\right)^2 \tag{23}$$

where p'_f is the edge-to-edge distance between adjacent strips and $d_{\min}$ is the minimum transverse dimension of the element; when the fibres are wrapped at an angle α_f with respect to the element axis, the angle efficiency factor, k_α, is:

$$k_\alpha = \frac{1}{1 + (\tan \alpha_f)^2} \tag{24}$$

Finally, the reduced design strain is:

$$\varepsilon_{fd,rid} = \min\{\eta_a \varepsilon_{fk}/\gamma_f;\ 0.004\} \tag{25}$$

where ε_{fk} is the FRP characteristic strain, and η_a and γ_f are the environment conversion factor and the partial factor of the FRP strengthening, respectively.

5 FRP Strengthening in Seismic Zones

Composite materials can be used effectively to seismically retrofit reinforced concrete structures. The objective is that of strengthening buildings that do not meet the safety requirements defined by the current seismic Codes under the design seismic action, with respect to one or more limit states.

Once a preliminary seismic assessment is performed on the existing structure, the strengthening intervention is designed based on its outcomes. The entire process goes through the following steps: (a) identification of safety requirements, (b) definition of protection levels (which yield the intensity of the seismic action), (c) choice of analysis methods, (d) choice of verification criteria, (e) assessment of the seismic safety, (f) definition of the material properties to use in the safety verifications.

Regarding the criteria for selecting the FRP strengthening method, it is widely recognized that stiffness irregularities cannot be solved by applying FRPs. In fact an intervention performed with FRP is classified as a selective technique, since strength irregularities can be adjusted by strengthening a selected number of elements. However, attention should be paid to ensure that the global ductility is not reduced.

The design of a strengthening intervention with FRP should include the following activities: (a) justification of the intervention type, (b) selection of techniques and/or materials, (c) preliminary design of the strengthening intervention, (d) structural analysis of the upgraded structure.

As mentioned above, from the seismic standpoint, FRP strengthening is regarded as a selective intervention technique, aiming at: (a) increasing the flexural capacity of deficient members through the application of composites with the fibres placed parallel to the element axis, (b) increasing the shear strength through the application of composites with the fibres placed transversely to the element axis, (c) increasing the ductility (or the chord rotation capacity) of critical zones of beams and columns through FRP wrapping (confinement), (d) improving the efficiency of lap splice zones, through FRP wrapping, (e) preventing buckling of longitudinal rebars under compression through FRP wrapping, (f) increasing the tensile strength of the panels of partially confined beam-column joints through the application of composites with the fibres placed along the principal tensile stress direction.

In general, the inspiring principles of the intervention strategies should be the followings: (a) all potential brittle failure mechanisms should be avoided, (b) all potential "soft story" collapse mechanisms should be eliminated, and (c) the global deformation capacity of the structure should be enhanced, either by (c1) increasing the ductility of the potential plastic hinge zones without changing their position, or, (c2) relocating the potential plastic hinge zones by flexure-strengthening the columns, with the aim of transforming the frame structure into a high dissipation mechanism with strong columns and weak beams.

For principle (a), as well-known, brittle failure mechanisms such as shear in beams and structural joints, lap splicing, and bar buckling should be avoided. For shear, the same criteria apply as for the non-seismic case, with the exception that side bonding is not allowed and FRP strips/sheets should only be applied orthogonally to the element axis. When avoiding potential brittle failure mechanisms, the relative strengthening modalities are quite straightforward. The most common case is that of potential shear failure either in beams or structural joints: in this case a strengthening of the regions of the structural member where shear mechanisms take place should be designed. More peculiar cases are those of longitudinal bars lap splices and buckling: in the former case, due to either bond degradation in splices or insufficient splice length, the relevant regions of potential plastic hinge formation should be adequately confined through an FRP wrapping; in the latter case, the strengthening intervention should consist in confining the potential plastic hinge zones where the existing transverse reinforcement cannot prevent the bars post-elastic buckling.

For principle (b), specific consideration should be given to potential "soft story" collapse mechanism, which can occur in the absence of walls, due to the simultaneous formation of plastic hinges at top and bottom of all columns at a certain story. In such cases, the strengthening intervention should aim at increasing the flexural capacity of these zones, with the objective of inhibiting the hinges formation.

For principle (c), when all possible brittle and soft story mechanisms are prevented, one could ascertain the extent to which the structure could exploit its ductility. This can be done, for example, through a nonlinear pushover analysis, included in the most modern seismic Codes. Usually, one is requested to check if the structure can actually exhibit a given ductility: this is expressed either by a pre-selected behaviour factor or by an attained target displacement obtained from the displacement spectrum. Such analysis allows to identify all those elements whose local collapse prevents the structure from exploiting its global ductility and from reaching the target displacement.

At this stage, one could face two different situations: (c1) the number of local collapses is not significant, or (c2) the number of local collapses is significant.

In the former case (c1), it is necessary to increase the deformation capacity of only those elements that collapse before the global target displacement is attained. The deformation capacity of beams and columns can be measured by the chord rotation q, that is, the rotation of the chord connecting the element end section with the contra-flexure section (shear span). Generally, the plastic deformation capacity is controlled by the compressive behaviour of concrete. An intervention of FRP-confinement on such elements (usually columns) increases the ultimate compressive strain of concrete, thus determining a ductility increase of the element.

In the latter case (c2), local collapses are so numerous that a different strategy should be pursued: the request of ductility should be spread over a larger number of elements. This can be achieved by relocating all potential plastic hinges in the columns to the framing beams, according to the capacity design criterion, which implies the elimination of all potential plastic hinges in columns. In "weak column-strong beam" situations, typical of frame structures designed for gravity loads only, the columns sections are under-designed both in size and reinforcement. In such cases, it is necessary to increase their flexural strength with the objective of changing the structure into a "strong column-weak beam" situation. It should be noted that, pursuing this strategy implies an increase of shear demand on columns due to the flexural capacity increase. It is therefore necessary to perform the required shear verifications and to increase the shear strength in order to avoid brittle failure modes.

As a matter of fact, the evaluation of the deformation capacity of FRP-strengthened existing RC elements under cyclic loads, has been a primary research for the past two decades; as a result, a relatively large number of analytical models describing the "axial load - bending - shear" cyclic response of RC structural members with FRP have been proposed, together with empirical formulas derived from experimental observations. However, such large number of available models and related research work denote also the difficulties that exist in finding a unified and undisputed assessing/design approach, which should include both a mechanics-based view of all FRP-strengthening techniques and a reliability-based framework. This stems from the relatively limited accuracy shown by some of the proposed models, as well as from the difficulty in extrapolating results from a limited number of experimental tests that, by their nature, cannot cover the full range of peculiarities of the response of FRP-strengthened RC elements under cyclic actions.

References

1. CNR-DT 200 R1/2013: Instructions for design, execution and control of strengthening interventions through fibre-reinforced composites. Consiglio Nazionale delle Ricerche (CNR), Roma, Italy. 2014.
2. EN. Eurocode 8: design of structures for earthquake resistance. Brussel: European Committee for Standardization; 1998. p. 2004.
3. Pellegrino C, Sena-Cruz J (eds) Design procedures for the use of composites in strengthening of reinforced concrete structures. RILEM State-of-the-Art Report 19, Springer, The Netherlands (2016).
4. Petrone F, Monti G. FRP-RC beam in shear: mechanical model and assessment procedure for pseudo-ductile behavior. Polymers. 2014;6(7):2051–64.

Other Selected References

Bond

5. Alam MS, Kanakubo T, Yasojima A. Shear-peeling bond strength between continuous fiber sheet and concrete. ACI Struct J. 2012;109(1):75–82.
6. Bilotta A, Faella C, Martinelli E, Nigro E. Indirect identification method for bilinear bond-law relationship. J. Compos Constr ASCE. 2011;. doi:10.1061/(ASCE)CC.1943-5614.0000253).
7. Bilotta A, Faella C, Martinelli E, Nigro E. Indirect identification method of bilinear interface laws for FRP bonded on a concrete substrate. J Compos Constr. 2012;16:171–184. ISSN: 1090-0268. doi:10.1061/(ASCE)CC.1943-5614.0000253.
8. Bizindavyi L, Neale KW. Transfer lengths and bond strengths for composites bonded to concrete. J Compos Constr ASCE. 1999;3:153–60.
9. Bizindavyi L, Neale KW, Erki MA. Experimental Investigation of Bonded Fiber Reinforced Polymer-Concrete Joints under Cyclic Loading. J Compos Constr ASCE. 2003;7(2):127–34.
10. Ceroni F, Pecce M. Evaluation of bond Strength in concrete element externally reinforced with CFRP sheets and anchoring devices. J Compos Constr ASCE. 2010;14(5):521–30.
11. Chajes MJ, Finch WW, Januszka TF, Thomson TA. Bond and force transfer of composite material plates bonded to concrete. ACI Struct J. 1996;93(2):208–17.
12. Chen JF, Teng JG. Anchorage strength models for FRP and steel plates bonded to concrete. ASCE J Struct Eng. 2001;127(7):784–91.
13. De Lorenzis L, Miller B, Nanni A. Bond of fiber-reinforced polymer laminates to concrete. ACI Mater J. 2001;98:256–64.
14. Ferracuti B, Savoia M, Mazzotti C. Interface law for FRP-concrete delamination. Compos Struct. 2007;80(4):523–31.
15. Ko H, Sato Y. Bond stress-slip relationship between FRP sheet and concrete under cyclic load. J Compos Constr ASCE. 2007;11(4):419–26.
16. Mazzotti C, Savoia M, Ferracuti B. An experimental study on delamination of FRP plates bonded to concrete. Constr Build Mater. 2008;22:1409–21.
17. Nakaba K, Kanakubo T, Furuta T, Yoshizawa H. Bond behaviour between fiber-reinforced polymer laminates and concrete. ACI Struct J. 2001;98(3):359–67.
18. Oller E, Cobo Del Arco D, Marì Bernat AR. Design proposal to avoid peeling failure in FRP-strengthened reinforced concrete beams. J Compos Constr. 2009;13(5):384–93.
19. Smith ST, Teng JG. FRP-strengthened RC beams-II: assessment of debonding strength models. Eng Struct. 2002;24(4):397–417.

Flexure and Shear Strengthening

20. Bousselham A, Chaallal O. Behaviour of reinforced concrete T-beams strengthened in shear with carbon fibre-reinforced polymer—An experimental study. ACI Struct J. 2006;103(3):339–47.
21. Bousselham A, Chaallal O. Mechanisms of shear resistance of concrete beams strengthened in shear with externally bonded FRP. J Compos Constr. 2008;12(5):302–14.
22. Bukhari IA, Vollum RL, Ahmad S, Sagaseta J. Shear strengthening of reinforced concrete. Magazine of Concrete Research. 2010;62(1):65–77.
23. Chen JF, Teng JG. Shear capacity of FRP-strengthened RC beams: FRP debonding. Constr Build Mater. 2003;17:27–41.
24. Carolin A, Taljsten B. Theoretical study on strengthening for increased shear bearing capacity. J Compos Constr. 2005;9(6):497–506.
25. Chen GM, Teng JG, Chen JF. Shear strength for FRP-strengthened RC beams with adverse FRP-steel interaction. J Compos Constr. 2013;17(1):50–66.
26. Chen GM, Teng JG, Chen JF, Rosenboom OA. Interaction between steel stirrups and shear-strengthening FRP strips in RC beams. J Compos Constr. 2010;14(5):498–509.
27. Chen JF, Teng JG. Shear capacity of FRP-strengthened RC beams: FRP rupture. J Struct Eng. 2003;129(5):615–25.
28. D'Antino T, Pellegrino C, Salomoni V, Mazzucco G. Shear behavior of RC structural members strengthened with FRP materials: a three dimensional numerical approach. ACI SP286-05, p. 69–84, 2012.
29. Khalifa A, Gold WJ, Nanni A, Abdel Aziz MI. Contribution of externally bonded FRP to shear capacity of RC flexural members. J Compos Constr. 1998;2(4):195–202.
30. Lima J, Barros J. Reliability analysis of shear strengthening externally bonded FRP models. Proc Inst Civ Eng (ICE) Struct. Build. 2011;164(1):43–56.
31. Modifi A, Chaallal O. Shear strengthening of RC beams with EB FRP: Influencing factors and conceptual debonding model. J Compos Constr. 2011;15(5):62–74.
32. Monti G, Liotta MA. Tests and design equations for FRP-strengthening in shear. Constr Build Mater. 2007;21:799–809.
33. Pellegrino C, Modena C. FRP shear strengthening of RC beams with transverse steel reinforcement. J Compos Constr. 2002;6(2):104–11.
34. Pellegrino C, Vasic M. Assessment of design procedures for the use of externally bonded FRP composites in shear strengthening of reinforced concrete beams. Compos B Eng. 2013;45(1):727–41.
35. Petrone F, Monti G. FRP-RC beam in shear: Mechanical model and assessment procedure for pseudo-ductile behavior. Polymers. 2014;6:2051–64.
36. Regan PE. Research on shear: A benefit to humanity or a waste of time? Structural Engineering. 1993;71:337–47.
37. Sas G, Täljsten B, Barros J, Lima J, Carolin A. Are available models reliable for predicting the FRP contribution to the shear resistance of RC beams? J Compos Constr. 2009;13(6):514–34.

Confinement

38. Carey SA, Harries KA. Axial behaviour and modeling of confined small, medium, and large-scale circular sections with carbon fiber-reinforced polymer jackets. ACI Struct J. 2005;102(4):596–604.
39. Chaallal O, Shahawy M, Hassa M. Performance of axially loaded short rectangular columns strengthened with carbon FRP wrapping. J Compos Constr. 2003;7(3):200–8.
40. De Lorenzis L, Tepfers R. Comparative study of models on confinement of concrete cylinders with fiber reinforced polymer composites. ASCE J Compos Constr. 2003;7(3):219–37.

41. Fahmy MFM, Wu Z. Evaluating and proposing models of circular concrete columns confined with different FRP composites. Compos B. 2010;41(3):199–213.
42. Girgin ZC. Modified failure criterion to predict ultimate strength of circular columns confined by different materials. ACI Struct J. 2009;106(6):800–9.
43. Harajli MH, Dagher F. Seismic strengthening of bond-critical regions in rectangular reinforced concrete columns using fiber-reinforced polymer wraps. ACI Struct J. 2008;105 (1):68–77.
44. Harajli MH. Axial stress–strain relationship for FRP confined circular and rectangular concrete columns. Cem Concr Compos. 2006;28:938–48.
45. Harajli MH, Hantouche E, Soudki K. Stress-strain model for fiber-reinforced polymer jacketed concrete columns. ACI Struct J. 2006;103(5):672–80.
46. Ilki A, Kumbasar N. Compressive behaviour of carbon fibre composite jacketed concrete with circular and non circular cross-section. J Earthq Eng. 2003;7(3):381–406.
47. Ilki A, Peker O, Karamuk E, Demir C, Kumbasar N. FRP retrofit of low and medium strength circular and rectangular reinforced concrete columns. ASCE J Mater Civ Eng. 2008;20 (2):169–88.
48. Jiang T, Teng JG. Analysis-oriented models for FRP-confined concrete: a comparative assessment. Eng Struct. 2007;29(11):2968–86.
49. Lignola GP, Nardone F, Prota A, Manfredi G. Analytical model for the effective strain in FRP-wrapped circular RC columns. Compos Part B. 2012;43(8):3208–18.
50. Lignola GP, Nardone F, Prota A, De Luca A, Nanni A. Analysis of RC hollow columns strengthened with GFRP. ASCE J Compos Constr. 2011;15(4):545–56.
51. Lin JL, Liao CI. Compressive strength of reinforced concrete column confined by composite material. Compos Struct. 2004;65:239–50.
52. Matthys S, Toutanji H, Audenaert K, Taerwe L. Axial load behavior of large-scale columns confined with fiber-reinforced polymer composites. ACI Structural Journal. 2005;102(2):258–67.
53. Mirmiran A, Shahawy M, Samaan M, El Echary H, Mastrapa JC, Pico O. Effect of column parameters on FRP-confined concrete. ASCE J Compos Constr. 1998;2(4):175–85.
54. Ozbakkaloglu T, Lim JC, Vincent T. FRP-confined concrete in circular sections: review and assessment of stress–strain models. Eng Struct. 2013;49:1068–88.
55. Pessiki S, Harries KA, Kestner JT, Sause R, Ricles JM. Axial behaviour of reinforced concrete columns confined with FRP jackets. ASCE J Compos Constr. 2001;5(4):237–45.
56. Realfonzo R, Napoli A. Concrete confined by FRP systems: confinement efficiency and design strength models. Compos B. 2011;42(4):736–55.
57. Spoelstra MR, Monti G. FRP-confined concrete model. ASCE J Compos Constr. 1999;3 (3):143–50.
58. Tan KH. Strength enhancement of rectangular RC columns using FRP. ASCE J Compos Constr. 2002;6(3):175–83.
59. Teng JG, Lam L. Behaviour and modeling of fiber reinforced polymer-confined concrete. ASCE J Struct Eng. 2004;130(11):1713–23.
60. Wang LM, Wu YF. Effect of corner radius on the performance of CFRP-confined square concrete columns: test. Eng Struct. 2008;30(2):493–505.
61. Wu YF, Wang LM. Unified strength model for square and circular concrete columns confined by external jacket. ASCE J Struct Eng. 2009;135(3):253–61.

Structural Repair and Strengthening of RC Elements with Concrete Jacketing

H. Rodrigues, P.M. Pradhan, A. Furtado, P. Rocha and N. Vila-Pouca

1 Introduction

Structural intervention in reinforced concrete (RC) buildings is an important topic for structural engineers, and recent years have shown an increase in interest from the technical and scientific community seeking information regarding an adequate technique for repairing or strengthening existing structures. Typically, the cause for these interventions can be related to accidents (e.g. fire, collisions, explosions), naturals hazards (e.g. extreme winds, earthquakes) or may occur due to design and

H. Rodrigues (✉)
RISCO, School of Technology and Management, Polytechnic Institute of Leiria, 4163-2411-901 Leiria, Portugal
e-mail: hugo.f.rodrigues@ipleiria.pt

P.M. Pradhan
Department of Civil and Geomatics Engineering, School of Engineering, Kathmandu University, Dhulikhel, Nepal
e-mail: prachand@ku.edu.np; prachand.pradhan@gmail.com

A. Furtado · N. Vila-Pouca
CONSTRUCT-LESE, Civil Engineering Department, Faculty of Engineering, University of Porto, 4200-465 Porto, Portugal
e-mail: afurtado@fe.up.pt

N. Vila-Pouca
e-mail: nelsonvp@fe.up.pt

P. Rocha
Department of Civil Engineering, School of Technology and Management, Polytechnic Institute of Viana do Castelo, Apartado 574, 4901-908 Viana do Castelo, Portugal
e-mail: procha@estg.ipvc.pt

P. Rocha
CONSTRUCT-LESE, Faculty of Engineering (FEUP), University of Porto, 4200-465 Porto, Portugal

A. Costa et al. (eds.), *Strengthening and Retrofitting of Existing Structures*, Building Pathology and Rehabilitation 9, https://doi.org/10.1007/978-981-10-5858-5_8

construction problems or even due to the change of the functionality of the building or in rehabilitation of the demand for new code requirements [1].

Several techniques are usually proposed for the repair and retrofitting of RC elements; the possibility of using RC jackets is usually considered, traditionally involving the addition of a thick layer of reinforced concrete in the form of a jacket with cast-in-place concrete or shotcrete, using longitudinal reinforcement and transverse ties. Recently, a new technique based on the use of thin jackets made with high performance fibre-reinforced concrete has been developed [2] with the advantage of reducing the thickness of the new concrete layer.

This technique has been used in the strengthening of slabs and beams on the bottom and, in some cases also on the top, used especially to increase the flexural and shear strength, stiffness and ductility of the elements, and in many cases in the strengthening of columns, especially in the cases of seismic strengthening of existing buildings.

This study offers a brief overview of the experimental tests conducted by other workers; the main conclusions will be presented as well as a brief literature review of numerical modelling of the behaviour of RC elements retrofitted with the RC jacketing technique. Thus, the construction procedures related to RC jacketing will be described for repair or for strengthening purposes. Finally, two case studies will be presented to evaluate the efficiency of RC jacketing to improve the original seismic performance of a soft-storey building and a university.

2 Experimental Tests Using RC Jacketing Technique

Some experimental tests can be found that were performed on original and damaged RC columns retrofitted with RC jacketing to improve their original capacity, or to restore them, in the case of damaged ones. Rodriguez et al. [3] noted that RC jacketing is labour-intensive compared to other jacketing methods; however, the results are satisfactory. Ramírez [4] tested 10 repair methods and concluded that concrete jackets are easy to construct and are the most interesting cost-efficient method; the obtained ratio of failure load between the repaired column and the original one is greater than 1. Júlio et al. [5] tested seven RC columns, six using RC jacketing under different interface adherence conditions between the original section and the additional one, concluding that a monolithic behaviour can be achieved without increasing significantly the roughness of the interface surface by designing RC jackets' thickness to be less than 17.5% of the original column width for cyclic monotonic or cyclic loading [5, 6]. These workers also verified that even if the strengthening operation was performed with or without an axial load applied, it did not significantly influence the results. The resistance and stiffness of the retrofitted specimens were higher than those of the original; however, the transverse reinforcement strain was higher in the original. Using this technique to repair damaged RC columns could restore some 80–90% of the original column capacity and 50–90% of their stiffness. Some jacketing procedures are presented in [1].

Konstantinos et al. tested three alternative methods of concrete jacketing [7], i.e. welding the jacket stirrup ends together, placing steel dowels across the interface between the original column and the jacket in combination with welding the jacket stirrup ends together, and connecting the longitudinal reinforcement bars of the original column to those of the jacket. Concrete jacketing increased considerably the columns' strength and stiffness, while placing CFRPs considerably increases their ductility. Krainskyi et al. [8] tested 10 strengthened RC columns with the same design but under different loadings equal to certain column strength limits, and observed that, by doubling the columns' cross sections, an increase in capacity of around 290% was achieved.

The RC jacketing technique was also tested to repair and/or strengthen existing non-damaged and damaged beams and beam-column joints to the original specimen state [9–16]; a good performance was achieved through the increase of the strength capacity, energy dissipation and displacement ductility of the retrofitted compared to the original specimens. RC jacketing can also change the failure mode of the specimen, depending on the retrofit design criteria.

3 Numerical Modelling of RC Jacketing

Different modelling approaches can be adopted to represent the behaviour of RC jacketing, since the simplified approach was based on fibre models until the use of finite element (FEM) models. There are some studies of numerical modelling that is calibrated with experimental tests [17–20]. Lampropoulos et al. [20] demonstrated that monotonic FE analysis with appropriate assumptions can simulate both monotonic and cyclic loading conditions to a reasonable degree of accuracy. According to the results of this study, simulation of the interface between the old and the new concrete is vital and cannot be ignored by simply considering a perfect bond at the interface. In the case of strengthened RC columns subjected to cyclic loading, strength degradation at the interface has to be included and can be effectively modelled by reducing the coefficients of friction and adhesion using a proposed formula. Finally, the effect of jacket concrete shrinkage is simulated, leading to a reduced maximum load and stiffness of strengthened columns. The results showed that RC jacketing increases significantly the element stiffness and strength.

Others have tested this technique to improve the seismic behaviour of existing buildings [21, 22]. Chaulagain et al. observed that, with RC column jacketing, the building had a significantly increased deformation capacity. RC jacketing was confirmed to be a very effective strengthening technique, leading to uniformly distributed values of strength and stiffness of the strengthened column that are considerably higher than those of the original column.

4 Construction Procedure of RC Jacketing

Different procedures can be adopted for the execution of RC jacketing, depending on whether it is used for repair or strengthening purposes [1]. If strengthening is intended for a damaged or deteriorated element, this process must be preceded first by a repair and then the strengthening process. In this case, the first step is to remove the concrete from the deteriorated zones, which can be made by any method that causes micro-cracking of the substrate, which should then be followed by the sand-blasting or water demolition techniques (Phase 0). Regarding non-damaged elements, there is no need to improve the roughness of the interface surface, except for short RC columns, where water demolition or sand blasting or similar techniques should be used. The improvement of the surface roughness can be obtained through the application of epoxy resin in the case of original undamaged elements. For the situation described before, that the water demolition or sand blasting methods are sufficient to guarantee an efficient surface roughness (Phase 1). The element should be temporary shoring to guarantee that the RC jacket will resist a part of the total load, and not just the incremental one (Phase 2). Subsequently holes have to be executed on the adjacent structural RC members (columns, beams or footing) for the anchoring of the longitudinal reinforcement that can be efficiently anchored with the application of two-component epoxy resin, after having been properly cleaned. The holes must be drilled to allow longitudinal bars to pass, in order to achieve continuity between the floors (Phase 3). Additional attention should be given to the position of the longitudinal reinforcement and distributed uniformly for the entire section. Others [1] have suggested that, if the objective is only to increase the shear and ductility capacity, continuity is not needed, and gaps should be provided before. It is recommended to adopt half the spacing of the original transverse reinforcement to guarantee a monolithic behaviour when subjected to cyclic loading (Phase 4). The application of steel connectors can be used to improve the level of strength and stiffness of the short RC columns. Finally, concrete with self-compacting characteristics, high strength and high durability should be applied (Phase 5). Figure 1 shows each phase of the construction procedure regarding the execution of RC jacketing.

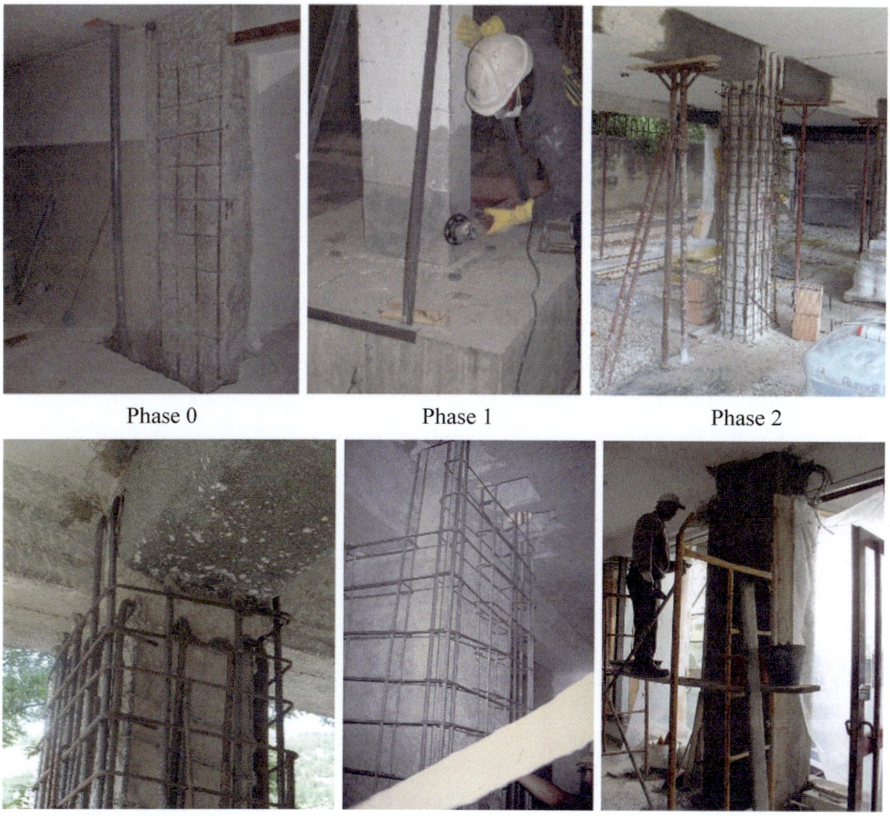

Phase 0 Phase 1 Phase 2

Phase 3 Phase 4 Phase 5

Fig. 1 Construction process of RC column jacketing

5 Case Study: Application of RC Jacketing to a Soft-Storey Building

5.1 Introduction

The study of seismic vulnerability of existing buildings in urban areas with moderate/high seismic risk is of extreme importance to evaluate their safety according to the recently proposed international codes and recommendations. One of the most common architectural trends observed since the 1950s is the construction of RC buildings characterised by a particular modern architectural style influenced by Le Corbusier. This architecture is characterised by the absence of infilled masonry walls in the ground floors of commercial buildings, car parks, or even pedestrian crossings. However, this architectural solution could introduce a critical stiffness of vertical irregularity, enhancing the formation of soft-storey mechanisms, one of the most common mechanisms that could cause buildings to collapse when subjected to earthquakes. In fact, the stiffness contribution of the infilled panels should not be neglected: it is necessary to fix these irregularities with the correct design of strengthening solutions in order to achieve satisfactory performance during future earthquakes.

One of the strengthening techniques that has been proposed is RC column jacketing (RCJ) of the soft-storeys with a stiffness-based design that allows one to achieve the stiffness deficit observed in these storeys without having to change the building's original architecture or even to evacuate the building during the application of this technique. For the application of this technique, it is necessary to guarantee the correct connection with the existing element through preliminary surface picking of the original section before the placing of the jacketing reinforcement and concreting. The strengthened columns must be linked to the adjacent beams and columns with the reinforcement adopted, and the correct anchor length and shear capacity of the foundations must be evaluated according to the new structural configuration of the building prior to strengthening.

In this context, we now present a case study of an existing RC building in Portugal with the behaviour potentially governed by the soft-storey mechanism; the effectiveness of the RCJ technique to eliminate/fix the original behaviour and to improve the building's seismic performance will be tested. The strengthened building results will be compared to those of the original, deducing information about the structural efficiency.

5.2 *Building Description and Design of the Ground Floor RCJ Columns*

The building under study is located in Lisbon and is characterised by not having masonry infill walls on the ground floor (Fig. 2a, b). The block plan is rectangular: 11.1 m in width and 47.4 m in length (Fig. 2b), and the building has 8 storeys plus the ground-floor column's height, making a total of 27.40 m. The main structural system (12 parallel plane frames) restricts the architecture. The layout of the units in the building block (six duplex apartments) was defined in accordance with the structural system. The distance between the frame's axes is 3.80 m. Each frame is supported by two columns and has one cantilever beam on each side with a span of 2.80 m, resulting in 13 modules. To simulate the structural behaviour of the building, we used the computer software SeismoStruct [23], which contemplates some important issues like the non-linear behaviour of RC elements and the influence of the masonry infill walls on the global seismic response of the building. The building was analysed in both principal directions by a 3D model (Fig. 2b).

When the member is considered to be of insufficient strength, an RC jacket may be used to enhance stiffness, strength and ductility. This is one of the most commonly applied methods of repairing and strengthening an RC member. Concrete

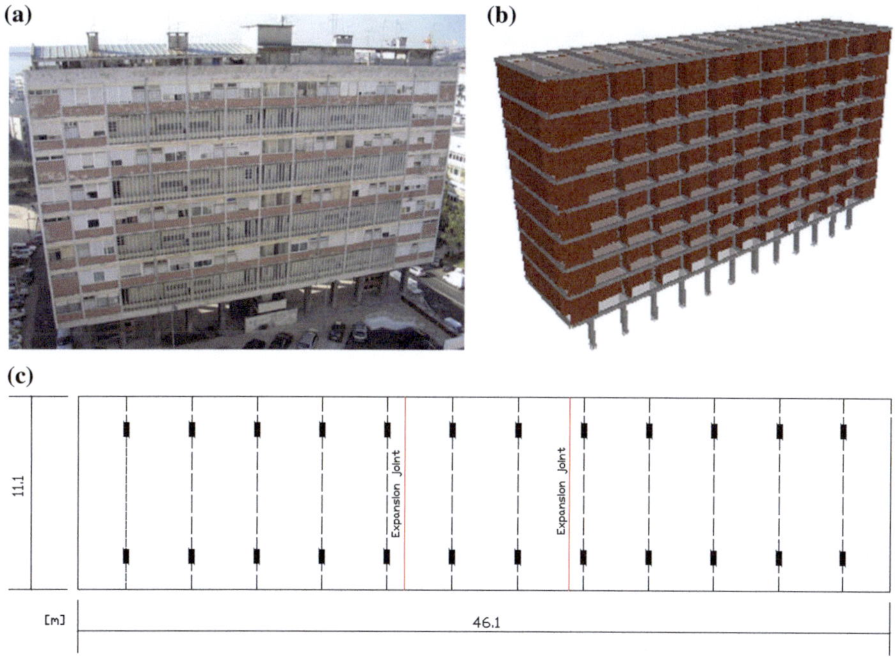

Fig. 2 General view of the building under study: **a** front view, **b** numerical model, **c** building plant

jackets can accommodate longitudinal and transverse reinforcement to increase the flexural and shear strength, enhance the deformation capacity, and improve the strength of deficient-splices [24].

One of the strengthening techniques proposed for the building is RC column jacketing (RCJ) of the ground floor columns. SeismoStruct [23] has the possibility of considering RC jacketing columns, by considering the initial section and the upgraded section (with the consideration of the disposition of the longitudinal reinforcement bars and confinement provided by the transverse reinforcement). The software allows for the use of elements with lumped plasticity (with fixed length, the so-called plastic-hinge). Fibre discretisation was adopted to represent the behaviour at the section level, where each fibre is associated with a uniaxial stress-strain law. The sectional moment-curvature state of the beam and column elements was then obtained through the integration of the nonlinear uniaxial stress-strain response of the individual fibres into which the section was subdivided. The numerical model takes into account the jacketing material's mass and the stiffness during the analysis. The plant disposition and the new sections of these columns designed according to the Eurocode 2 [25] (Sect. 6, 7, 8 and 9) and Eurocode 8 (Sect. 5) [26] are illustrated in Fig. 3.

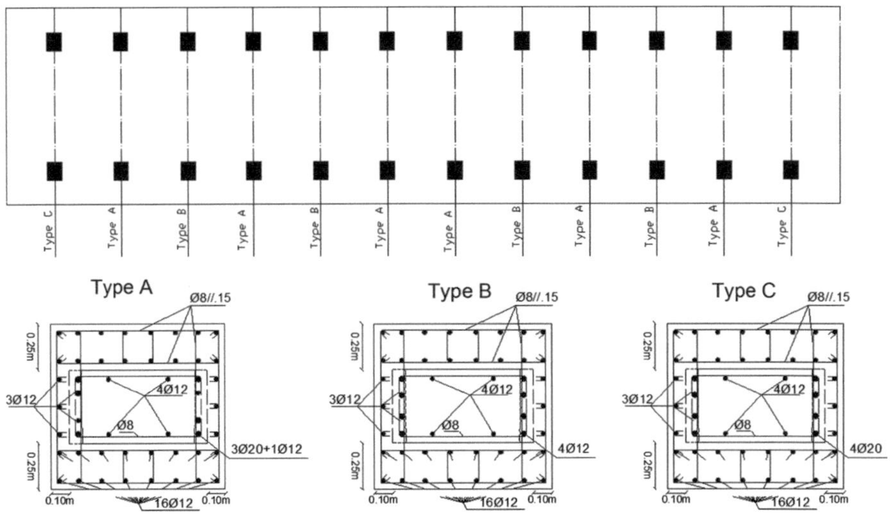

Fig. 3 Plant disposition and cross section of the new strengthened ground-floor columns using the RC jacketing (RCJ) technique

5.3 *Evaluation of RC Jacketing Efficiency*

The efficiency of the RCJ technique was evaluated by subjecting the numerical model to an artificial earthquake that was generated for a medium/high risk scenario in southern Europe [27] for different return periods. The obtained results allow us to assess the seismic safety according to the hazard levels proposed by the international recommendations VISION-2000 [28] and FEMA-356 [29] to evaluate the building safety. Another global drift limit was used, namely the Gobarah proposal [30] recommended for non-ductile structures, which is the case for the rehabilitation/strengthening of existing buildings. From the non-linear dynamic analyses, the maximum first-storey drift and maximum inter-storey and upper-storey drift was determined for each peak ground acceleration (Fig. 4). We also studied whether the introduction of RCJ increased the maximum upper-storey drifts, particularly whether any damage occurred on the IM walls when the strengthened building is subjected to seismic activity. The definition of limit states for infills can be directly related to the inter-storey drift demand. Based on the strut model, Magenes and Pampanin [31] proposed an empirical evaluation for the damage level of the infills that corresponds to a certain limit state, depending on the axial deformation. The FEMA-306 [32] and FEMA-307 [33] documents also provide reference values of inter-storey drift for RC buildings with masonry infill walls.

According to the results, RCJ significantly reduces the maximum first-storey drift about by 35–50% for pga above 0.4 g at least in the longitudinal direction, and by 20–40% in the transverse direction. As a consequence of the maximum

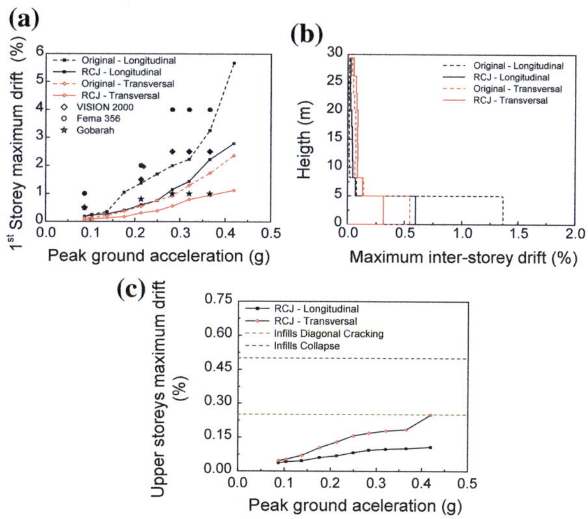

Fig. 4 Non-linear dynamic analysis results: **a** maximum First-storey drift, **b** maximum Inter-storey drift, **c** upper-storey maximum drift

inter-storey drift envelope, it can be concluded that the soft-storey mechanism was fixed, and that no significant damage occurred in the upper-storey infill masonry walls.

Globally, it can be concluded that the correct design and implementation of the RCJ technique to improve soft-storey buildings and, in particular, to fix/eliminate such mechanisms can be achieved by the application of this technique.

6 Case Study: Kathmandu University

6.1 Introduction

Though the term “retrofitting” is not new to the Structural Engineering arena, it became quite popular in Nepal after the earthquake of magnitude 7.8 (Moment magnitude) on 25 April 2015 marking the epicentre at Barpak, Gorkha at 11:56 a. m., Nepalese Standard Time. Another earthquake of magnitude 7.3 was triggered on 12 May 2015, with its epicentre at Dolakha. These two epicentres are located at the west and east of Kathmandu within less than 100 km. A series of aftershocks greater than magnitude 4 hit the country since then, further gradually deteriorating the existing structures of Nepal. The psychological trauma of people who were forced to take shelter in open spaces was made even worse by the onset of winter. Rehabilitation of the structures started after the loss of more than 8500 lives [34] and damage to a huge number of dwellings. The lives of those taking shelter in tents and other temporary means became miserable as winter started. Nevertheless, people started to repair and retrofit their houses on their own, with the limited experience that prevailed in the country.

6.2 Present Scenario

After acquiring technical suggestions from civil and structural engineers, the population started to reform their dwellings with techniques limited by their resources. The majority of buildings, however, were merely repaired in the name of retrofitting, possibly either due to ignorance or due to their adverse economic conditions. Thus, it is expected that another strong earthquake could lead to another devastating situation to Nepal.

Retrofitting actually means strengthening the buildings to such an extent that they should be able to resist future large earthquakes. Just repairing is not the solution; moreover, it may lead to a catastrophic scenario in the future. In Nepal, there are various types of buildings; the majority is of reinforced concrete, brick masonry load-bearing structures, stone masonry houses, adobe etc. The majority of adobe and stone masonry buildings collapsed, while many reinforced concrete and other types of buildings were also not spared. Likewise, heritage buildings and temples could not survive the major earthquake of 25 April 2015. The damage and collapse of many buildings were due to many reasons, of which the most important were the poor quality controls in construction materials, poor design and non-engineered construction methods, and vulnerable buildings due to deficient characteristics from an earthquake-resistant point of view. Many structures were damaged due to their old age and maintenance deficiencies. Although the Gorkha and Dolkha earthquake epicentres were quite far from Kathmandu, the capital of Nepal, the structures lying in Kathmandu were also affected. Kathmandu, Bhaktapur and Lalitpur, the three major cities in the valley, suffered extensive damage and casualties, in addition to the collapse of some major monuments and temples. The devastation was visible in Sindhupalchowk, Nuwakot, Rashuwa, Dolkha, Kathmandu Valley, Gorkha and many districts from Gorkha to Dolkha. In general, the collapsed reinforced concrete buildings and stone masonry buildings were deficient in earthquake-resistant design principles. Besides, many buildings did not follow basic codal requirements. The damage due to the short column effect, soft storey, pounding effect, slopes and reclaimed land, was later realised in most of the buildings. Some buildings were also damaged due to alterations by building owners after the initial design.

Many reinforced concrete structures whose column and beam joints needed detailed retrofitting design have simply been repaired by general cement-slurry grouts or by the addition of steel angle frames. These steel frames are welded to the longitudinal bars of the damaged beams and columns in the name of retrofitting. Eventually, it is understood that if the frames are rectangular, the rectangular steel frames added as supports in this way will deform into a parallelogram shape when the earthquake force hits them again. The stability of the repaired buildings remains a question of safety.

6.3 Retrofitting at Kathmandu University

Very few buildings have been retrofitted considering detailed retrofitting design in the country. Some of the buildings of Kathmandu University suffered minor and major damage. A few of the buildings needed detailed retrofitting design after preliminary rapid visual assessment. The library block at Dhulikhel (Fig. 5) and the Management building located at Balkumari required detailed study, as they were damaged to a large extent (Fig. 6). The staff quarter at Dhulikhel was also studied in depth, even though it suffered only minor damage, primarily of brick masonry infill walls. Mostly, the columns were found to be weaker compared to the beams; thus, jacketing retrofitting works were applied to the damaged columns. Some beams that were considered vulnerable were retrofitted with FRP wrapping and infills that were severely affected were pressure-grouted with cement-slurry with admixtures, further strengthened by chicken wire mesh and shotcretes.

Indian Code IS 15988: 2013 [35] was adopted for the retrofitting design of columns, while the ACI 31-95 code was considered for FRP wrap calculations.

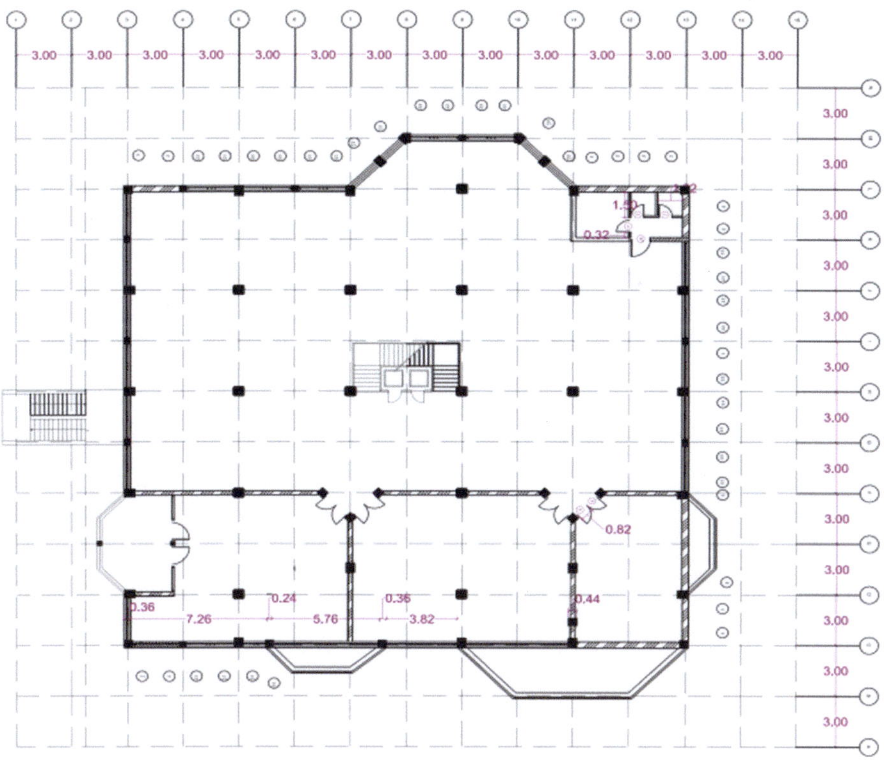

Fig. 5 First Floor Plan of Kathmandu University Central Library

Fig. 6 Damage observed in Kathmandu University Central Library (Dhulikhel) after the major earthquake

Since the majority of columns were found to be inadequate in size as well as in terms of reinforcements, steel and concrete jacketing works were chosen.
As per IS 15988: 2013, certain criteria must be followed:

Step 1—Preliminary visual evaluation: It is a quick procedure to establish the actual structural layout and assess its characteristics that may affect seismic vulnerability. This is an approximate procedure based on conservative parameters to identify the potential risk of a building, and may be used to screen buildings for detailed evaluation. It is primarily based on observed damage characteristics in previous earthquakes, coupled with some simple calculations. First, a site visit is carried out to check for visible deficiencies like configurations of load path, distressed geometry, weak storeys, soft storeys, vertical discontinuities, mass, torsion, short columns and pounding effects.

Step 2—Column retrofitting: The retrofitting of deficient columns is essential to prevent storey collapse. Hence, it is more important to retrofit columns compared to beams. The columns are retrofitted to increase their flexural and shear strengths, to increase the deformation capacity near the beam-column joints, and to strengthening the regions of faulty splicing of longitudinal bars. The columns in an open ground storey or next to openings should be prioritised for retrofitting. The retrofitting strategy is based on the strong column-weak beam principle of seismic design. Concrete jacketing involves the addition of a layer of concrete (Fig. 7), longitudinal bars and closely spaced lateral ties (Fig. 8). As shown in Fig. 8,

Fig. 7 Concrete jacketing in columns of Kathmandu University Central Library (Dhulikhel)

Fig. 8 Concrete jacketing of a damaged column

the jacket increases both the flexural and shear strength of the column. If the thickness of the jacket is less than 100 mm on each side of column, the effectiveness will be lower in terms of stiffness.

To increase the flexural strength, additional longitudinal bars need to be anchored to the foundation and should continue through the floor slab. Usually, the required bars are placed at the corners so as to avoid intercepting the beams that frame into the column. A tie cannot be made of a single bar due to the obstruction in placing, so it may be constructed of two bars properly anchored to the new longitudinal bars. It is preferred to have a 135° hook with adequate extension at the ends of the bars. Since the thickness of the jacket is at least 100 mm, casting micro-concrete or the use of shotcrete are preferred to conventional concrete.

To ensure the composite action of the existing and the new concrete, the options for preparing the surface of the existing concrete are by chiselling, roughening with a wire brush, or using bonding chemicals. If the jacket is only partially around the existing column, existing bars can be exposed at a few locations, then welded to the Z- or U-shaped bent bars to the new bars.

The minimum specifications for concrete jacketing are:

1. The compressive strength of the new concrete must be at least 5 MPa greater than that of the existing concrete (IS 15988: 2013).
2. For columns where extra longitudinal bars are not required for additional flexural capacity, a minimum of 12 mm diameter bars in the four corners and ties of 8 mm diameter must be provided.
3. The minimum dimensions of the jacket should be 100 mm.
4. The minimum diameter of the ties should not be less than 8 mm and should not be less than one-third of the diameter of the longitudinal bars. The angle of the end of the ties should be 135°.
5. The centre-to-centre spacing of ties should not exceed 200 mm preferably. Close to the beam column joints, or a height of one-quarter of the clear height of the column, the spacing should not exceed 100 mm.

When a building is acted upon by earthquake forces, its structural components may become damaged due to lateral forces. It is always better to retrofit a structure after any major shock is resisted by it. Retrofitting of columns becomes essential, as they are the major components that contribute towards resisting the lateral deformation of buildings. During analysis, whether the existing foundations are safe or not must be studied; accordingly, strengthening of foundations may also be

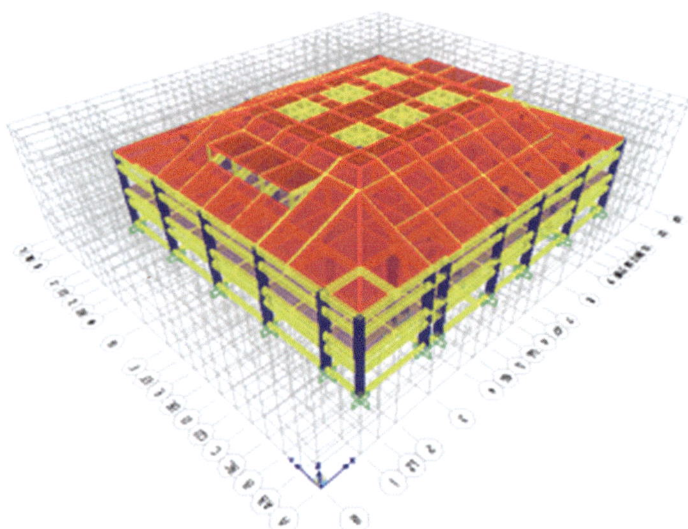

Fig. 9 3D model of Kathmandu University Central Library—Identification of failed columns

undertaken. The safe bearing capacity of soil must be known in order to check the foundation safety.

First, the structure was modelled in computer software as shown in Fig. 9, by Pradhan et al. [36] as per the built drawings, and the analysis was performed to identify the columns that fail. The failing columns were then noted.

The cross section of the columns that failed is required to be enlarged (this applies also to the case of retrofitting beams), so that they can resist potential lateral forces. A minimum jacket of 100 mm thickness around the column is mandatory. The new model prepared after assigning the columns of increased cross section is

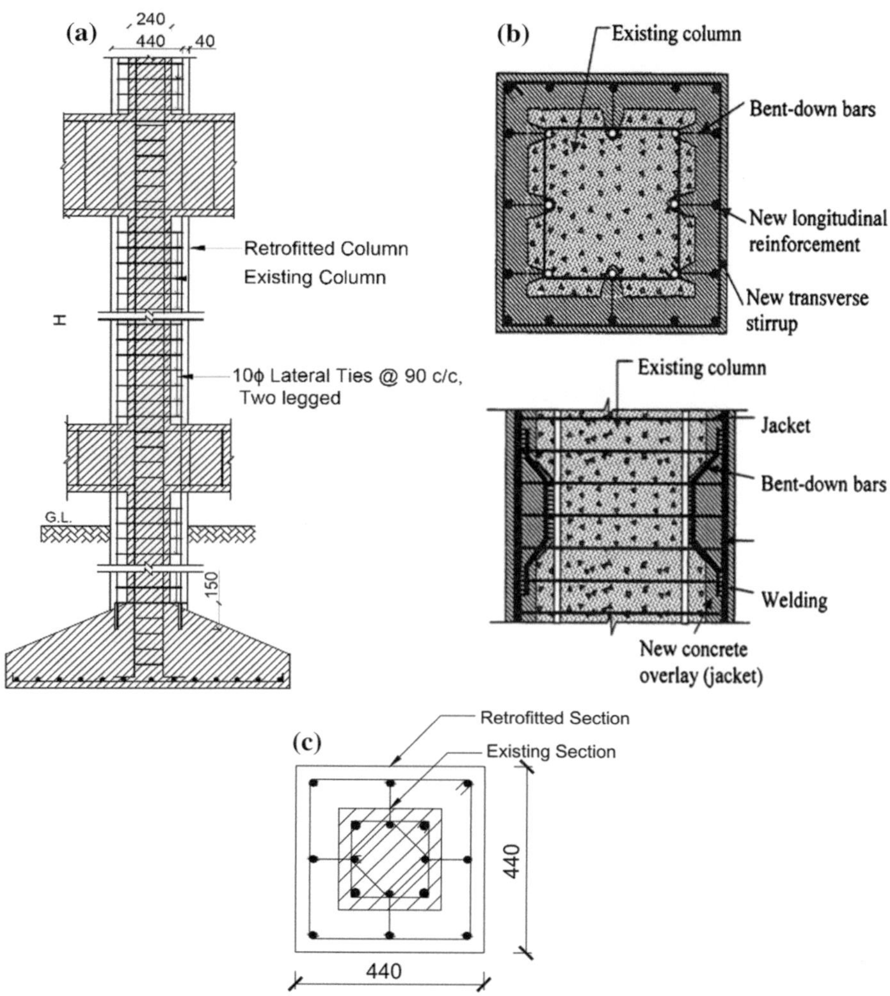

Fig. 10 **a** Longitudinal section of retrofitted column, **b** reinforcement bar and connector arrangement, **c** detail showing placement of connectors

again analysed to check if any further failure has occurred; the increasing in column size is continued until all columns and beams become safe.

After a few trials of analyses with increased column sizes, none of the elements will fail; then, the area of steel required is obtained from computer analysis software like SAP2000. The original area of steel (obtained as per built drawing details) is deducted from the area obtained from software analysis (i.e. the required steel area). Figure 10 shows the longitudinal section of a retrofitted column. Some partial factor of safety is provided for both added concrete section and steel, thus additional concrete area as well as steel must be provided as per the retrofitting code [35]. In the study, almost all the ground floor columns were found deficient in flexure and were suggested for retrofitting as shown in Fig. 10 [36]. Many columns on the upper floors required retrofitting as well.

7 Final Comments

The proper characterisation and evaluation of the existing structural elements, namely through the analysis of their current state, loading conditions and a clear definition of the intended objectives, allows one to design optimised retrofitting solutions that will improve the element's behaviour. It can be concluded that the RC jacketing technique is very effective and, when well executed, increases the elements' stiffness, strength and ductility. Aspects such as the interface between the new and the existing element section should be given special attention, to guarantee that the surface roughness will lead to a monolithic behaviour of the element when subjected to lateral loading.

Finally, the efficiency of the RC jacketing technique to reduce the seismic vulnerability of the soft-storey building was analysed: RC jacketing of all the ground-storey columns significantly reduced the soft-storey mechanism and improved the buildings' strength capacity.

Acknowledgements Part of this work reports to research financially supported by Project POCI-01-0145-FEDER-007457—CONSTRUCT—Institute of R&D in Structures and Construction, funded by FEDER funds through COMPETE2020—Programa Operacional Competitividade e Internacionalização (POCI) and by national funds through Fundação para a Ciência e a Tecnologia (FCT).

References

1. Júlio E, Branco F, Silva V. Structural rehabilitation of columns with reinforced concrete jacketing. Prog Struct Mat Eng. 2003;5:29–37.
2. Martinola G, Meda A, Plizzari G, Rinaldi Z. Strengthening and repair of RC beams with fiber reinforced concrete. Cement Concr Compos. 2010;32:731–9.

3. Rodriguez M, Park R. Seismic load tests on reinforced contrete columns strengthened by jacketing. ACI Struct J. 1994;91:150–9.
4. Ten Ramírez J. concrete column repair methods. Constr Build Mater. 1996;10:195–202.
5. Júlio E, Branco F. Reinforced concrete jacketing—interface influence on cyclic loading response. ACI Struct J. 2008; 105(4):1–7. https://www.concrete.org/publications/international concreteabstractsportal.aspx?m=details&ID=19861.
6. Júlio E, Branco F, Silva V. Reinforced concrete jacketing—interface influence on monotonic loading response. ACI Struct J. 2005; 252–57.
7. Konstantinos G, Vandoros S, Dritsos E. Concrete jacket construction detail effectiveness when strengthening RC columns. Constr Build Mater. 2008;22:264–76.
8. Krainskyi P, Blikharskiy Z, Khmil R. Experimental investigation of reinforced concrete columns strengthened by jacketing. J Multidiscip Eng Sci Technol (JMEST). 2015; 2:1959–1963.
9. Bindhu K, Mohana N, Sivakumar S. New reinforcement detailing for concrete jacketing of nonductile exterior beam column joints. J Perform Constr Facil. 2014; 30(1) (February 2016). https://doi.org/10.1061/(ASCE)CF.1943-5509.0000700.
10. Alcocer S. RC frame connections rehabilitated by jacketing. J Struct Eng. 1993; 119(5) (May 1993). https://doi.org/10.1061/(ASCE)0733-9445(1993)119:5(1413).
11. Cheong H, MacAlevey N. Experimental behavior of jacketed reinforced concrete beams. J Struct Eng. 2000; 126.
12. Tsonos A. Effectiveness of CFRP-jackets and RC-jackets in post-earthquake and pre-earthquake retrofitting of beam-column subassemblages. Eng Struct. 2008;30:777–93.
13. Ruiz-Pinilla J, Pallarés F, Gimenez P, Calderón P. Experimental tests on retrofitted RC beam-column joints underdesigned to seismic loads. General approach. Eng Struct. 2014;59:702–14.
14. Altun F. An experimental study of the jacketed reinforced-concrete beams under bending. Constr Build Mater. 2004;18:611–8.
15. Tahsiri H, Sedehi O, Khaloo A, Raisi E. Experimental study of RC jacketed and CFRP strengthened RC beams. Constr Build Mater. 2015;95:476–85.
16. Chalioris C, Pourzitidis C. Rehabilitation of shear-damaged reinforced concrete beams using self-compacting concrete jacketing. International Scholarly Research Network. 2012.
17. Thermou G, Papanikolaou V, Kappos A. Cyclic response of R/C jacketed columns including modelling of the interface behaviour. Presented at the 15th World Conference on Earthquake Engineering, Lisboa, 2012.
18. Nishad C, Anto J. Non linear analysis of jacketed reinforced concrete column. Int J Eng Trends Technol (IJETT). 2015; 28:369–373.
19. Siddappa KYG, Chandra S, Anushree. Finite element analysis of jacketed reinforced concrete columns subjected to axial and uniaxial load. Int J Res Eng Technol. 2015; 4:132–138.
20. Lampropoulos A, Dristsos S. Modeling of RC columns strengthened with RC jackets. Earthq Eng Struct Dyn. 2011;40:1689–705.
21. Chaulagain H, Rodrigues H, Spacone E, Varum H. Assessment of seismic strengthening solutions for existing low-rise RC buildings in Nepal. Earthq Struct. 2015;8:511–39.
22. Furtado A, Rodrigues H, Varum H, Costa A. Evaluation of different strengthening techniques' efficiency for a soft storey building. Eur J Environ Civ Eng. 2015; 1–18.
23. SeismoSoft. SeismoStruc—a computer program for static and dynamic nonlinear analysis of framed structures; 2004. http://www.seismosoft.com.
24. Varum H. Seismic assessment, strengthening and repair of existing buildings. Aveiro: Departamento de Engenharia Civil, Universidade de Aveiro; 2003.
25. EC2. Design of concrete structure, part 1-1: general rules and rules for buildings. European standard EN 1992-1-1. European Committee for Standardization (CEN), Brussels; 2004.
26. Eurocode 8. Design of structures for earthquake resistance—Part 1–3: Strengthening and repair of buildings—European prEN 1998-1-3, B. European Committee for Standardization, Belgium; 2003.

27. Rodrigues H, Varum H, Costa A. A non-linear masonry infill macro-model to represent the global behaviour of buildings under cyclic loading. Int J Mech Mater Des. 2008;4:123–35.
28. SEAOC-Vision2000. Performance based seismic engineering of buildings, vols. I and II: Conceptual framework. Ed: Structural Engineers Association of California, Sacramento (CA); 1995.
29. FEMA356. Prestandard and commentary for the seismic rehabilitation of buildings. Ed: Federal Emergency Management Agency, Washington (DC); 2000.
30. Gobarah A. On drift limits associated with different damage levels. In: Proceedings of the international workshop Bled Slovenia; 2004, pp 321–332.
31. Magenes G, Pampanin S. Seismic response for gravity-load designed frame systems with masonry infills. Presented at the 13th World Conference on Earthquake Engineering (13WCEE), Vancouver, BC, Canada; 2004.
32. Evaluation of Earthquake Damage Concrete and Masonry Wall Buildings—Basic Procedures Manual Federal Emergency Management Agency; 1999.
33. Evaluation of Earthquake Damage Concrete and Masonry Wall Buildings—Technical Resources Federal Emergency Management Agency; 1999.
34. http://earthquake.usgs.gov/learn/topics/Nepal_Slides.pdf; 2015.
35. IS 15988. Seismic Evaluation and Strengthening of Existing Reinforced Concrete Buildings — Guidelines, Bureau of Indain Standards, New Delhi; 2013.
36. Pradhan PM, Adhikari R, Dahal A, Shrestha A, Subedi DL, Thapa S, Kharel P. Retrofitting Design of Kathmandu University Library Building after Gorkha Earthquake 2015. Lowland Technology International (Accepted for Special Issue on "Nepal Earthquake and Disasters"), vol 18, No 2; 2016.

Strengthening of RC Bridges

Pedro Delgado and Andreas Kappos

1 Introduction

As stated in several reports on recent earthquakes, bridges not properly designed for seismic resistance are damaged under strong earthquakes, and the consequences of this damage can be more dramatic than for normal building structures. Even for "moderate magnitude" earthquakes, damage in some existing bridges and viaducts has been very grave, on several occasions causing their partial destruction, and in some cases total collapse, with corresponding heavy costs. In most cases, the bridge safety is governed by pier performance. Previous studies have addressed the seismic performance of both solid and hollow piers.

Many studies have been carried out on RC solid columns aiming to assess the effectiveness of retrofit solutions with Carbon Fibre Reinforced Polymers (CFRPs), addressing the increase in strength (bending and shear) and ductility. Carbon fibre jackets (most commonly by applying CFRP wrapped sheets) have been used in strengthening/retrofitting several types of columns, from circular to rectangular cross sections, under the influence of different loading actions, as is the case of the

P. Delgado (✉)
Department of Civil Engineering, School of Technology and Management, Polytechnic Institute of Viana do Castelo, Apartado 574, 4901-908 Viana do Castelo, Portugal
e-mail: pdelgado@estg.ipvc.pt

P. Delgado
CONSTRUCT-LESE, Faculty of Engineering (FEUP), University of Porto, 4200-465 Porto, Portugal

A. Kappos
Research Centre for Civil Engineering Structures, City University of London, Northampton Square, London EC1V 0HB, UK
e-mail: Andreas.Kappos.1@city.ac.uk

A. Costa et al. (eds.), *Strengthening and Retrofitting of Existing Structures*,
Building Pathology and Rehabilitation 9, https://doi.org/10.1007/978-981-10-5858-5_9

following works (among several others): Seible et al. [1], Gergely et al. [2], Parvin et al. [3], Santarosa et al. [4], Teng et al. [5], Hadi [6].

On the other hand, studies concerning the retrofit of RC hollow section piers are more scarce, nevertheless, some works are found in the literature, namely the FRP retrofit of hollow section bridge piers: Ogata and Osada [7], Cheng et al. [8], Mo et al. [9], Pavese et al. [10], Yeh and Mo [11], Tsionis and Pinto [12], Lignola et al. [13], Delgado et al. [14]. From these studies, in some cases, the improved performance was attributed to the change of shear failure mode to a flexural one, combined with the confinement effect.

Despite the substantial amount of studies on pier strengthening, the number of retrofitted bridges subjected to strong ground motions is still quite small to allow drawing meaningful conclusions. Therefore, it is quite important to perform analytical studies on the effect of bridge pier retrofitting, mainly concerning the derivation of analytical fragility curves for as-built and retrofitted bridges. In fact, the fragility curves are extremely useful to evaluate the efficiency of different seismic retrofit strategies and to identify the optimum solution compared to the as-built condition for different levels of seismic hazard.

2 Typical Bridge Piers Damage

It is during the occurrence of earthquakes that the deficiencies caused by poor behaviour of structures are highly evidenced and, therefore, lessons can be drawn for the design aspects that allow providing them with a good seismic behaviour.

In this section, the general framework of the seismic behaviour of bridges is set, with special emphasis on aspects that most directly influence the seismic vulnerability. Thus, the most important aspects of some of the earthquakes that have occurred in the past, namely the most significant damage attributed to deficiencies in the structural behaviour.

The failure of the Cypress viaduct [15], which led to the collapse of its upper deck, was one of the most serious failures that occurred during the earthquake of

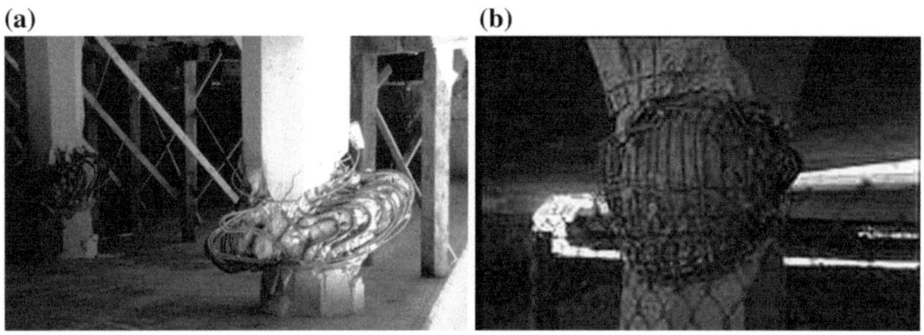

Fig. 1 Northridge earthquake [16]: **a** insufficient bending ductility and **b** shear failure

Loma Prieta, mainly due to failure that occurred in the joints which connect the piers and beams.

During the Northridge earthquake [16], poor flexural ductility available in plastic hinges and insufficient shear capacity were responsible for the collapse of several piers, as illustrated in Fig. 1. This type of failure is due to inadequate transverse reinforcement, leading to insufficient concrete confinement and, thus, causing the buckling of the longitudinal rebars.

Reinforced concrete bridges in the Kobe Line, Hanshin Expressway, were not provided with appropriate specifications for the cyclical behaviour, leading to extensive damage and the collapse of some of them. The drastic collapse of the piers and the consequent viaduct fall, was due to the premature interruption of the longitudinal reinforcement [17].

In the Taiwan earthquake of 1999, about 20 bridges located along the fault near Juahan were severely damaged, many of them with simply supported multiple spans [18]. This fact conducted to deformations and forces that probably exceeded those expected due to the seismic action.

The existing bridges of Wenchuan, China, were severely affected due to the magnitude of the earthquake of 2008 [19, 20]. The total number of damaged and collapsed bridges due to the large earthquake stands over 576, including very recent bridges, meaning thereby heavy economic losses. In the research work and inspection after the earthquake, conducted by the Chinese Ministry of Communications, it is stated that the number of bridges that survived with little damage was very low (<40%) and the number of severely damaged bridges or even collapsed was very high (14%), taking into account that these infrastructures are very important for rescue teams in the post-earthquake.

3 Strengthening Techniques and Strategies

3.1 Solid Piers

Solid piers are often damaged after an earthquake due to inadequate strength or ductility, and the strengthening of these piers can be performed with jacketing of different material (reinforced concrete, steel, or FRP). For increasing strength capacity the entire pier height needs to be retrofitted while for ductility enhancement the strengthening technique can be applied at the plastic hinge region.

Before performing the retrofit, if the pier needs to be repaired the following steps should be carried out: (1) Delimitation of the repairing area (usually the plastic hinge region); (2) Removal and cleaning of the damaged concrete; (3) Alignment and replacement of the longitudinal reinforcement bars (if the bars had buckled or failed it needed to be replaced in order to ensure the alignment); (4) Application of formwork and new concrete (usually Microbeton, a pre-mixed micro concrete, modified with special additives to reduce shrinkage in the plastic and hydraulic phase).

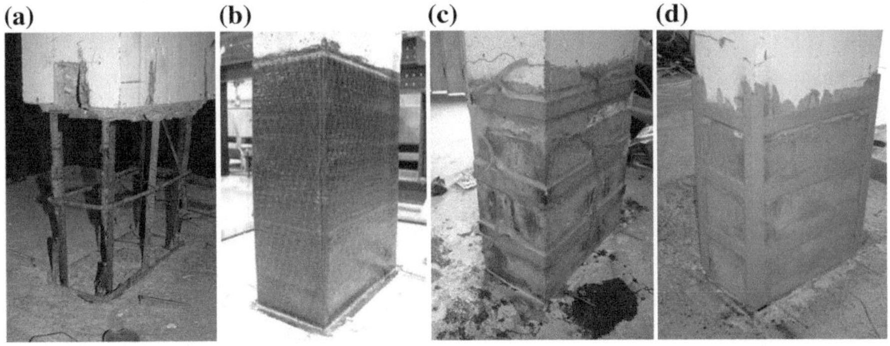

Fig. 2 Lap spliced zone (**a**); retrofitted column with CFRP sheet jacket (**b**); steel plates (**c**); and Steel plates connected by equal leg angle steel profiles (**d**)

After the repair stage, the selected technique and strategy of retrofitted can be applied, with three different suggestions illustrated in Fig. 2: CFRP jacket; steel plates; and steel plates connected by equal legs angles steel profiles.

In order to design the retrofit jackets, Priestley et al. [15] approach is suggested to calculate the thickness of the jacket. Inelastic deformation capacity of flexural plastic hinge regions can be increased by improving confinement of the pier concrete with the jacketing system.

3.2 *Hollow Piers*

The main steps for the repaired and retrofitted (after suffering shear damage) hollow piers are the following: (1) delimitation of the repair area; (2) removal and cleaning of the damaged concrete; (3) alignment or replacement of the longitudinal rebars; (4) application of formwork and new concrete (usually Microbeton, a pre-mixed micro concrete, modified with special additives to reduce shrinkage in the plastic and hydraulic phase); (5) outer retrofit with jacketing (with a given technique and material). To provide a general idea of the pier damage and of the retrofit process, Fig. 3 shows the piers during repair and after CFRP sheet jacketing. The CFRP sheet properties (supplied by manufacturer) are as follows: Elastic modulus, $E_j = 240.000$ MPa; Ultimate strain, $\varepsilon_{ju} = 0.0155$; Ultimate strength, $f_{ju} = 3800$ MPa and Layer thickness: $t_{j1} = 0.117$ mm.

In order to design the outer shear retrofit with CFRP jackets, the methodology suggested by Priestley et al. [15] was adopted to evaluate the thickness of the rectangular hollow pier jacket for increasing the shear strength above the maximum flexural force while keeping the initial section conditions. According to this methodology the shear strength can be conveyed by Eq. (1):

Fig. 3 Hollow piers before and after the shear retrofitting with CFRP sheets

$$V_d = V_c + V_s + V_p + V_{sj} \tag{1}$$

where V_c, V_s and V_p are the shear force components accounting, respectively, for the nominal strength of concrete, the transverse reinforcement shear resisting mechanism and the axial compression force; the term V_{sj} corresponds to the possible retrofit contribution with CFRP or metal jackets and can be estimated according to Eq. (2):

$$V_{sj} = \frac{A_j}{s} f_j \cdot h \cdot \cot \theta \tag{2}$$

where h is the overall pier section dimension parallel to the applied shear force, f_j is the adopted design jacket stress, A_j is the transverse section area of the jacket sheets spaced at a distance of s and inclined at the angle of θ relative to the member axis. Therefore, Eq. (1) can be applied to estimate the number of CFRP sheet layers required to increase the shear capacity of the pier.

3.3 *Experimental Test on Hollow Cross Section Piers with Shear Strengthening*

To study the behaviour of hollow cross section piers with shear strengthening, experimental and numerical tests were carried out and one specimen (PO2-N6) was selected for presentation herein. In this case, one strip layer of CFRP sheet was used, with 0.177 mm thick, 100 mm wide and spaced at 100 mm along the pier height in order to increase the shear capacity, following the strategy referred in the previous section. It was decided to leave a 100 mm distance at the base in order to analyse the available ductility of the pier if shear collapse mechanism was avoided.

The specimens were tested under cyclic loading with increasing values of drift intensity. The axial load of 250 kN was considered, that corresponds to a normalized axial force of 0.05. The test setup characteristics and more detailed results are available in previous reports [14, 21, 22].

When 3.0% drift was achieved, corresponding to 45 mm of top pier displacement, the second strip of CFRP (counting for the bottom) broke up in the northeast corner. In the subsequent cycles, broke up the 3°, 4°, 5° and 6° fibre strips, in ascending order, in the same corner as the second strip, Fig. 4. After the abrupt rupture of the fibres, the pier loses the concrete confinement, causing the desegregation of the material on the east side wall pier, as can be seen in Fig. 4a. Finally, after the experimental test has been stopped, high deterioration of the concrete and some buckling phenomena of longitudinal rebars were observed [23]. Although the failure mechanism of the pier PO2-N6-R1 was achieved by shear mechanism, it can be observed from Fig. 5 that the maximum strength increased about 50% in comparison with the original pier (PO2-N6). Therefore, a higher bending contribution was obtained in the pier behaviour (in comparison with the original pier), despite the shear deformation effects that occurred during the test. But in terms of displacement, this pier retrofit just has reached the ductility between 3 and 4, due to CFRP strips failed to accommodate the pier web deformation and broken.

The numerical simulations were carried out using two different methodologies: (i) fibre model and (ii) damage model. The fibre models are based in a finite element discretization with non-linear behaviour distributed along the element length and cross-sectional area, while the damage model is supported on refined finite element (FE) meshes, with high complexity and detail levels in the constitutive laws defined for both concrete and steel. The concrete is simulated with a continuum damage model where several applications for bridges with hollow section piers can be found at Faria et al. [24].

east side south side west side

Fig. 4 Final damage of pier PO2-N6-R1, for 3.0% drift

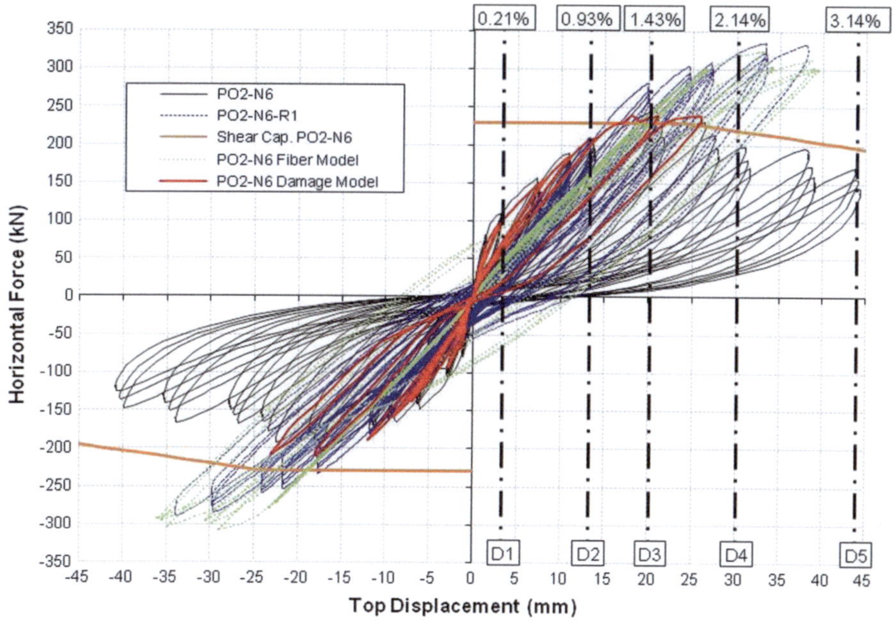

Fig. 5 Experimental and analytical results—comparison of PO2-N6 versus PO2-N6-R1

Analytical results of pier PO2-N6, with and without retrofit, are also shown in Fig. 5. To evaluate the Shear Capacity line, the methodology suggested by Priestley et al. [15] was adopted.

The difference between the results of the adopted fibre model [25] and the experimental test of pier PO2-N6 is due to a limitation of the numerical model to simulate the shear effects. Namely, it assumes that the pier is governed by bending behaviour, which leads to a maximum forces achieved similar to those obtained in the experimental test of the retrofitted pier (PO2-N6-R1), once it is capable to explore a higher bending component due to the CFRP strips retrofit. This numerical model does not consider the deformation and stiffness degradation from shear mechanisms, which is quite relevant for the pier behaviour when inclined cracks occurs in its webs. This inability of the model to consider the shear effect leads to maximum forces (in Fig. 5) close to values expected for flexural capacity (about 320 kN).

The deformation and strength mechanisms associated to shear are related to the concrete tensile behaviour [22]. The damage model [24] makes use of "effective stress tensors" decomposed in compression and tensile stresses that allows exploring the deformation and stiffness degradation by shear. As can be seen in Fig. 5, the damage model results show a good approximation to the experimental test of pier PO2-N6, with a quite accurate simulation of the shear effects.

3.3.1 Concluding Remarks

Experimental tests were performed in hollow cross section piers with and without shear strengthening, and the maximum force applied to the retrofitted pier increased about 50% in comparison with the original pier, but without significantly improvement on the ductility. Regarding the numerical analysis, with the more refined model, the damage model, it was possible to simulate the complex shear behaviour of this type of hollow cross section pier without retrofit, while the fibre model [25] captured the experimental behaviour of the retrofitted pier (PO2-N6-R1), achieving similar maximum forces once it is mainly governed by bending behaviour.

4 Effect of Strengthening on Seismic Performance of Bridges

The seismic performance of retrofitted bridges was first tested during the 1987 Whittier Narrows and 1989 Loma Prieta earthquakes that caused damage to a number of around 1000 California bridges that had been retrofitted against unseating by the addition of cable or rod restrainers after the 1971 Sylmar (San Fernando) earthquake [26, 27]. Some of these restrainers broke, but more concerning was the fact that, while unseating was prevented in most cases, other severe damage patterns emerged, mainly failures of concrete piers. This led to a second Caltrans retrofitting programme that included 1155 bridges; the focus of that programme was column retrofitting through jacketing (mostly with steel jackets). During the 1994 Northridge earthquake, which struck i.e. more than 200 retrofitted bridges, only one retrofitted (with restrainers) bridge suffered significant damage [26]. Overall the retrofitted bridges, particularly those of the 2nd phase of the Caltrans programme, performed well, suffering no damage or only minor damage (in the joints at the locations of deck hinges and abutments). Since then, a number of retrofitted bridges world-wide have been subjected to strong ground motions, but the size of the sample is still quite small to permit drawing meaningful conclusions and, even more, to derive empirical fragility curves for retrofitted bridges. In fact all studies of such bridges are primarily analytical; the analysis-based approach to seismic assessment of retrofitted bridges is presented in the following.

4.1 Modelling of RC Piers with Different Types of Jackets and Fragility Curves for Bridges with Strengthened Piers

The most comprehensive way to describe the seismic vulnerability of a structure is through a set of fragility curves, for a number of damage states; these can also serve as a valuable guide for designing a retrofitting scheme. Fragility curves for retrofitted bridges have been proposed for evaluating the efficiency of different retrofit measures and strategies with regard to the bridge seismic performance [28, 29] and identify the optimum solution compared to the as-built condition for different levels of seismic hazard. Recently, a new methodology for the derivation of bridge-specific fragility curves was proposed, considering the effect of varying structural and geometric component properties on capacity and demand estimation [30]. The method is suitable for application to inventories of existing bridges, using an ad hoc developed software, and is presented in the remainder of this section. It can be applied to retrofitted bridges with different structural systems [31], i.e. it is not restricted to specific bridge configurations or retrofitting schemes. In this method bridge piers, abutments, bearings and foundations are selected as the most important components for both as-built and retrofitted bridges, as far as performance under seismic actions is concerned.

The first step of the methodology is to define the capacity of critical components, based on the results of inelastic analysis, considering different failure modes. Component limit states (minor, moderate, major and collapse), are qualitatively described based on experimentally observed damage patterns, and quantified defining threshold limit state values in terms of a global engineering demand parameter (displacement), dependent on component geometry, reinforcement, material properties, and loading. Retrofitted component threshold limit state displacement values are correlated to as-built relevant threshold values in order to identify the effect of each retrofit measure on component capacity. Uncertainty in capacity, namely in damage threshold definition, is estimated using reduced Monte Carlo (Latin Hypercube Sampling) simulation for each component.

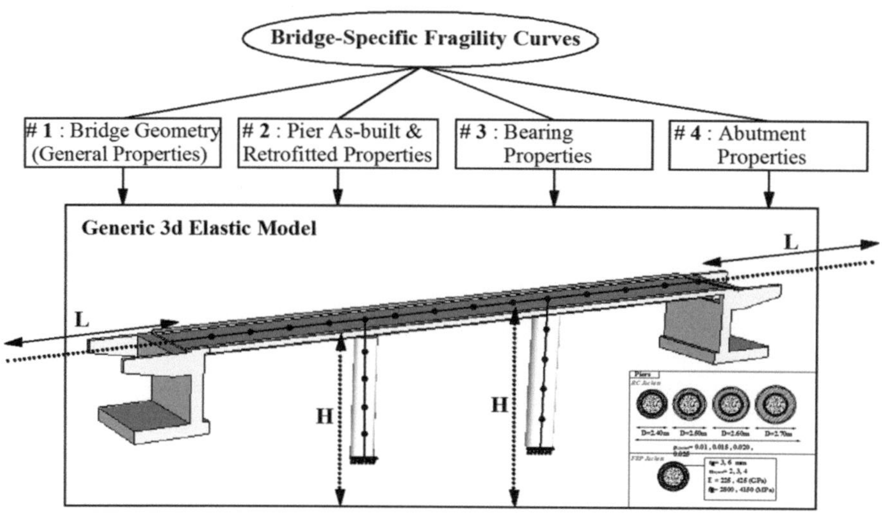

Fig. 6 Bridge fragility assessment: input data for setting-up a simplified 3D model

The second step is the development of a simplified 3D model (linear springs for bearings and abutments, see Fig. 6), providing general properties of as-built bridge components and retrofit measures, in order to estimate seismic demand on each component. Seismic analysis is performed at bridge level, with a view to establishing the correlation among different components during the evaluation of bridge performance for different levels of earthquake intensity. However, since the main target of the methodology is its applicability to a large bridge inventory for evaluating the retrofit effectiveness, simple elastic response spectrum analysis is used for the demand calculation.

To account for inelastic performance of the bridge system under seismic actions, the uncertainty in demand is calculated using nonlinear response history analysis for selected accelerograms that generally vary with seismic intensity level (Modified Incremental Dynamic Analysis). Representative bridges of each category (according to the classification scheme described in [30] are selected and Monte Carlo simulation with Latin Hypercube sampling is performed to quantify the uncertainty in seismic demand; in a practical context, uncertainty in demand is assumed to be the same for bridges classified in the same category. Having defined capacity and demand at a component level, bridge fragility is calculated assuming a series connection between components (except for limit state 4—'collapse' for which bearings are not deemed as critical). Since the demand calculation is based on the results of an elastic model, the effect of gap closure should be accounted for, considering two models (open and closed gap), retaining the results of the first model until gap closure.

The procedure for deriving the bridge-specific fragility curves is described in detail in [30]. In the following only the aspects related to retrofitted bridges will be presented, along with an illustrative case study.

4.1.1 Retrofitted Component Limit State Threshold Values and Related Uncertainty

Different parameters related to geometry, reinforcement, materials, and loading affect the available strength and ductility, and hence the seismic performance of piers. The effect of different pier shapes (cylindrical, hollow cylindrical, wall, rectangular, hollow rectangular), dimensions, material properties and constitutive laws, longitudinal and transverse reinforcement ratio and finally axial load, on the limit state threshold values of as-built bridge piers was evaluated. Regarding pier retrofit, two widely used in Europe retrofit techniques for strengthening and confinement (passive or active) were examined, namely RC jackets and CFRP jackets. Retrofitted piers with RC jackets of different thickness, reinforcement ratio and material properties and CFRP jackets with different layer number, thickness, elastic modulus and jacket strength, have different threshold limit state values. The considered limit states 1–4 are shown in Table 1 for FRP-jacketed columns (similar definitions apply to RC jacketing [31]). A database of retrofitted components was compiled (the case of cylindrical piers is presented herein) and empirical

Table 1 Limit states for FRP jacketed bridge piers

Limit state	Threshold values of curvature (φ)	Qualitative performance description
LS 1—minor/slight damage	$\boldsymbol{\varphi_1}$: φ_y	Microcracking of concrete and shifting of aggregates
LS 2—moderate damage	$\boldsymbol{\varphi_2}$: min (φ: $\varepsilon_c > 0.004$, φ: $\varepsilon_s \geq 0.015$)	Spalling of the cover concrete, strength may continue to increase
LS 3—major/extensive damage	$\boldsymbol{\varphi_3}$: min (φ: $\varepsilon_c \leq 0.004 + 1.4 \cdot \rho_w \cdot \frac{f_{yw}}{f_{cc}}$, φ: $\varepsilon_s \geq 0.06$)	First hoop fracture, buckling of longitudinal reinforcement, initiation of crushing of concrete core
LS 4—failure/collapse	$\boldsymbol{\varphi_4}$: min (φ: $M < 0.65 \cdot M_{\max}$)	60–70% of the ultimate load, patches of white began to show, plastic flow of resin, FRP rupture

relationships were derived for the quantification of damage in displacement terms, considering different failure modes of retrofitted components and correlating the retrofitted to as-built threshold values in order to assess the effect of each retrofit measure on component capacity.

Different properties for all retrofitted and as-built parameters were selected and moment versus curvature analysis of all possible combinations was performed using appropriate software [32]. Threshold values for the different limit states are initially defined in terms of curvature as depicted in Table 1, related to material strain limits. Using the advanced least squares method (robust fit), empirical relationships for the estimation of yield and ultimate moment and curvature values for the retrofitted columns (secant stiffness at yield $EI_{eff} = M_y/\varphi_y$) were derived for each different pier type (see Fig. 7, referring to cylindrical piers). An example of empirical relationship is given in Eq. (3) referring to curvature values for a CFRP-jacketed column

$$\varphi_{FRPj}/\varphi_{core} = \beta_0 + \beta_1 \cdot (D_{FRP,j}/D_{core}) + \beta_2 \cdot (E_{FRP,j}/E_{c,core}) + \beta_3 \cdot (f_{jFRP,j}/f_{c,core}) + \beta_4 \cdot (\rho_{fFRP,j}/\rho_{w,core}) \tag{3}$$

Values of β_i coefficients for yield and ultimate curvatures can be found in [33].

To define bridge capacity and threshold limit state values in displacement terms, an inelastic lumped plasticity model of the pier is analysed (bilinear moment-curvature curve) for all possible combinations of core/retrofitted section properties, and the displacement of the cantilever top (component control point) at the time step that the deformation of the plastic hinge exceeds threshold limit state values (φ_1, φ_2, φ_3, φ_4 in Table 1) is recorded. Shear failure mode is also considered, since the shear demand at each step is compared with the ultimate shear capacity and the displacement value associated with LS4 is recorded and compared with the one derived considering flexural failure; reduction in the concrete contribution (V_c) with increasing curvature ductility is accounted for.

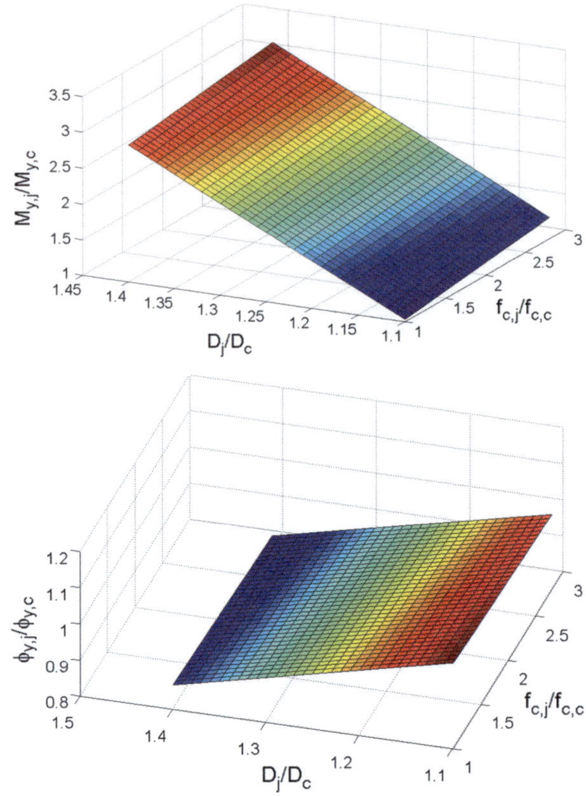

Fig. 7 $M_{y,j}/M_{y,c}$ and $\varphi_{y,j}/\varphi_{y,c}$ diagrams for RC jacketed cylindrical bridge piers

Practically all possible combinations of section properties and pier heights for common bridge piers are considered (height range 5–20 m for RC jacketed piers, resulting in $\sim$157,500 analyses) and threshold limit state values in displacement terms (δ_1, δ_2, δ_3, δ_4) are obtained, correlated with the relevant threshold displacement values for the core. Therefore, the threshold limit state values of the retrofitted pier can be easily defined if the corresponding values of the as-built component are known (see Eq. 3), whereas the effect of the retrofit measure and different retrofit properties can be easily evaluated. Analysis results are processed using the advanced least squares method and empirical relationships for threshold δ_i values are provided for the case of reinforced concrete and FRP jacket; as an example, the thresholds for FRP jacketed columns are given by:

$$(\delta_{FRPj}/H)/(\delta_{core}/H) = \beta_0 + \beta_1 \cdot (D_{FRP,j}/D_{core}) + \beta_2 \cdot (E_{FRP,j}/E_{c,core}) + \beta_3 \cdot (f_{jFRP,j}/f_{c,core}) + \beta_4 \cdot (\rho_{fFRP,j}/\rho_{w,core}) \tag{4}$$

where the β_i coefficients are different for each limit state (all values given in [33]).

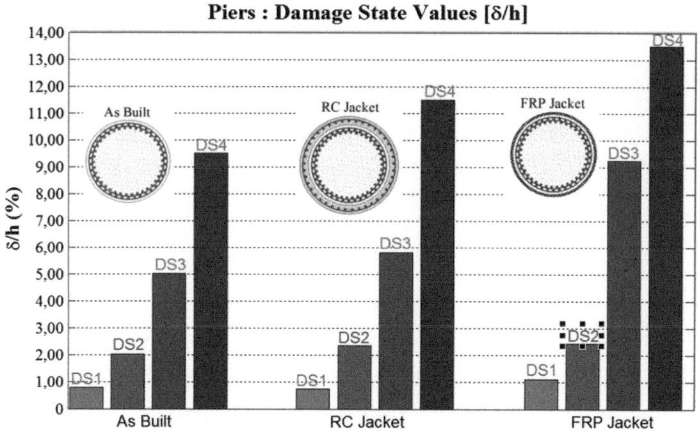

Fig. 8 Drift for all limit states and retrofit measures in cylindrical piers

Figure 8 presents the drifts (at the column top), estimated by applying empirical relationships such as (4) to all cylindrical columns retrofitted using different retrofit techniques; they are average values derived from all inelastic analyses performed. It is seen that both RC and FRP jackets improve the seismic performance, increasing the threshold limit state values. The fact that FRP jacketing results in larger increase in LS thresholds, compared to reinforced concrete jacketing of cylindrical piers, is additionally related to the fact that threshold limit state values are dependent on the ultimate concrete strain value ε_{cu}, which is significantly increased using FRP confinement. Moreover, it should be noted that RC jackets, apart from providing strengthening and confinement, result in an increase in initial stiffness, and eventually, as a rule, in higher input seismic forces.

Uncertainty in capacity was considered, adopting distributions of the random variables available in the literature [31, 33]. Latin Hypercube sampling was used with 100 realizations of each retrofitted member. Analysis results were processed and dispersion values for capacity (β_c) were estimated for each retrofit technique (RC and FRP jackets).

The definitions of the limit state thresholds for abutments and bearings (common elastomeric, and lead-rubber, bearings), i.e. for the other critical components of the bridge, are not affected by the retrofit scheme and are not presented herein; relevant information can be found in [30, 33].

4.1.2 Case Study

A typical overpass of Egnatia Motorway (N. Greece) was assessed using the previously described methodology. The bridge (Fig. 9) has cylindrical piers of 2 m diameter and approximately 9 m height, monolithically connected to the deck

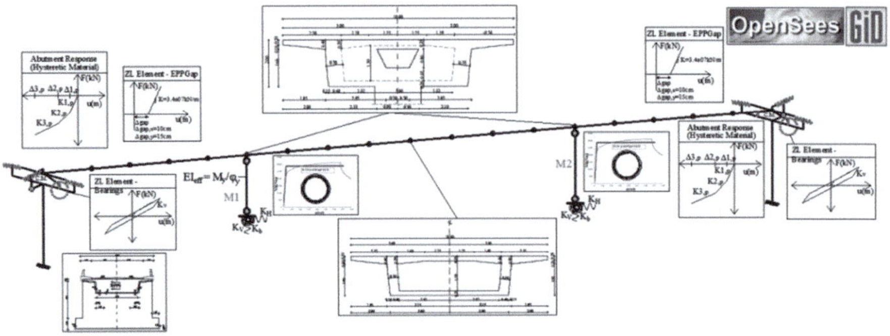

Fig. 9 3D model of the case-study bridge

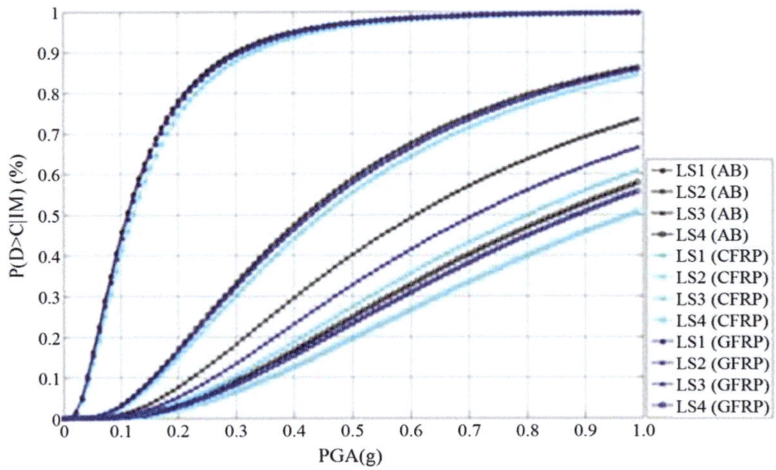

Fig. 10 Fragility curves for as-built and retrofitted bridge (transverse direction)

(a prestressed concrete box girder), and it is retrofitted using four different strategies: (i) RC jackets; (ii) FRP jackets; (iii) high-damping elastomeric bearings; (iv) lead rubber bearings. For all these retrofit schemes LS threshold values are available in the database of retrofitted components described in the previous section. The thresholds in global terms (δ_i) are correlated with the displacements developed in the considered bridge by carrying out standard elastic response spectrum analysis of the spine model shown in Fig. 9 for various levels or earthquake intensity. Fragility curves were finally derived using the series connection assumption (critical component for each LS defines the LS threshold for the entire bridge). The uncertainty associated with each LS was also derived considering the most critical component (in this case the piers); in a practical context it is not calculated for the specific bridge but is taken from the database for the category wherein the bridge falls (in this case the studied bridge was also the 'representative' bridge for the category 'box girders connected monolithically to single-column piers'). Using the properties of the overpass of Fig. 9, different bridge models were generated, with different number of spans [31].

Fragility curves for the as-built (AB) and retrofitted (with RC or CFRP jacketing of the piers) cases are depicted in Fig. 10. The strengthening techniques applied aimed at producing the same pier ductility, i.e. the common criterion was performance-based rather than cost-based; clearly a cost-based approach can also be adopted. Since the first limit state is associated with bearing damage, the beneficial effect of retrofitting is evident for the higher damage states only, i.e. for limit states 3 and 4, as shown in the Fig. 10. It has to be noted that if the strengthened component is not the critical one for a certain LS, the corresponding threshold can be lower than that for the as-built bridge; this illustrates the importance of a performance-based approach to retrofitting (different retrofit schemes are appropriate for different performance objectives). Further examples of fragility curves for retrofitted bridges can be found in [33].

4.1.3 Concluding Remarks

There is a clear need for collecting more information on the seismic performance of retrofitted bridges; this is expected to happen in the coming years as earthquakes strike areas like the US and Japan where a substantial number of retrofitted bridges exist. In the meantime, analytical approaches are the only viable option for estimating the seismic performance of bridges where one or more retrofitting techniques were applied. The component-based methodology for deriving bridge-specific fragility curves presented herein is arguably the most appropriate one for retrofitted bridges, as it makes feasible the exploration of several alternative schemes with a view to identifying the optimal one.

5 Conclusions

This chapter presented some strategies for the strengthening of RC bridges and the respective benefits to their structural behaviour.

For the hollow pier tested with and without shear retrofitting, and the maximum force applied to the retrofitted pier increased about 50% in comparison with the as-built pier, but without significant improvement in the available ductility. Numerical analysis of this pier allows to simulate the complex shear behaviour of this type of hollow cross section without retrofit, with the damage model which takes into account the deformation and strength mechanisms associated to shear behaviour, while the retrofitted pier behaviour, mainly governed by bending mechanism, was satisfactorily captured by the fibre model, achieving similar maximum forces.

Analytical studies were conducted for estimating the seismic performance of bridges where several retrofitting techniques were applied. The component-based methodology for deriving bridge-specific fragility curves presented in Sect. 4 is perhaps the most appropriate one for retrofitted bridges, since it allows the evaluation of several alternative schemes for achieving the optimal strengthening solution.

Acknowledgements Part of this work reports to research financially supported by Project POCI-01-0145-FEDER-007457—CONSTRUCT—Institute of R&D in Structures and Construction, funded by FEDER funds through COMPETE2020—Programa Operacional Competitividade e Internacionalização (POCI) and by national funds through Fundação para a Ciência e a Tecnologia (FCT).

References

1. Seible F, Priestley M, Hegemier G, Innamorato D. Seismic retrofit of RC columns with continuous carbon fiber jackets. J Compos Constr. 1997;1:52.
2. Gergely I, Pantelides C, Nuismer R, Reaveley L. Bridge pier retrofit using fiber-reinforced plastic composites. J Compos Constr. 1998;2:165.
3. Parvin A, Wang W. Behavior of FRP jacketed concrete columns under eccentric loading. J Compos Constr. 2001;5:146.
4. Santarosa D, Campos Filho A, Beber A, Camplagnolo J. Concrete columns confined with CFRP sheets. In: JG T, editor. International conference on FRP composites in civil engineering. Hong Kong, China; 2001.
5. Teng J, Chen J, Smith S, Lam L. FRP strengthened RC structures. New York: Wiley; 2001.
6. Hadi MNS. Behaviour of FRP wrapped normal strength concrete columns under eccentric loading. Compos Struct. 2006;72:503–11.
7. Ogata T, Osada K. Seismic retrofitting of expressway bridges in Japan. Cem Concr Compos. 2000;22:17–27.
8. Cheng CT, Yang JC, Yeh YK, Chen SE. Seismic performance of repaired hollow-bridge piers. Constr Build Mater. 2003;17:339–51.
9. Mo YL, Yeh YK, Hsieh DM. Seismic retrofit of hollow rectangular bridge columns. ASCE J Compos Constr. 2004;8:43–51.
10. Pavese A, Bolognini D, Peloso S. FRP seismic retrofit of RC square hollow section bridge piers. J Earthq Eng. 2004;8:225–50.
11. Yeh YK, Mo YL. Shear retrofit of hollow bridge piers with carbon fiber-reinforced polymer sheets. J Compos Constr. 2005;9:327–36.
12. Tsionis G, Pinto AV. Numerical analysis of RC bridge piers with rectangular hollow cross-section retrofitted with FRP jackets. J Earthq Eng. 2007;11:607–30.
13. Lignola GP, Prota A, Manfredi G, Cosenza E. Non-linear modeling of RC rectangular hollow piers confined with CFRP. Compos Struct. 2009;88:56–64.
14. Delgado P, Arêde A, Vila Pouca N, Rocha P, Costa A, Delgado R. Retrofit of RC hollow piers with CFRP sheets. Compos Struct. 2012;94(4):1280–7.
15. Priestley M, Seible F, Calvi G. Seismic design and retrofit of bridges. New York: Wiley; 1996.
16. Oliveira CS, Azevedo J, Delgado R, Costa AG, Costa AC. O Sismo de Northridge, Los Angeles, 17 de Janeiro de 1994—Ensinamentos para Portugal. Edição Conjunta dos IC-IST/IC-FEUP: Lisboa; 1995 (in Portuguese).

17. EASY. Earthquake Engineering Slide Information System; 1997. http://www.ikpir.fgg.uni-lj.si/easy/.
18. NISEE. National Information Service for Earthquake Engineering, University of California, Berkeley; 2006. http://nisee.berkeley.edu/taiwan/.
19. EERI. Learning from Earthquakes: The Wenchuan, Sichuan Province, China, Earthquake of May 12, 2008, EERI—Special Earthquake Report—October, 2008.
20. ISSN. General introduction of engineering damage of Wenchuan Ms 8.0 earthquake. J Earthq Eng Vib; 2008.
21. Delgado R, Delgado P, Vila Pouca N, Arêde A, Rocha P, Costa A. Shear effects on hollow section piers under seismic actions: experimental and numerical analysis. Bull Earthq Eng. 2009;7:377–89.
22. Delgado P, Monteiro A, Arêde A, Vila Pouca N, Delgado R, Costa A. Numerical simulations of RC hollow piers under horizontal cyclic loading. J Earthq Eng. 2011;15(6):833–49.
23. Delgado P, Sousa F, Rocha P, Arêde A, Vila Pouca N, Costa A, Delgado R. Experimental and numerical analysis on the cyclic behavior of bridge piers with and without CFRP retrofit. In: 5th ECCOMAS thematic conference on computational methods in structural dynamics and Earthquake engineering. COMPDYN 2015.
24. Faria R, Oliver J, Cervera M. A strain-based plastic viscous-damage model for massive concrete structures. Int J Solids Struct. 1998;35(14):1533–58.
25. Seismosoft. SeismoStruct. computer program for static and dynamic nonlinear analysis of framed structures. http://www.seismosoft.com.
26. Yashinsky M. Performance of bridge seismic retrofits during Northridge earthquake. J Bridge Eng ASCE. 1998;3(1):1–14.
27. Federal Highway Administration (USA): Seismic retrofitting manual for highway structures: part 1—Bridges. Publication no. FHWA-HRT-06-032, McLean, VA; 2006.
28. Kim S-H, Shinozuka M. Development of fragility curves of bridges retrofitted by column jacketing. Probab Eng Mech. 2004;19(1–2):105–12.
29. Padgett JE, DesRoches R. Methodology for the development of analytical fragility curves for retrofitted bridges. Earthq Eng Struct Dyn. 2008;37(8):1157–74.
30. Stefanidou SP, Kappos AJ. Methodology for the development of bridge-specific fragility curves. Earthq Eng Struct Dyn. 2017;46(1):73–93.
31. Stefanidou SP, Kappos AJ. Efficiency of different retrofit measures in reducing seismic fragility of bridges. In: 16th European Bridge conference, Edinburgh, June 2015, Paper no. 1477.
32. Papanikolaou VK. Analysis of arbitrary composite sections in biaxial bending and axial load. Comput Struct. 2012;98–99:33–54.
33. Stefanidou SP. Structure-specific Fragility Curves for As-Built and Retrofitted Bridges. Ph.D. thesis, Civil Engineering Department, Aristotle University of Thessaloniki, Greece; 2016.

Strengthening of Masonry Bridges

Cristina Costa, António Arêde and Aníbal Costa

1 Introduction

The structural condition assessment of masonry arch bridges has been assuming increased importance in the last decades, for which, at least, the two following major reasons can be pointed out. On the one hand, both at regional and country levels, this type of clay brick or stone bridges includes important pieces of high value heritage constructions that must be safeguarded and preserved for upcoming generations. On the other hand, bridge owners and stakeholders are increasingly interested in establishing suitable safety criteria and management plans, for which both the identification of limit conditions for traffic loading and the definition of adequate intervention proposals are issues of major concern to reduce damage and detrimental effects caused by road and railway traffic.

The longevity of thousands masonry arch bridges in current operation is itself an evidence allowing attesting the good performance of these constructions, which is

C. Costa (✉)
Department of Engineering, Tomar School of Technology,
Polytechnic Institute of Tomar, 2300-313 Tomar, Portugal
e-mail: c.costa@ipt.pt

C. Costa
CONSTRUCT-LESE, Faculty of Engineering (FEUP),
University of Porto, 4200-465 Porto, Portugal

A. Arêde
CONSTRUCT-LESE, Civil Engineering Department, Faculty of Engineering,
University of Porto, 4200-465 Porto, Portugal
e-mail: aarede@fe.up.pt

A. Costa
RISCO, Department of Civil Engineering, University of Aveiro,
Campus Universitário de Santiago, 3810-193 Aveiro, Portugal
e-mail: agc@ua.pt

A. Costa et al. (eds.), *Strengthening and Retrofitting of Existing Structures*,
Building Pathology and Rehabilitation 9, https://doi.org/10.1007/978-981-10-5858-5_10

mainly due to the type of the materials used (e.g. stone masonry and infill material) and also to the high robustness of the built solutions. Stone bridges, in particular, are examples of sustainability compared to modern reinforced concrete and steel bridges, generally requiring higher maintenance costs.

In such context, several experimental and numerical studies have been presented worldwide, particularly in the eight editions of the International Conference on Arch Bridges [1–8]. The study of stone arch bridges was also addressed in research projects, namely "Sustainable Bridges—Assessment for Future Traffic Demands and Longer Lives" [9] (developed within EU FP6) and "Improving Assessment, Optimisation of Maintenance and Development of Database for Masonry Arch Bridges" [10] (developed by the International Union of Railways–UIC). Other working groups, such as the Construction Industry Research and Information Association (CIRIA), have also contributed to a better organization of data available on this subject [11].

The European project reported in [9], which involved about 220,000 bridges in Europe, concluded that *circa* 41% are arched bridges, of which 35% are over 100 years old and 62% are small span bridges.

Concerning the railway network, the large number of existing and in operation masonry arch bridges across Europe, also justifies the need to study this type of bridges. According to data reported in UIC [10], which several European countries have contributed to, about 60% of railway bridges are either arched ones or culverts; out of these, 80% have spans shorter than 5 m and 70% are aged between 100 and 150 years. For example, in Italy, Portugal and France there are about 56,888, 11,746 and 78,000 such cases, respectively, amounting to 95, 90 and 77% of the total existing railway bridges in each country.

Conservation plans should be drawn up in order to protect the structural integrity and safety of this type of bridges, ensuring performance and preserving heritage value. For this purpose, several interventions can be implemented, namely: (i) preventive; (ii) remedial and (iii) strengthening measures. The choice of the solution to be adopted must be made according to the degree of structural deterioration; it can involve structural replacement, consolidation or stabilisation, resorting either to traditional techniques (using only original-like materials and techniques) or modern innovative techniques, the latter typically adopting potentially more effective solutions relying on cutting-edge materials and/or equipment.

The characterization of the structural response resulting from most common loading (self-weight, support settlements, traffic loads and induced vibrations) allows relating structural damages, defects and failure modes with the bridge behaviour in the longitudinal and transverse directions; the former is governed by arch hinge mechanisms, while the latter is mainly influenced by interactions between infill material, spandrel walls and arches. Damage and degradation present in masonry arch bridges result from complex processes involving several material deterioration phenomena (physical, chemical, mechanical and erosive), due to time decay and inadequate maintenance plans and/or rehabilitation interventions. Due to their nature, these defects are generally spread throughout the whole construction and less influenced by the interdependence between equilibrium mechanisms and load transmission or the structural behaviour of a particular structural element.

Concerning strengthening interventions, it is well known that passive structural reinforcement solutions are only activated for loads higher than those acting in the structure prior to the intervention or for deferred deformations. By contrast, active strengthening techniques can be also an option (e.g. prestressing), which are likely to require modifications of the bridge conditions and generally mobilize immediate structural reaction and inherent change the state of equilibrium and deformability; although possible, the latter are seldom adopted in stone arch bridges.

The strengthening and/or rehabilitation solutions most commonly adopted to tackle the structural problems are found to be mainly based on the use of transverse ties and longitudinal anchors, reinforced concrete elements, as well as injections of mortar or cement grout in the infill.

Considering that many stone bridges are part of historic heritage, the respect for their cultural and historic values requires interventions aimed at improving and/or restoring structural safety governed by several specific criteria broadly disseminated in international charters [12]. However, such interventions should be adopted only when there are prior or possible future problems, likely to be identified in several stages, namely anamnesis, diagnosis, therapy and control, as suggested in [12]. Moreover, interventions should be the least possible intrusive (principle of minimum intervention) and the characteristics of materials used in the intervention shall ensure: (i) mechanical-structural compatibility, to minimize stiffness and strength changes in the original structure; (ii) physical-chemical compatibility, to prevent new pathologies; (iii) durability, to minimize future intervention needs and (iv) reversibility, to allow removing new elements inserted during the intervention at the end of their lifetime, or if they show signs of inappropriateness, without causing damage to the original materials.

In this framework, this chapter aims at describing the most common strengthening techniques of masonry arch bridges and their effects on the structural system. It refers to cases where some of such strengthening solutions were implemented to correct structural deficiencies in the longitudinal and transversal bridge directions. Some general strengthening solutions are also addressed, for cases where the increase of load carrying capacity is required, particularly due to traffic condition modifications.

As part of this study, a survey was made on some intervention works carried out on stone masonry arch bridges. A few references are made to bridges intervened in Portugal by IP–"Infraestruturas de Portugal", the national official body for Infrastructure Management, by IGESPAR (the Portuguese Architectural Heritage Management Institute) and by other national authorities, in order to identify the most commonly used solutions in the country. Other interventions carried out in this type of bridges are also referred, based on publications resulting from projects developed by UIC [13], as part of the assessment, safety and maintenance of masonry arch bridges, by the Transport Research Laboratory (TRL) [14, 15], and on other works published in further bibliographic references.

2 Structural Behaviour, Frequent Damages and Degradation

2.1 Basics of Structural Behaviour

The structural system of these bridges essentially comprises two types of constituent materials: masonry and infill material. Stone masonry is used to form the main components, namely arches, spandrels and piers. The infill material, placed in between the spandrels and above the arches, is usually made up of soil granular material, ideally with sparse granulometry, but other types of materials can be found. These materials are generally heterogeneous and anisotropic, in some cases containing discontinuities with complex behaviour; usually they have very low, or virtually null, tensile strength, while shear strength is much dependent on the friction conditions, cohesion and normal stresses installed.

The structural behaviour of stone masonry arch bridges is mainly controlled by the strength mechanisms and interaction between structural elements, as well as the mechanical behaviour of constituent materials; all these issues contribute for load transmission due to weight and traffic, foundation settlements and for the response to traffic induced vibrations.

Concerning the global structural response under the most common loading (self-weight and traffic) two main mechanisms are activated, according with the distinct bridge behaviour in the longitudinal and the transverse directions. Under heavy loading, likely to approach failure conditions, the former is characterized by arch hinge mechanisms resulting from transmission of vertical loads across the infill (see Fig. 1a); the latter, mainly influenced by the interaction between infill, spandrels and arches, is related with the transmission of horizontal pressures and the way how they are reacted (see Fig. 1b, [16]).

Damage and degradation of stone masonry arch bridges result from mixed processes causing several deterioration phenomena. These can be related with time decay of material characteristics, poor maintenance, strength mechanisms, interaction between structural elements related with load transmission, traffic induced vibrations and supports' settlements. Typical occurrences and their direct causes are briefly overviewed in the next paragraphs.

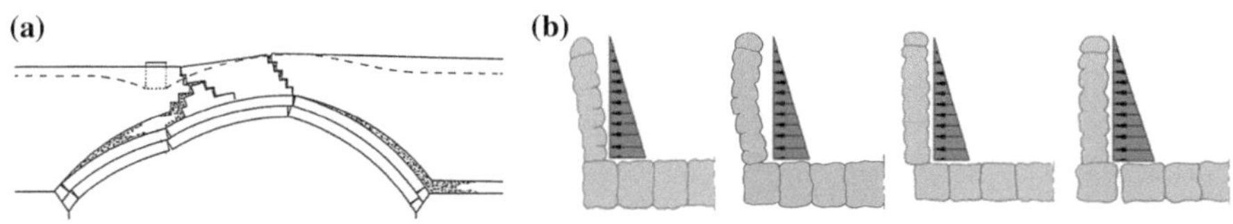

Fig. 1 Failure modes: **a** Longitudinal direction, **b** Transversal direction (picture credits [16])

2.2 Structural Damages

Structural damage, resulting from the behaviour of the bridge structure and its materials, often appears in the form of cracking, sliding, crushing and excessive deformation of specific structural elements.

In fact, for instance, the identification of theoretical positions of hinges characterising arch failure mechanisms, allow associating the existence of transverse cracks with the bridge structural response in the longitudinal direction, since it can be a sign of lack of material strength capacity under excessive load or occurrence of foundation settlements.

The interaction between different elements of the bridge, considering both the longitudinal and transverse direction response, is also a cause of structural damage which may influence the overall behaviour of the bridge.

As mentioned above, cracks in transverse joints can be plausibly associated with the development of hinge mechanisms in the arch.

In turn, arch longitudinal cracks, along the spandrel-arch connection (Fig. 2), are likely related with the structural response in the transverse direction, typically influenced by the interaction between arches, spandrels walls and infill material. In fact, vertical loads lead to the development of horizontal pressures in the infill which are responsible by out-of-plane effects in spandrel walls, tensile cracking in arches' stones and sliding in longitudinal interfaces between arches and spandrels (Fig. 1b), [16]). Consequently, due to such longitudinal cracks, the arch tends to perform independently from the spandrel walls and, therefore, the favourable effect of the latter for the arch behaviour (which conveys into the spandrel-arch system stiffness) gradually disappears in the central area. In this way, the formation of longitudinal hinge mechanisms in the central area becomes more facilitated and the effectiveness of the load transfer system decreases.

Fig. 2 Longitudinal cracks due to the bridge structural performance in the transverse direction

2.3 *Material Degradation*

Agents of material deterioration such as aggressive water, soluble salts, atmosphere gases, temperature, wind, ice and living organisms, may lead to: (i) stone decay by erosion, dissolution and disaggregation; (ii) occurrence of efflorescence, incrustations and black films, moisture stains, fluid water run-off, biological deposits and vegetation on exterior surfaces, and (iii) loss of joint and infill material (see Fig. 3). These are physical, chemical and erosive damages resulting from time passage and lack-of or non-adequate maintenance plans and/or rehabilitation interventions. These types of damage usually do not depend on the bridge structural behaviour and may appear in a generalized way anywhere in the bridge.

As a first preservation step, measures should be taken to eliminate the causes of deterioration and to implement maintenance plans to prevent damage. Therefore, cleaning the surfaces to eliminate biological and atmospheric contaminations, as well as maintenance of waterproofing and drainage systems, should be part of regular maintenance procedures.

Fig. 3 Physical, chemical and environmental damages: **a** Dissolution, **b** Granular disintegration, **c** Vegetation, black films, moss and lichens, **d** Lack of mortar and block erosion

3 Interventions for Damage Prevention and Rehabilitation

3.1 Preventive Measures

Preventive interventions are pro-active actions that can be taken to avoid or minimize the occurrence of structural problems. These measures include current maintenance, which can be performed during routine inspections, and/or specific works that may require more specialized workmanship.

Routine maintenance is important to reduce the rate of deterioration in masonry bridges. This type of intervention requires few resources, in terms of workmanship and equipment, but can have a significant impact on the conservation cost of such infrastructures.

These measures may involve cleaning degraded material, vegetation, biological colonization and other undesirable substances in bridges and its surroundings (e.g. slopes), using appropriate methods. Clearing the drainage system is particularly relevant due to the serious implications arising from water infiltration into the structure, namely the degradation of infill material and joint mortar. Also, the water course should be periodically cleaned, regularized and protected to ensure maintenance of proper water flow and to prevent sudden increase of the water level, typical of flood events, which can cause significant damage in bridges, especially in piers and breakwaters. As an example, Fig. 4 illustrates the case of the Ronfos Bridge (Portugal), before and after intervention made to repair the damage caused in breakwaters due to poor shallow water flow.

The application of protective systems, such as superficial treatment coatings on masonry (particularly in buttresses and/or parapets) and the inclusion of impermeable membranes underneath the deck, may be crucial to decrease the deterioration rate and to prevent further damage.

Fig. 4 Shallow water flow under Ronfos Bridge (Portugal): **a**, **b** Damaged breakwaters before remedial works, **c** After the intervention

Sometimes, it is necessary to implement load and speed limits for vehicles in order to avoid excessive overloads and to slow down the rate of degradation of the bridge.

3.2 Remedial and Rehabilitation Measures

Remedial measures are interventions aiming at restoring the structural integrity and the whole construction to its original conditions. These measures require appropriate means and skilled labour and should only be carried out when deemed necessary according to recommendations drawn from major or special inspections. The works included here involve both routine repairs, generally not requiring the preparation of specific design or more extensive interventions.

3.2.1 Joint Re-pointing

Joint re-pointing consists in injecting mortar into the joints where the material is deteriorated or missing. This technique should be used as an integrant good practice of maintenance programs because mortar pointing reduces water penetration and promotes the increase of the contact area between the stones which, in turn, contributes to reduce contact stresses in the joints, thus improving the structure's durability.

Technically it is simple and not very expensive, but it requires due care concerning the type of mortar to be adopted. It is recommended to be a lime based hydraulic mortar that is mechanically, chemically and aesthetically compatible with the existing material. Current cement-based mortars should be avoided since they are recognized to be chemically non-compatible with stone units which can become subjected to accelerated degradation due to the high levels of soluble salts present in Portland-type cement. Moreover, the use of mortar stronger than masonry units can lead to units' cracking and undesirably modify the original masonry resistant mechanism based on weak joints and strong units. Therefore, mortar composition and mechanical properties should allow the masonry units to move while maintaining the cohesion of stone elements.

Figure 5 shows the case of the São Lázaro bridge (nearby Porto, Portugal) where joints have been re-pointed with bastard mortar containing lime, cement and aggregate. The procedure for this method usually includes the following steps: (i) removing the degraded joint material with a suitable cleaning method (e.g. water jet) that does not cause damage to the masonry units; (ii) washing the open joints with low pressure water to remove loose material and moisten the surfaces where the mortar is to be applied; (iii) filling the joints with mortar compatible with the existing material; (iv) stabilising and finishing the joint by applying mortar with a pointing trowel to obtain a concave surface of the joint.

Fig. 5 São Lázaro bridge. General view of exterior surfaces: **a** Before and **b** After joint re-pointing

3.2.2 Replacement of Masonry Units

Replacement of masonry units is used when it is need to add missing units or when existing blocks clearly exhibit loss of material. Replaced units should be similar to existing ones and inserted in the structure by filling the joints with compatible mortar. Compatibility of materials should consider mechanical, chemical and aesthetic characteristics.

3.2.3 Replacement of Infill Material

Replacement of infill material can be a suitable option when it is found to have degraded mechanical characteristics, namely loss of thin particles washed-out due to water percolation inside the bridge structure. In such cases, infill material shear strength reduces which implies increase of horizontal pressure in spandrel walls. Another motivation for the use of this technique can be the reduction of dead loads due to infill, by replacing the existing material with a lighter one.

In bridges with historical interest, this technique may be particularly useful to stabilize movements of spandrel walls with no influence on the overall aesthetics of the structure.

The replacement of infill materials should be done very carefully due to the modifications introduced in the load-bearing capacity of the structure. In particular, the material must be inserted uniformly and symmetrically around the arch axis, to prevent damage caused by lack of equilibrium in the involved structure elements.

Although this operation it-self preserves the original structural concept, it is often found in association with arch strengthening using reinforcing straps or concrete slabs. Some cases of such can be found in rehabilitation interventions in historical Portuguese bridges carried out in the 1940s and 1950s, such as the Lagoncinha bridge [17] or the Ponte de Lima bridge [18]. In the French Marillais bridge case, which exhibited settlement of the central support pier, the rehabilitation consisted in replacing the infill with concrete, complemented with steel bar anchoring in the damaged zone (further addressed in latter section). The central pier was internally filled with concrete and the same was made in the abutment area, though with lightweight concrete. A concrete slab has been built in the deck area and the whole waterproofing and drainage system reconditioned [19].

By contrast, in the recent case of Esmoriz bridge (Portugal) [20], which was found highly deteriorated, stone arrangements were made and the infill material was replaced by a new well graded granular material, but no other less compatible and/or irreversible interventions were made.

3.2.4 Grout Injection

Grout injection using fluid grout consists in filling empty cavities, voids and cracks in the infill and the masonry of arches and spandrels with the aim to improve the material continuity and load transfer mechanisms between the structural elements thus benefiting the global bridge structural behaviour in both longitudinal and transverse directions.

The application of this method requires prior cleaning and re-pointing of masonry joints. The injection of grout in voids, cavities and cracks must be performed following bottom-up sequence, starting from the lowest levels (if possible), at low controlled pressure or by gravity into holes previously drilled in the bridge components. These holes serve not only to inject grout but also to place drain pipes, which allow controlling the execution process and ensure that all voids are filled. The holes should be drilled carefully with adequate equipment to prevent possible damage in masonry and sealed after work completion (Fig. 6).

As for the type of grout to be used, it is necessary to consider the compatibility with existing materials. The grout should be composed of a lime-based mortar, fluid enough to penetrate inside the structure. Cementitious mortars should be avoided because of their incompatibility with the stone-like material and because, unlike

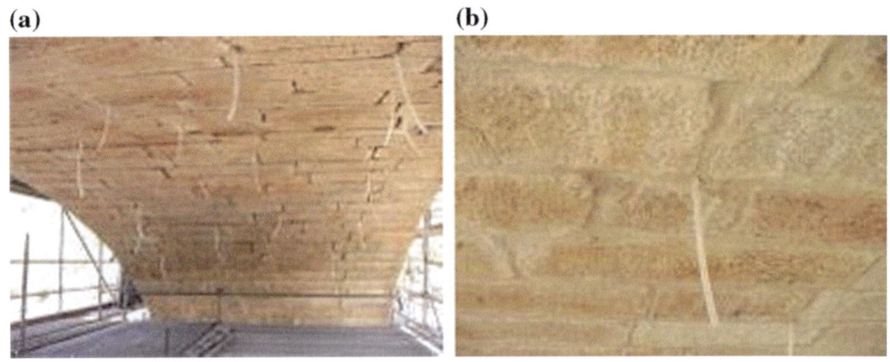

Fig. 6 Remondes bridge: **a** Before and **b** After repointing of joints [28]

lime-based mortars, do not allow water evaporation out of the structure. When crack sealing is sought, specially formulated resins can be injected.

Grout can be injected into all masonry elements of the bridge, foundations included. The repair of these elements, however, involves very specialized work, which may require particular technical expertise. Specifically in foundations, mortar injection is also quite suitable to make masonry more impermeable, by increasing the consistency of its constituent material in order to reduce internal circulation of water.

For example, this reinforcement technique was used simultaneously with transverse steel ties rods (further addressed in Sect. 4) connecting opposite spandrel walls in some Portuguese bridges: Segura, Real, Formigosa, Pedrinha, Caninhas, Sancheira and Remondes, [21–28]. In the latter case, Remondes bridge, injections were made using a 3.5 bar pressure chamber and low pressure pumps with injection details to achieve perfect control of flow rates and injection pressures (between 0.2 and 0.3 MPa), in order to avoid damaging the structure (Fig. 6).

This solution has the advantage of not affecting the bridge external look, while allowing works to be carried out without interrupting traffic flow.

The main disadvantages of this solution concern the high level of uncertainty in the quality of work performed and the fact that filling may not have been carried out effectively with voids not being completely filled.

3.2.5 Pavement Repair

The solutions that involve the repair of the paving stones are intended to keep the bridge operation through one of the following options (Fig. 7): (i) repairing the original bridge pavement, reusing the masonry stones in good condition and adding new similar slabs, where necessary; (ii) covering the original pavement with an additional layer usually made of bituminous material, taking due care to keep operational the drainage system which is often negatively affected by the height increase of the pavement and, (iii) replacing the original pavement by a new one, made either of bituminous material or stone units.

Fig. 7 Pavement repair: **a** Replacement of masonry stones, **b** Bituminous pavement overlay and **c**, **d** Replacement pavement

Regardless the adopted procedure, deck drainage and waterproofing (preferably underneath the deck) should be assessed and ensured, in order to prevent the infiltration of water into the structure.

3.2.6 Rehabilitation of the Drainage System

In order to provide adequate drainage conditions in the bridge and prevent or reduce water infiltrations into the structure, the drainage system rehabilitation may include: deck profile modification, introduction and/or repairing of ducts and opening of gutters through the mortar, deviating the water flow to any existing drainage systems in the area (Fig. 8). This intervention should be carried out in association with deck waterproofing.

Drainage systems should be sized according to the average rainfall in the region, in order to allow water to flow with the least impact possible on acting loads and structure operation. These systems should be designed so as to allow regular cleaning and routine inspections.

In the case of the gargoyles (or gutters), which are designed to allow water to exit from the inside of the structure, it is necessary to prevent the loss of the fine infill particles through an appropriate filtering.

3.2.7 Dismantling and Reconstruction

The dismantling and reconstruction option aims at restoring the original geometry of the bridge, and hence its original operating conditions. Therefore, although it can be seen as the most severe rehabilitation intervention, rather than "intrusive", it is fairly legitimate since it focuses on keeping built heritage, by preserving it for upcoming generations, provided the materials and techniques adopted are compatible with the original concept of the structure and the whole construction details.

This is a complex and time-consuming process used to replace elements partially/fully collapsed or severely affected by cracking or deformation, in bridges

(a) (b)

Fig. 8 Drainage system: **a** Sinkhole and **b** Gargoyles

risking collapse and even possible disappearance. In such context, for structures with historic value, it can be deemed as an appropriate option, economically sustainable when the original materials and construction techniques can be used with advantages in terms of durability and reliability.

This option, however, has some limitations such as the degradation of original materials and the possible lack of skilled labour to reproduce old construction techniques; in addition, the demands inherent to load carrying capacity increase and lane widening, might not be compatible with this rehabilitation option, since it risks to enforce mixing other complementary interventions not respecting the heritage authenticity preservation issues.

One such case, for example, is the intervention carried out in the 1950's in the southern part of the Portuguese Lagoncinha bridge [17], wherein, beyond dismantling and reconstruction using original materials and techniques, new materials and elements were added, namely cementitious concrete to replace the original infill material and also reinforced concrete slab underneath the pavement. However, recently and in line with heritage conservation charters, similar rehabilitation options were adopted in the Esmoriz [19] and Goimil [29] bridges, Portugal.

The Esmoriz bridge exhibited serious structural damage such as longitudinal cracking in the arch intrados, pavement deformation, subsidence and crushing of foundations, widespread damage caused by moisture, vegetation and biological pollution and loss of mortar in the joints (see Fig. 9a). Before dismantling the structure, paving stones and spandrel walls have been numbered with erasable ink and this numbering has been recorded through photographs taken *in situ*. After placing the falsework shoring (see Fig. 9b), the arch blocks were readjusted keeping the original bridge stones that were in good conditions. Cracked and damaged stones were replaced by other similar granite stones cut in the appropriate way (Fig. 9c, d). The reconstruction of spandrel walls was made ensuring appropriate interlocking of stone blocks and the infill material was replaced by well-graded and compacted granular material.

Fig. 9 Esmoriz Bridge: **a** View of the arch intrados before the intervention, **b** Falsework, **c** View of the arch extrados after infill removal and spandrel walls' dismantling, **d** View of the arch intrados after reconstruction

Fig. 10 Goimil Bridge: **a** View of the arch intrados before the intervention, **b** Falsework and general view of the arch extrados after infill removal and spandrel walls' dismantling, **c** View of the bridge after reconstruction

After the main structure reconstruction, the granite parapets have been restored, the deck was waterproofed and the drainage system was reviewed. All the block joints were replaced and finished with commercial ready-mixed lime-based mortar.

The intervention in the Goimil Bridge followed similar procedures as described in previous paragraphs for the Esmoriz Bridge; in none of them, no other new materials and techniques were adopted. Some images of the intervention phases in the Goimil Bridge are shown in Fig. 10.

4 Reinforcement Solutions for Bridges with Frequent Damages

Reinforcement measures refer to the introduction of new elements in the original structure, in order to account for deterioration or changes in the operating conditions (e.g. increase in road traffic). These measures should be adopted considering the bridge structural behaviour, both before and after intervention works, the existing anomalies and their possible causes, having in mind the architectural and historic value of old masonry bridges.

Therefore, wherever possible, methods and solutions should be adopted so as not to affect the aesthetic quality, the physical and functional integrity of the intervened bridges.

The interventions' purposes should be clear, according to the existing problems and the intended lifespan of the structure. The latter must be set according to the loads acting on the structure and based of complementary studies, namely geotechnical and hydraulic investigations, as well as traffic surveys. Considering these conditions, the structural safety assessment of the bridge should be made in order to adopt the most effective measure(s). The next section will describe some of the key strengthening techniques for masonry arch bridges.

4.1 Strengthening Solutions for Longitudinal Behaviour

General Comments

The damage related with poor longitudinal behaviour is mainly motivated by weak load carrying capacity or excessive loads, causing first compression and high deformation due to degradation of the structural system strength and/or stiffness and ultimately leading to transverse joint opening between blocks located next to hinges' positions that characterise arch hinge mechanisms.

Where structural degradation is related to excessive load caused by the volume of traffic, a decrease in traffic flows should be suggested in order to reduce service loads and prevent further degradation.

Where joints between blocks are degraded due to loss of mortar, joint repointing should be made to improve structural behaviour of the bridge because it contributes for more adequate load transfer. This is a minimally intrusive and necessary measure for the great majority of situations which result from poor maintenance.

Common arch strengthening solutions involve increasing the arch thickness by inserting a concrete slab on the arch intrados or extrados, which may be reinforced with steel or composite materials.

In order to add new concrete elements, on-site moulding, prefabrication or concrete injection techniques may be used. This type of reinforcement is often adopted when the arch has low strength capacity due to its reduced thickness or material degradation, but it must be emphasized that it is a very heavy and intrusive reinforcement solution, clearly not aligned with the recommendations above referred [12].

By contrast, the option of acting on the load degradation mechanism by reinforcing the pavement allows less intrusive solutions.

Several reinforcement solutions include transverse and radial anchoring, connecting the arch to the spandrel walls and (possibly) to the infill, as well as the application of steel bars or composite materials on the arch intrados or extrados.

Finally, it should be noted that consolidation of the transverse behaviour also improves the bridge's longitudinal behaviour, since the infill material becomes more confined and, therefore, with improved shear strength. Where passive pressures tend to develop in the infill, they are potentially more effective, thus better opposing possible arch extrados deformation towards the infill; in turn, the infill active pressures onto the arch tend to decrease, leading to the reduction of arch deformation outwards the infill.

Reinforcement at Intrados Versus Extrados

The option of placing the longitudinal reinforcement at the intrados or extrados has implications on the structural behaviour of the arch. In this respect, it is worth recalling that arch mechanisms can develop when more than three hinges are installed: typically four hinges when acted by non-symmetrical load relative the arch crown, e.g. a concentrated force at or near the quarter span, or five hinges when subjected to symmetrical or distributed loading along the arch span.

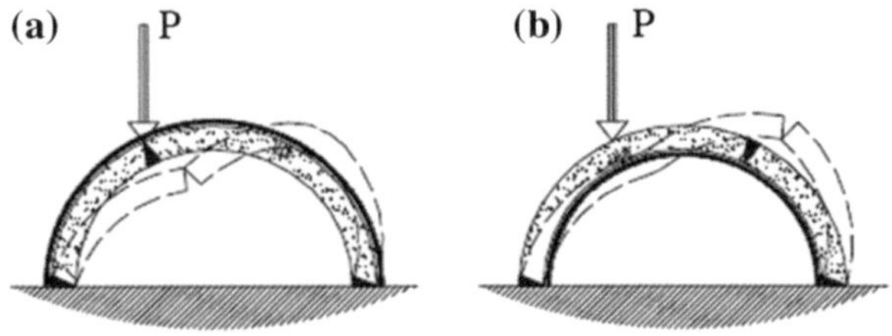

Fig. 11 Longitudinal reinforcement of arches: **a** In the extrados and **b** In the intrados (picture credits [30])

Taking for reference the most commonly critical situation of a non-reinforced arch loaded by a concentrated force at/near the quarter span, hinges are likely to develop by joint opening at the supports, in the arch intrados underneath the force location and in the extrados in the opposite quarter span zone. Therefore, should the force increase far enough, the failure mechanism can be installed and the force limit can be estimated by plastic limit analysis methods.

By contrast, when the arch is reinforced, at either the extrados or the intrados, by some kind of tensile resistant material or system connected to the arch, it is apparent that hinges are not free to develop, because where joints are expected to open, there will be reinforcement to resist tensile stresses as illustrated in Fig. 11.

Assuming that the reinforcement is not effective in arch springings, hinges will still develop by joint opening at the supports. Along the arch span, the location of a third hinge (which transforms the arch into an "isostatic" three-hinge system) will depend on where the reinforcement is installed and where the load is applied.

Considering non-symmetrical concentrated force action, when the arch is reinforced at extrados (Fig. 11a), the third hinge is still free to open in the arch intrados underneath the force location (e.g. at the left quarter span), but the fourth hinge, that would tend to form at extrados in the opposite quarter arch zone (thus, the right one), will require increased load due to reinforcement strength which is mobilized there. This means that kinematic hinge mechanisms are prevented to develop because extrados joints are restricted from opening and, therefore, the load-bearing capacity of the arch is increased.

For the same loading conditions, but with arch intrados reinforcement (Fig. 11b), the third hinge is not free to form in the intrados below the load, due to the reinforcement strength, but it can develop by joint opening in the extrados in the opposite quarter span zone. Again, the additional fourth hinge that would form underneath the load (by joint opening in the intrados), will mobilize increased load to reach the arch load-bearing capacity, but it is clear that it is a more vulnerable situation due to the risk of reinforcement detachment from the arch.

Considering the above mentioned, in particular for the case of non-symmetrical loading relative to the arch crown, it is clear that both extrados and intrados arch reinforcement (not extended to the arch supports) allow free development of enough hinges to transform the arch into an isostatic three-hinge system. Beyond that, both reinforcement types are suitable to mobilize increased strength of the arch.

Taking into account the presence of the infill and spandrel walls on the extrados, which have a favourable effect for the arch strength, it appears that intrados reinforcement is likely to yield more effective solutions provided due care is taken to prevent degradation and detachment in the connection between the arch and the reinforcement, in order not to loose the reinforcing efficiency.

Concerning the extrados reinforcement, workability issues are quite important since it requires removal and replacement of large amounts of material (pavement, infill and in some cases, spandrels) and, consequently, the traffic interruption. The same applies when infrastructure facilities are buried along the bridge, such as supply lines, for which it may be necessary to disrupt their normal operation.

Still in this case, it is necessary to assess whether the infill has a decisive role in the arch stability during the works' development, when the material is removed, and check the need for temporary shoring to ensure the arch safety. Also, when the infill depth over the arch is not sufficient to accommodate the new thickness with the reinforcement, the change of the bridge longitudinal profile may be required.

By contrast, intrados strengthening does not imply constraints of bridge traffic nor interruptions of any kind of lifeline installed. However, the final appearance of the bridge is greatly affected and the free height under of the arch may become well reduced.

4.1.1 Addition of a Reinforced Concrete Layer to the Arch

Adding a layer of concrete to the arch has several disadvantages due to the large amount and type of new material involved. Therefore, this technique should be considered with caution, particularly because it is a very intrusive process, irreversible and does not easily allow future inspections in the retrofitted components (intrados or extrados).

In terms of waterproofing and drainage system, the introduction of a new element in the arch, providing an impermeable layer (such as a concrete cover), leads to changes on water percolation and capillarity, which may require appropriate drainage and efficient waterproofing to prevent the existence of water inside the structure.

Beyond the considerable increase of structural mass, material chemical degradation may occur, either in the existing or new materials, due to the high amounts of salts present in the concrete.

The addition of a reinforced concrete overlay on the extrados can behave as a composite material with the existing arch, or the latter can be considered as permanent formwork and the new overlay is designed to resist all actions transmitted to the arch.

For example, the strengthening system of the Sandro Gallo brickwork bridge in Venice comprised increasing the thickness of the existing masonry arch in its extrados with a new brick layer, reinforced concrete and composite material (CFRP); the foundation of the new layer was materialized by micro-piles. The intervention aimed to increase the load-bearing capacity of the bridge due to a change in the category according to the Italian rules to a higher requirement level [31].

Fig. 12 Downstream view of Remondes bridge after the arch reinforcement with shotcrete

The strengthening with shotcrete in the intrados consists in thickening the arch to increase its load bearing capacity. The new material is high-pressure sprayed and as a result adheres to surfaces and compacts the material already applied forming a new layer that usually is reinforced with rebars [15].

As main advantages, this process is fast and does not require formwork. However, their effectiveness is seriously compromised in the case of deterioration of the bond between the two layers, particularly vulnerable due to concrete retraction problems.

Figure 12 shows the final aspect of Remondes bridge over the River Sabor, Portugal, in which one of the arches was reinforced with shotcrete [28].

4.1.2 Strengthening of the Load Degradation System on the Deck

Adding a slab of reinforced concrete above the infill is indicated when it is necessary to improve the degradation system of loads applied on the deck.

This solution has the advantage of involving a small area of excavation, not affecting the bridge aesthetics. It is relatively simple intervention that requires only a brief interruption in traffic.

For example, the rehabilitation and strengthening of the Segura bridge, Portugal [22], included the application of a reinforced concrete slab as part of the pavement. Figure 13 shows two phases preceding the slab execution, namely the removal of the existing bituminous layer and the regularization of the surface for application of the slab [21].

A similar solution of pavement reinforcement was tested to failure in a single-arch bridge built in laboratory [32]. Results of this test, compared with those obtained from similar experimental campaigns in a bridge reinforced with a concrete overlay in the arch extrados and another bridge non-intervened, enabled identifying similar values of maximum load and deformation capacity in both reinforced bridges, both cases showing maximum strength about 3.7 times higher than the unreinforced solution.

Fig. 13 Segura bridge: **a** Bituminous removal and **b** Surface regularization for the application of a concrete slab on the pavement [21]

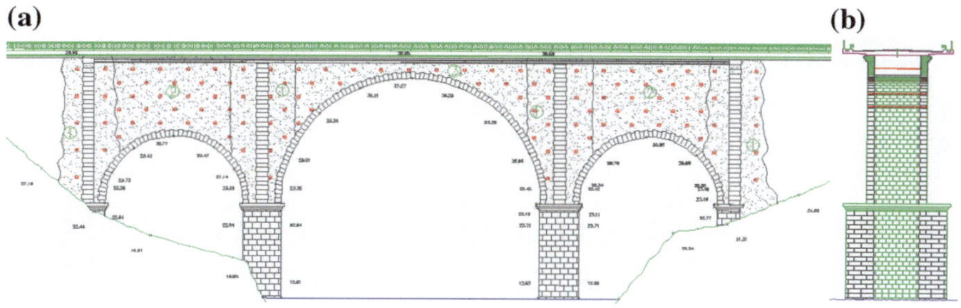

Fig. 14 Caninhas bridge: **a** Elevation, **b** Transverse cross-section [26]

The adaptation of the bridge to a wider cross-section generally involves the deck widening. The most common intervention involves the execution of a concrete slab in the pavement and the reinforcement of other existing parts of the bridge. Many stone masonry arch bridges have been adapted to new traffic demands, such as the Marillais bridge, France [20], the Sandro Gallo Bridge, Italy [31] and the Portuguese cases of Caninhas [26] (Fig. 14), Sancheira [27] or Real [23] bridges, among others mentioned in the literature [33–35].

Although making use of concrete as new material in existing masonry bridges, the above mentioned solutions do not strongly conflict with the original bridge structural system. In fact, the load bearing elements have their function preserved, particularly when the concrete slab is installed without deck widening. One important aspect to take into account is that, again, adequate drainage system must be ensured, together with proper impermeable isolation of the masonry components, in order to prevent water percolation coming from concrete zones down to the masonry infrastructures and to avoid aggressive salts' penetration therein.

Provided these non-structural issues are accounted for, the above described solutions are efficient to strength the bridge, not only in the longitudinal direction but also in the transverse one, for which additional measures might be required as described in a latter section.

4.1.3 Reinforcement with Rebars, Bars and Laminates on Arch Surface

Using steel rebars or composite material in longitudinal grooves on the arch intrados or extrados, as well as steel bars and laminates or sheets of composite materials fixed in the arch surface allow avoiding the addition of concrete.

In this case, the increment of the load-bearing capacity of the structure is achieved without changing geometric and mass of the structure, thus improving the strength, stiffness and deformation capacity as well as the load distribution by introducing current and durable materials.

Such a strengthening system, proposed by Bersche-Rolt Ltd [36] is constituted by steel rebars placed in the intrados complemented with radial and transverse anchors, the latter suitable for strengthening behaviour in the transverse direction. Thus, it is a global strengthening solution of the arch also which improves operating conditions in both the longitudinal and transverse direction. Figure 15 shows a schematic elevation view of the reinforcement and one phase of placing the longitudinal bars in the arch intrados grooves.

Garrity [37] also presented a similar strengthening solution of the arch using stainless steel rebars glued near the surface in grooves and holes.

In the same line, Foraboschi [38], Melbourne and Tomor [39] tested the performance of composite bars applied on the intrados of an arch built in laboratory. Failure mode by detachment of the strengthening was identified on the arch after being subjected to load tests.

The use of composite materials on the external faces of the arch intrados was also studied in the laboratory by Baratta and Corbi [40] in stone masonry arches with dry joints by applying horizontal displacements in one of the supports to simulate horizontal settlements and consequent arch decompression. From the evolution of mobilized strength with the vertical displacement at the arch models' crown (with and without reinforcement), the load capacity was found largely increased, but the behaviour became brittle and the collapse occurred due to strengthening detachment in the strengthened model.

(a) (b)

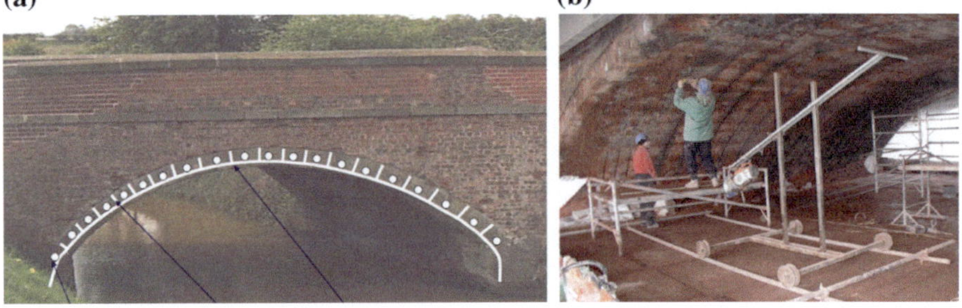

Fig. 15 Strengthening solution for bridges: **a** Elevation scheme and **b** View of the longitudinal strengthening at the arch intrados [36]

Jurina and Mazzoleni [30] proposed an arch strengthening solution in the extrados which prevents the development of the hinge mechanisms by placing rebars in the arch longitudinal direction, which are anchored or grounded in the arch abutments; this was complemented with transverse strengthening by flexible carbon fibre composite materials in the intrados or transverse steel ties.

4.1.4 Radial and Secant Anchors

The application of anchors is a strengthening technique widely used in old masonry structures consisting of a steel bar, enclosed in a hole previously made in the masonry and sealed by means of grout (adherent anchor).

The arch can be strengthened in the longitudinal direction by means of radial anchors from the intrados connecting the arch to the spandrel walls. This was used in the Portuguese cases of Donim bridge, along with transverse ties as shown in Fig. 16a [41], and in the Pedrinha bridge [25].

For multi-ring arches this technique is also appropriate to connect the various rings. The strength system, proposed by Bersche-Rolt Ltd [36] and already illustrated in Fig. 15, includes the use of radial anchors for this purpose. Figure 16b shows the phase of radial drilling through the arch intrados for installing the adherent anchors [36] consisting of stainless steel bars sealed with cement grout.

Sumon [42] describes the results of five load tests made in the TRL–Transport Research Laboratory, for assessing the effectiveness of three different reinforcement techniques in multi-ring arches with identical characteristics. One arch specimen was reinforced with shotcrete sprayed in the intrados and the other with a concrete belt in the upper surface; two other arches were built with ring separation, one of them tested to failure without strengthening and the other reinforced with a stainless steel mesh applied from the intrados so as to connect the different layers; finally, another test was conducted in a regular unreinforced arch (standard arch).

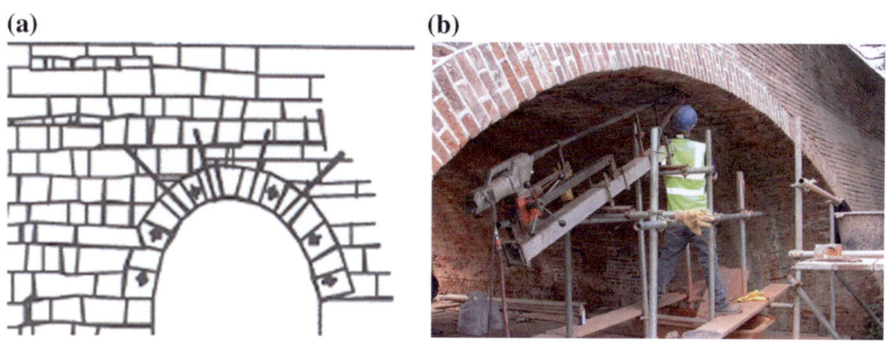

Fig. 16 Radial and transverse anchors: **a** Donim bridge strengthening scheme [41], **b** Drilling phase for anchor installation [36]

The results for the arch built with ring-separation showed lower load bearing capacity than the standard arch and the use of the strengthening technique with stainless steel anchors allowed 14% increase in the load bearing capacity compared to the previous test. By applying a shotcrete layer in the intrados, the load capacity increased 3.9 times relative to the standard (unreinforced) arch, but brittle failure and reduced deformation capacity was observed. The introduction of an extrados concrete layer increased the load capacity by a factor of 2.9 without losing deformation capacity; the formation of an hinge in the arch was observed for 78% of the maximum applied load, after which the reinforcing concrete layer was exclusively responsible for supporting further load increase.

Another type of arch strengthening in the longitudinal direction can be made resorting to secant anchors by adopting the Archtec system [43] developed by a partnership formed by the companies Cintec International, Rockfield Software and Gifford. This system, widely used for reinforcing bridges' in the United Kingdom [44], comprises the arch strengthening with stainless steel anchors placed inside holes drilled down from the pavement and diagonally arranged in the longitudinal direction with a secant profile across the arch, as schematically illustrated in Fig. 17a; the corresponding drilling phase from the upper surface is shown in Fig. 17b.

The basic idea of this reinforcement system consists of placing the reinforcement in critical positions in order to restrict the development of arch hinge mechanisms. Results of load testing in laboratory built bridges evidenced that this type of reinforcement increases the bridge load bearing capacity by a factor of approximately 3.0 compared with unreinforced cases, but with reduction of deformation capacity in the reinforced arch.

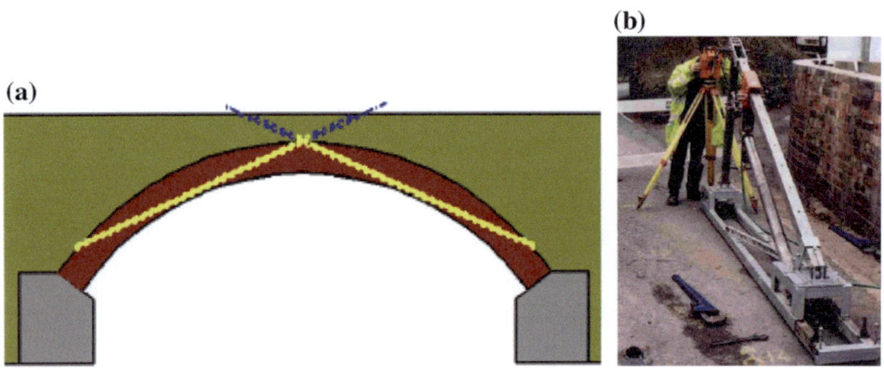

Fig. 17 Archtec system: **a** Schematisation of the strengthening and **b** Drilling equipment from the pavement [44]

4.2 *Strengthening Solutions for the Transverse Direction*

As above mentioned, damages and faults caused by deficient behaviour in the transverse direction result from the interaction between spandrel walls and the arch, leading to longitudinal tensile cracks in the arch under the spandrel inner face, slippage between the spandrel base and arch extrados, longitudinal cracks concentrated along the arch longitudinal axis, deformation and rotation of spandrel walls with inherent decompression of infill material and pavement subsidence.

Such poor performance in the transverse direction, responsible by loss of transversal stiffness, also affects the longitudinal direction behaviour and contributes for reducing the restrictions to the development of the longitudinal hinges' mechanisms of the arch.

There are several techniques to repair these damages and to consolidate the transverse direction performance. Amongst the most common, there can be referred: the use of steel ties placed across the arches' voussoirs and spandrels (distributed or concentrated on the arches' zones); the addition of a concrete layer over the bridge inner facings (i.e. spandrel walls and arches' extrados); and strengthening the infill material resorting to mortar (e.g. cement grout) injections or replacement of the existing infill material.

4.2.1 Transverse Ties

Transverse ties anchored in stone masonry can be adopted either to restore the original arch shape (configuring active reinforcement) or to prevent and/or restrain deformation in the transverse direction possibly occurring in future cases of increased out-of-plane deflections, by resisting higher horizontal impulses transmitted by the spandrel walls and, consequently, the resultant internal forces (thus, conveying a situation of passive ties).

Steel ties can be made of ordinary reinforcing rebars threaded just at the ends or in its entire length, placed in holes previously made in the masonry structure, filled or not with low shrinkage mortar, materializing bonded or unbounded ties, respectively.

This solution has the advantage of having little effect on traffic and does not involve excavation of infill material, resulting in a low cost solution. Concerning solution drawbacks, difficult drilling operation can be an issue and corrosion of ties and/or anchor plates may rise and become visible, thus requiring appropriate measures.

This solution was used both in the Donim bridge, Portugal (Fig. 18a), to control the longitudinal joint opening detected in the arch intrados [41], and in the Segura bridge, Portugal [22], wherein longitudinal cracks in the arches were also found.

Besides other longitudinal strengthening elements already mentioned before, the strengthening system proposed by Bersche-Rolt Ltd [36] also comprises bonded transverse steel ties which are then surrounded by low-shrinkage cement grout

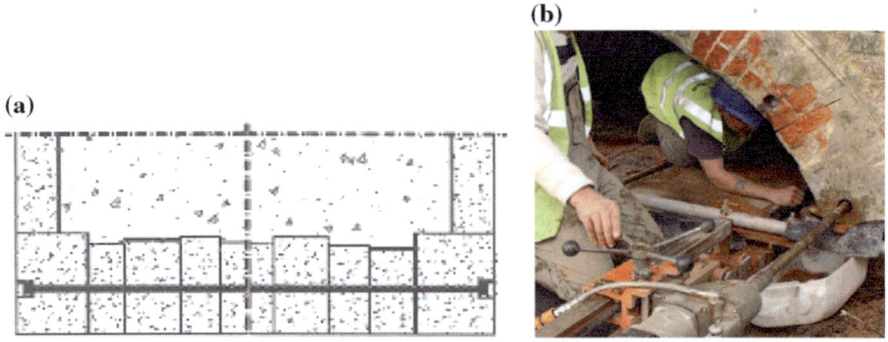

Fig. 18 Transverse ties in the arch: a Scheme of Donim bridge strengthening [41], b Drilling operation for installing the Bersche Rolt Ltd [36] transverse strengthening system

Fig. 19 Transverse ties in the bridge over the River Sul [46]

injected after completion of the tie installation in the drilled hole. Figure 18b illustrates the phase of arch perforation to install the transverse tie therein.

Since the traffic induced vibration is normally more intense in the arch than in the spandrel, this transverse strengthening solution may lead to concentration of internal forces in the spandrel-arch connection which are likely to enforce propagation of initial cracks to other locations. Therefore, associated with the transverse ties, this system usually includes also radial anchors connecting the arches to the spandrels and/or transverse anchors distributed throughout the area of the spandrel walls.

Strengthening with transverse ties distributed in the spandrel walls' area is indicated when out-of-plane walls' deformations are likely to occur. Additionally, this reinforcement provides important confinement of the infill material, which improves its stiffness and further prevents arch hinge mechanisms from developing, therefore increasing also the arch strength.

This reinforcement solution was used in several Portuguese bridges, namely: Segura [22], Real [23], Formigosa [24], Pedrinha [25], Caninhas (see Fig. 14) [26], Sancheira [27], Areosa [45] and Remondes [28].

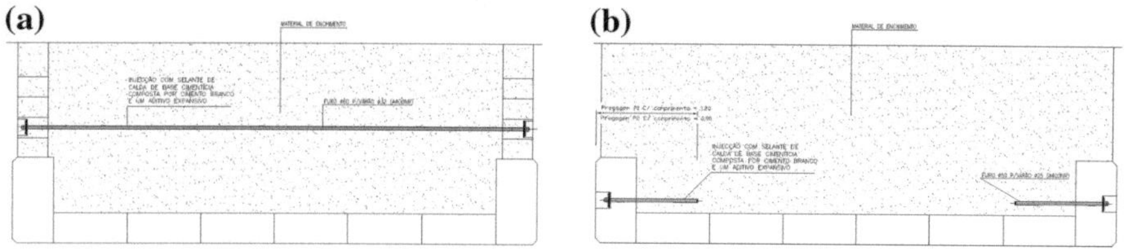

Fig. 20 Schematic representation of the Canharda bridge structural reinforcement. Transverse, **a** Ties connecting spandrels and **b** Anchors [47]

Transverse ties applied in spandrel walls, near the arch extrados, were also used in the roadway bridge over the River Sul (Portugal), as illustrated in Fig. 19, in order to solve the problem of longitudinal joint opening in the arch aligned with the internal faces of spandrel walls [46].

A similar reinforcement system comprising transverse ties was also used in Canharda railway bridge (Portugal) in the spandrels close to the arch extrados (see Fig. 20a) together with anchors to connect the arch and the infill (see Fig. 20b) [47].

Melbourne et al. [48] conducted a test campaign on three bridges built in laboratory constituted by multi-ring arches: one bridge without defects and two others built without some voussoirs of the longitudinal arch ring under the spandrels; one of the two latter bridges was reinforced with transverse ties in the spandrels. The models exhibited collapse by arch hinge mechanisms with separation between the spandrel walls and the arch, as well as rings' separation. The study concluded that the transverse direction performance became improved because ties contributed for preventing the transverse displacement of spandrel walls. However, this strengthening system was found to have little influence in the development of the load bearing mechanism of this type of bridges.

4.2.2 Vertical Anchors in the Spandrel Walls

The reinforcement of the spandrel walls through the use of vertical anchor as schematically shown in Fig. 21, is intended to improve the behaviour against the out-of-plane collapse, by providing a tensile strength system in the inner face of the wall. This type of solution is also used to strengthen the bridge side guards (or parapets), often damaged due to vehicle impacts.

Hobbs et al. [49] also presented some proposals for strengthening the parapets resorting to other techniques based on composite materials and reinforced concrete elements. Similarly, Kiang Hwee and Patoary [50], Triantafillou [51] and Muszynski and Purcel [52] proposed techniques for strengthening spandrel walls using composite materials.

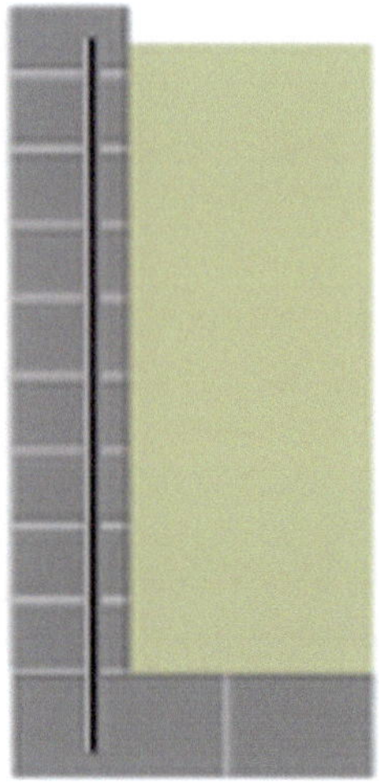

Fig. 21 Strengthening solutions proposed for the parapets, adapted from Bersche-Rolt Ltd [36]

4.3 *Foundation and Piers*

Current solutions found for foundation strengthening often involve first the soil improvement through injections, jet-grouting, adding piles and micro-piles.

Besides soil treatment, as the ultimate and overall bridge support, foundation structural elements (typically consisting of enlarged piers reaching the ground firm) may also require strengthening. This can be done by injection filling of internal voids, cavities or damaged zones and by increasing the foundation section with new enlargements typically made of reinforced concrete (surrounding the original/existing foundation) properly anchored to the existing materials. Of course, the chemical compatibility between new and original materials (stone) is not easy to ensure in this situation, but it must be recognized that its importance is clearly diminished because such intervention zones are normally submerged by the water flow.

The reinforcement of foundations using the micro-piles and grout injections can also be used to improve the infill area over the piers in order to direct the internal forces straight to the foundations. This case is usually associated with the construction of a concrete slab in the pavement, as adopted in the Sancheira [27] (Fig. 22) and Tavira [53] (Fig. 23) bridges, as well as the Marillais bridge case [20].

4.4 *Global Strengthening Solutions*

Global strengthening solutions include those acting on the improvement of the three-dimensional behaviour, involving arches, spandrels and infill strengthening, which can be done by adding new elements in the internal surfaces of the bridge.

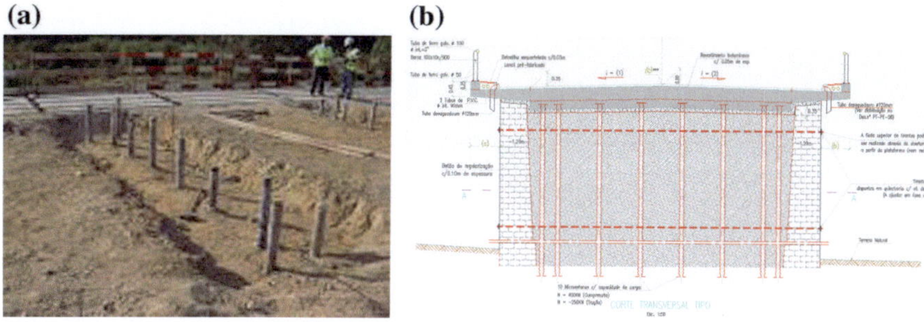

Fig. 22 Sancheira Bridge [27]: foundation and pier strengthening with micropiles. **a** Construction detail and **b** Cross section representation

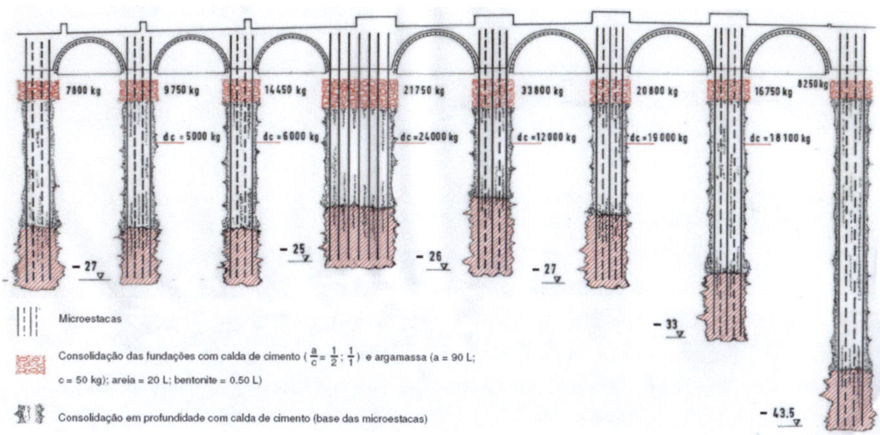

Fig. 23 Tavira Bridge [53]: schematic representation of soil, foundation elements and piers strengthening with micro-piles

One possibility to achieve this goal consists on adding a new concrete envelope inside the bridge over the arches and internal faces of the spandrel walls as adopted in the Tavira Bridge case (Fig. 24a) [53]. However, it is definitely a very intrusive technique because the masonry system is significantly modified and it has the non-negligible problem of adding a large amount of materials chemically incompatible with existing stone masonry.

A similar effect on the overall behaviour can be obtained using standard steel elements attached to the extrados and spandrel walls as shown in Fig. 24b), which corresponds to the reinforcement solution used in one of the arches of the Donim Bridge [41], in this case without the disadvantage of adding new cementitious or other chemically incompatible materials.

Given the structural importance of the infill for the overall behaviour of masonry bridges, any kind of infill strengthening will act as a global strengthening solution. Therefore, infill injection with appropriate fluid mortar or infill replacement by a

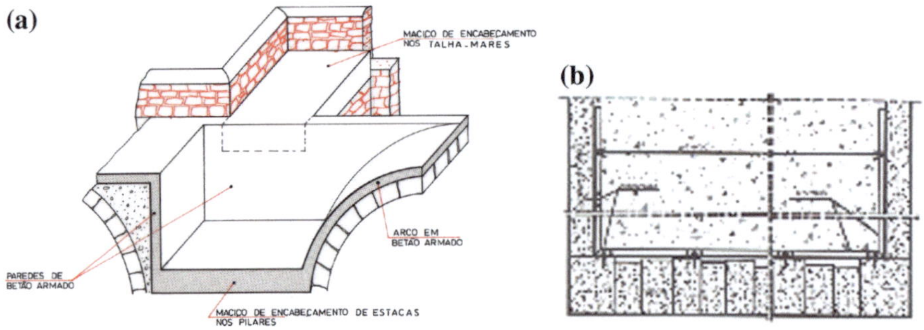

Fig. 24 Global strengthening in the internal paraments: **a** With the addition of a concrete layer [53] and **b** A frame of metallic profiles [41]

more suitable material (e.g. well graded granular soil type material) should be always an option to be considered, possibly in conjunction with other ones already referred for the main structural elements.

5 Final Remarks

Given the current scenario of interventions in this type of structures, the most widely used solutions for strengthening can be divided according to the type of materials and techniques used.

Essentially the following solutions can be highlighted: (i) addition of metallic elements (rebars, steel shapes, ties and anchors); (ii) addition of composite materials reinforced with fibres (rebars, laminates and flexible materials); (iii) addition of reinforced concrete elements, cast *in situ* or sprayed; (iv) mortar injections (mostly cement grout) and (v) replacement (or substitution) of the existing material.

For bridges of particular heritage interest, reconstruction or renovation techniques likely to change the structure characteristics and authenticity should not be considered. Although arguable, this is an increasingly accepted intervention principle which applies to bridges due to their importance, not only as part of infrastructures' network, but also (and mainly) as heritage constructions which must be preserved for upcoming generations. Therefore, in these cases, repair options should mainly include measures to prevent further deterioration where maintenance actions play a decisive role.

Acknowledgements Part of this work reports to research financially supported by Project POCI-01-0145-FEDER-007457—CONSTRUCT—Institute of R&D in Structures and Construction, funded by FEDER funds through COMPETE2020—Programa Operacional Competitividade e Internacionalização (POCI) and by national funds through Fundação para a Ciência e a Tecnologia (FCT).

References

1. Arch Bridges. In: Melbourne C, editor. Proceedings of the 1st international conference on arch bridges. 3–6 September 1995, Bolton, Thomas Telford, ISBN: 0727720481.
2. Arch Bridges: History, analysis, assessment, maintenance and repair. In: Sinopoli A, editor. Proceedings of the 2nd international arch bridge conference. 6–9 October 1998, Venice: A.A. Balkema, ISBN: 9058090126.
3. Arch'01. In: Abdunur C, editor, Proceedings of the 3th international conference on arch bridges. 19–21 September 2001, Paris, Presses de l'école nationale des ponts et chaussées, ISBN: 2859783471.
4. Arch Bridges IV: advances in assessment, structural design and construction. In: Roca P, Molins C, editors. Proceedings of the 5th international conference on arch bridges. 12–14 September 2004, Barcelona: Center for Numerical Methods in Engineering (CIMNE), ISBN: 8495999633.
5. Arch'07. In: Lourenço PB, Oliveira D, Portela V, editors. Proceedings of the 5th international conference on arch bridges. 12–14 September 2007, Funchal, ISBN: 9728692315.
6. Arch'10. In: Chen, editors. Proceedings of the 6th international conference on arch bridges, 11–13 October 2010, Fuzhou, ISBN: 8371252655.
7. Arch'13. In: Radic J, Kuster M, Savor Z, editors. Proceedings of the 7th international conference on arch bridges, 2–4 October 2013, Trogir–Split, SECON-CSSE, ISBN: 9537621162.
8. Arch 2016: Arch bridges in culture. In: Biliszczuk J, Bien J, Hawryszkow P, Kaminski T, editors. Proceedings of the 8th international conference on arch bridges, 5–7 October 2016, Wroclow, EDITUS, ISBN: 8371252655.
9. Sustainable Bridges: sustainable bridges–assessment for future traffic demands and longer lives. Integrated project in the sixth framework programme on research, technological development and demonstration of the European Union, FP6-PLT-001653. 2007. http://www.sustainablebridges.net.
10. UIC: International union of railways: improving assessment, optimization of maintenance and development of database for masonry arch bridges. 2005. http://orisoft.pmmf.hu/masonry.
11. McKibbins L, Melbourne C, Sawar N, Gaillard C. Masonry arch bridges: condition appraisal and remedial treatment. London: CIRIA; 2006.
12. ICOMOS/ISCARSAH: Recommendations for the analysis, conservation and structural restoration of architectural heritage. 2005. http://www.icomos.org.
13. UIC: Assessment, reliability and maintenance of masonry arch bridges. In: Orban Z, editors. UIC masonry arch bridges study group, state-of-the-art research report. Paris: International Union of Railways, 2004.
14. Ashurst D. An asssessment of repair and strengthening tecniques for brick and stone masonry arch bridges. Report 284. TRL-Transport Research Laboratory, 1992.
15. Page J. A guide to repair and strengthening of masonry arch highway bridges. Report 204. TRL–Transport Research Laboratory, 1996.
16. Gilbert M. The behaviour of masonry arch bridges containing defects, PhD-Thesis, University of Manchester, 1993.
17. DGEMN: Ponte da Lagoncinha. Boletim n.º 87. Boletim da Direcção Geral dos Edifícios e Monumentos Nacionais, MOP, 1957.
18. DGEMN: Ponte sobre o rio Lima. Sistema de Informação para o Património Arquitectónico–IPA.00000424. 1995. http://www.monumentos.pt.
19. Fauchoux G, Abdunur C. Strengthening masonry arch bridges though balckfill replacement by concrete. In: Sinopoli A, editor. Second international conference on arch bridges. Venice: Balkema; 1998.
20. Costa A, Arêde A, Paupério E, Costa C. Ponte da Esmoriz. Relatório de inspeção estrutural, IC142-RVT218-A-NG306 (in Portuguese). IC-Instituto da Construção, FEUP, 2012.

21. Esteves P. Estudo de soluções de reforço em pontes em arco de alvenaria de pedra. Relatório de Seminário do 5º Ano da Licenciatura em Engenharia Civil. 2007. Unpublished internal document (in Portuguese), Instituto Politécnico de Tomar.
22. IEP: Ponte internacional de Segura sobre o rio Erges. Memória descritiva e justificativa do projecto de reabilitação e reforço estrutural. Unpublished internal document (in Portuguese), Instituto de Estradas de Portugal, 2004a.
23. IEP: Projecto de reabilitação e reforço estrutural da ponte do Real. Projecto de execução realizado pela Tecnofisil. Unpublished internal document (in Portuguese), Instituto de Estradas de Portugal, 2004b.
24. EP: Reabilitação e reforço estrutural da ponte da Formigosa. Projecto de execução. Unpublished internal document (in Portuguese), Estradas de Portugal, 2005.
25. IEP Reabilitação/reparação geral da Ponte Pedrinha. Projecto de execução realizado pela BETTAR, SA. Unpublished internal document (in Portuguese), Instituto de Estradas de Portugal, 2001.
26. IEP: Alargamento e reforço da Ponte de Caninhas. Projecto de execução realizado pela Consulplano, S.A.. Unpublished internal document (in Portuguese), Instituto de Estradas de Portugal, 2002.
27. ICERR: Reabilitação da Ponte da Sancheira. Projecto de execução realizado pela BETTAR, SA. Unpublished internal document (in Portuguese), Instituto para a Conservação e Exploração da Rede Rodoviária, 2001.
28. STAP: Relatório final da obra: Ponte de Remondes na EN 216. EP - Estradas de Portugal. Unpublished internal document (in Portuguese), STAP, 2005.
29. Costa A, Guedes J, Costa P. Relatório de Inspecção da Ponte D. Goimil. IC9-RVT28-A-NG36 (in Portuguese). NCREP, IC-Instituto da Construção, FEUP, 2003a.
30. Jurina L, Mazzoleni M. Ponti in muratura: verifiche statiche ed interventi di consolidamento con la tecnica dell'arco armato. Corso di aggiornamento e specializzazione: Ponti e viadotti: ispezioni visive e tecniche di risanamento, Bolzano, 2004.
31. Modena C, Valluzzi MR, da Porto F, Casarin F, Bettio C. Structural upgrading of a brick masonry arch bridge at the lido (Venice). In: Roca P, Molins C, editors. Fourth international conference on arch bridges, Barcelona, 2004.
32. Miri M, Hughes T. Increased load capacity of arch bridges using slab reinforced concrete. Arch Bridges IV, Barcelona, P. R. a. C. M., editors. 2004.
33. Healey T, Counsell J. Widening and strengthening of London's Kingston Bridge. In: Sinopoli A, editor. Second international conference on arch bridges. Venice: Balkema, 1998.
34. Healey T, Counsell J. Widening and strengthening of Kingston Bridge, London. In: Abdunur C, editors. Third international conference on arch bridges, Paris, 2001.
35. Martínez J, Castillo L. The adaptation of masonry arch bridges. The cerezo bridge experience. In: Roca P, Molins C, editors. Fourth international conference on arch bridges, Barcelona, 2004.
36. Bersche-Rolt Ltd: Masonry reinforcement for arch bridge. Abridged presentation on the bersche-rolt retrofitted system, electronically published at bridge owners forum web site 2004. 2005. http://www.bridgeforum.com/bof/subgroups/masarch/.
37. Garrity S. The Strengthening of single span masonry arch bridge using near-surface reinforcement. In: Abdunur C, editor. Third international conference on arch bridges, Paris, 2001.
38. Foraboschi P. Strengthening of masonry arches with fiber-reinforced polymer strips. J Compos Constr. 2004;8(3):191–202.
39. Melbourne C, Tomor A. Fatigue performance of composite and radial-pin reiforcement on multi-ring masonry arches. In: Roca P, Molins C, editors. Fourth international conference on arch bridges, Barcelona, 2004.
40. Baratta A, Corbi I. Consolidamento di Strutture Murarie con Fibre di Carbono. XXXII Convegno Nazionale dell'Associazione Italiana per L'Analisi delle Sollecitazioni, Salerno, 2003.

41. Oliveira DV, Lourenço PB. Repair of stone masonry arch bridge. In: Roca P, Molins C, editors. Fourth international conference on arch bridges, Barcelona, 2004.
42. Sumon S. Repair and strengthening of five full scale masonry arch bridges. In: Sinopoli A editor. Second international conference on arch bridges. Venice: Balkema, 1998.
43. Brookes CL, Mullet P. Service load testing, numerical simulation and strengthening masonry arch bridges. Arch Bridges IV, Barcelona, 2004.
44. Gifford and Partners Ltd.: Archtec—verification of structural analysis. Gifford and partners report by brookes CL, Electronically published at Bridge Owners Forum web site 2004. 2004. http://www.bridgeforum.com/bof/subgroups/masarch.
45. Martins F. Rehabilitation of the Minho line's Areosa and Afife masonry railway bridges. In: Lourenço PB, Roca P, editors. Historical constructions, Guimarães, 2001. p. 159–1164.
46. Guedes J, Paupério E, Costa P. Relatório de Inspecção à Ponte sobre o Rio Sul. OU4-RVT13-A-NG19 (in Portuguese). NCREP. IC-Instituto da Construção, FEUP, 2002.
47. Martins F. Beneficiação da Ponte Ferroviária de Canharda. JPEE - Jornadas Portuguesas de Engenharia de Estruturas (in Portuguese), Lisboa, 2006.
48. Melbourne C, Begimgil M, Gilbert M. The load test to collapse of a 5 m span brickwork arch bridge with tied spandrel walls. In: 1st international conference on arch bridges, Bolton, UK, 1995.
49. Hobbs B, Gilbert M, Molyneaux T. Effects of vehicle impact loading on masonry arch parapets. In: Sinopoli A, editor. Second international conference on arch bridges, Balkema, 1998. p. 281–7.
50. Kiang Hwee T, Patoary MKH. Strengthening of masonry walls against out-of-plane loads using fiber-reinforced polymer reinforcement. J Compos Constr. 2004;8(1):79–87.
51. Triantafillou TC. Strengthening of masonry structures using epoxy-bonded FRP laminates. J Compos Constr. 1998;2(2):96.
52. Muszynski LC, Purcell MR. Use of composite reinforcement to strengthen concrete and air-entrained concrete masonry walls against air blast. J Compos Constr. 2003;7(2):98–108.
53. Appleton J, Silva J. Reabilitação e reforço da ponte de Tavira. Revista Pedra Cal, 2002; 14:12–14 (in Portuguese), GECoRPA - Grémio das Empresas de Conservação e Restauro do Património Arquitectónico.

Strengthening and Retrofitting of Steel Bridges

José M. Jara, Manuel Jara, Bertha A. Olmos and Jamie E. Padgett

1 Steel Bridge Typologies

Steel is a common material used to build bridges. Even in concrete bridges, steel is used in joints, bearings, parapets and deck's systems, being one of the principal components in superstructures and substructures. Railroad and highway steel bridges have been built for centuries as a competitive material in terms of the structural cost for medium- and long-span length structures. However, its strength/weight ratio and excellent ductility properties make steel a suitable material for a wide range of bridge span lengths. The following sub-sections describe the typologies of steel bridges.

1.1 Steel Superstructures

Bridge superstructures are frequently built with steel elements. The superstructure type depends on the bridge use, span length, and aesthetic requirements,

J.M. Jara (✉) · M. Jara · B.A. Olmos
Faculty of Engineering, University of Michoacan, Morelia
Michoacán, Mexico
e-mail: jmjara@umich.mx

M. Jara
e-mail: mjara@umich.mx

B.A. Olmos
e-mail: bolmos@umich.mx

J.E. Padgett
Department of Civil and Environmental Engineering, Rice University,
Houston, TX, USA
e-mail: jamie.padgett@rice.edu

A. Costa et al. (eds.), *Strengthening and Retrofitting of Existing Structures*,
Building Pathology and Rehabilitation 9, https://doi.org/10.1007/978-981-10-5858-5_11

Fig. 1 Deck-type truss bridge

Fig. 2 Through-type truss bridges

among other variables. In all cases, reinforced concrete is the most common material used to construct the deck. Historic bridges use steel trusses or steel girders to support the deck. Figure 1 shows a typical deck-type truss bridge, where the deck is over the top chord of the truss.

In through-type truss bridges, the deck slab is located at the bottom chord plane. Figure 2 shows this type of bridge. Steel truss bridges have been used as a suitable solution for bridge span lengths that range from 30 m to 400 m. The selection of the truss type depends more on the space limitations beneath the superstructure than on structural issues. However, the top chord in truss bridges is subjected to compression, which is an additional difference between these two types of structures. In deck-type bridges, the compression chord can be inside of the deck slab, which avoids the buckling problem in these elements.

Arch steel bridges are also commonly used; Fig. 3 shows an arch bridge that conforms with I-type girders composed of steel-bolted plates. Although they have been used in small span lengths, their use is more attractive for span lengths that range from 150 m to 450 m.

Fig. 3 Arch-type bridge with pin bearings

Fig. 4 Arch-type bridges

The arch can be built by several pieces or by several open elements to increase the depth at the arch ends (Fig. 4). In any case, transverse elements (bracing) connect the arch girders to give stiffness and an integrated action in that direction.

The arch girder can also be built with box or tubular cross-sections (Fig. 5). If the span length increases, the arch rise must also increase; the hanger sections can be cables or laminated steel sections.

When the span length is longer, arch bridges can be created with truss elements. Figure 6 shows an arch bridge in which the arch is a truss structure with variable depth. This versatile structure uses the advantages of both systems and creates an aesthetic bridge structure.

Girder bridges are composed by slab decks supported on plate girders. They are common structures for medium span length bridges constructed in urban areas. These bridges can be simply supported or continuous structures (Fig. 7). In the first case, the girders are simply supported on steel or elastomeric bearings in each span. The bearing type and its geometric characteristics depend on the bridge age, span length and the bridge loads. Although similar bearings are used in continuous bridges, steel girders are continuous elements on bridge bents.

Fig. 5 Arch bridges with tubular- and box-type cross-sections

Fig. 6 Arch bridge with a truss structure

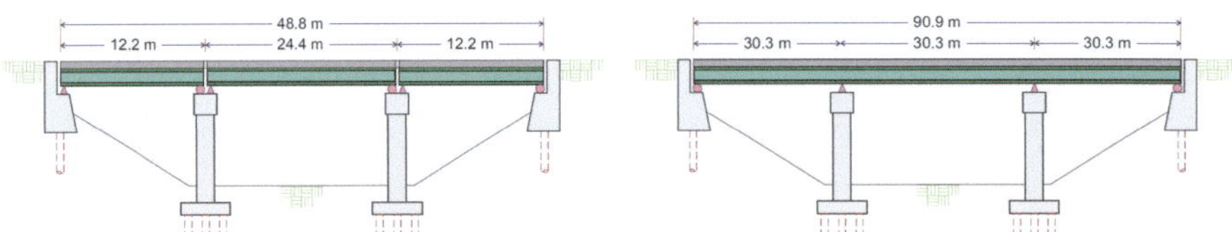

Fig. 7 Simply supported and continuous bridges

On the left, Fig. 8 displays a bridge built with bolted steel plates and an orthotropic deck system and on the right, an I-type girder superstructure with a reinforced concrete (RC) deck slab. In the left image, steel columns support the girders, which form a steel frame structure, and in the other image, the girders rest on rocker steel bearings on RC columns.

Figure 9 shows two typical steel girder bridges. The girders typically have webs with a high depth/thickness ratio that requires web stiffeners along the girder length to avoid local buckling of the plates.

Fig. 8 Steel girder bridges

Fig. 9 Steel girder bridges with web stiffeners

The superstructure of cable-stayed bridges mostly consist of steel girder elements. These bridges can reach a span longer than 1000 m, and the cables are generally fixed to either side of the tower in orthotropic steel decks. Figure 10 displays two cable-stayed bridges. In the left image, a single-span bridge with only

Fig. 10 Cable-stayed bridges

Fig. 11 Suspension bridge

one tower is shown, and in the right image, a three-span bridge with two A-shape towers is shown.

Suspension bridges are the bridge structures with the longest span lengths. These structures can also be found several centuries ago, and similar to cable-stayed bridges, suspension bridges also typically have decks composed by steel girders and truss structures. These bridges use steel cables; however, they are fixed to the ground at each end of the structure. Figure 11 shows an example of an old suspension bridge with a deck composed of steel plate girders.

Modern suspension bridges have lighter elements and can reach span lengths close to 2000 m. Figure 12 shows two suspension bridges. In both cases, the deck is constructed with steel elements.

Curved bridges are often constructed with steel superstructures. The malleability of the steel makes it a suitable material to build girders with this shape. Figure 13 shows two curved bridges, where a box-type girder is shown on the left, and an I-shaped girder is shown on the right.

Fig. 12 Suspension bridges

Fig. 13 Curved steel bridges

Fig. 14 Pin bearings in old steel bridges

Fig. 15 Elastomeric bearings in steel bridges

Bearings in steel bridges depend on several variables, such as span length, bridge age, and bridge location, among others. Figures 14, 15 and 16 present several of the most common bearings found in old and contemporary steel bridges.

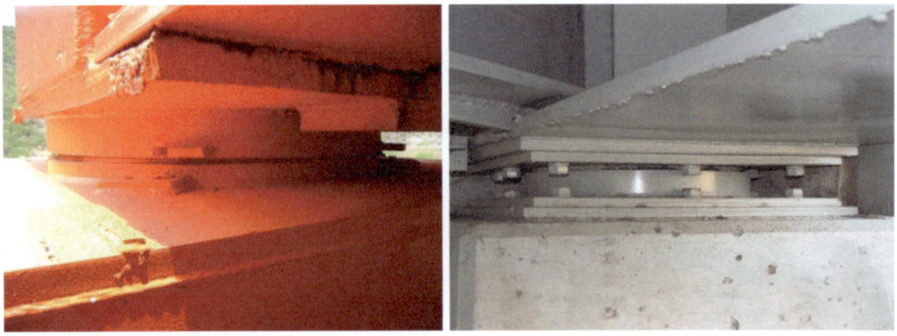

Fig. 16 Pot bearings in steel bridges

2 Superstructure Rehabilitation Techniques

The superstructure of a bridge is basically composed of girders and diaphragms. It supports the deck, directly transfers traffic loads, and connects the substructure elements among them. As a consequence of the age of bridges, steel girders and steel diaphragms must be rehabilitated to sustain the two primary design objectives: serviceability and strength. In general, there are three possibilities for repairing damaged structures: total replacement, partial replacement, and strengthen a structural element in place; in general, bridge rehabilitation is preferable over bridge replacement. A portion of the damage of the superstructure is induced by the presence of salt water running from the deck through the full depth of transverse floor cracks and longitudinal construction joints, which deteriorate the elements. It is important to seal deck cracks to prevent deck deterioration because this is a permanent treatment for steel girders and diaphragms. Deck deterioration is mostly due to environmental effects, application of chemicals and/or abrasives, and the impact of vehicular traffic. Bridge manuals recommend rehabilitation techniques, such as the use of deck overlays, water-proofing membranes to seal all deck cracks, and the method of payment for deck reparation [1, 2].

Life-cycle behaviour of steel bridges involves damage due to several factors, including overloads, exposure to aggressive environments, vehicle impacts, fatigue, hurricanes, and earthquakes, among others. Corrosion is often observed during physical inspections of bridges. Figure 17 shows the corrosion in a bearing (left) and in webs of a pedestrian truss bridge (right).

Extreme event loads also damage steel bridges. During strong earthquakes, axial forces on the piers can substantially increase. The 1995 Kobe earthquake damaged the bearings and pier columns and caused bridge structures to collapse [3]. Figure 18 shows buckling plates of bridge elements damaged during the Kobe earthquake.

Fig. 17 Corrosion in bridge structural elements

Fig. 18 Damage to structural elements caused by the 1995 Kobe earthquake

2.1 *Traditional Techniques*

Traditionally, the inclusion of steel plates or laminated sections by welding or bolting has been the most used rehabilitation technique. The retrofitted steel elements are attached to the tension flange; in the case of continuous steel girders, the flange is connected with the deck, which makes it more difficult, time consuming, and expensive regarding the rehabilitation process because the deck must be removed, which interrupts traffic. Despite the wide use of this technique, reports have revealed the development of fatigue cracks at the weld toe of the cover plate ends after being subjected to cyclic tensile loads at the connection angles, web gusset plates, and at the longitudinal stiffeners that could cause girder failure. To solve this problem, researchers have developed methodologies to rehabilitate fatigue cracks, such as the friction-type bolted splice plate connection, air-hammer peening along the weld toe, and a hybrid of the previous two methods, called the partial bolted splice connection [4, 5]. The Manual for Repair and Retrofit of Fatigued Cracks in Steel Bridges contains descriptions of the methodologies that can be used to solve these type of problems as well as examples [6].

Fatigue strength is critical in welded cover plates and has been the cause of failure in several bridge girders. Partial length cover plates are used frequently in the original design of slab-on-girder steel bridges due to the typical flexural moment distribution on these elements. This technique reduces the girder depth, weight, and cost of the superstructure. It is also an extremely common traditional retrofitting technique used to increase the flexural resistance of the superstructure elements in existing bridges.

Fatigue cracks have been detected at the ends of the cover plate, mostly in the case of tapered cover plates. Figure 19 shows a typical steel beam retrofitted by a tapered cover plate.

Certain existing non-damaged bridges must be retrofitted for fatigue due to the following reasons: (a) a longer service life is expected, (b) the live loads are higher than those assumed during the design process, and (c) the traffic volume over the structure is growing faster than expected. In such cases, the need for retrofitting using welded cover plates is common. Typically, the bolted splice plate connection at the ends of the cover plates is adopted to improve the flexural resistance and service life of the bridge girder. Figure 20 shows this methodology. Before the plate splice is installed, the plates and girder must be blast-cleaned.

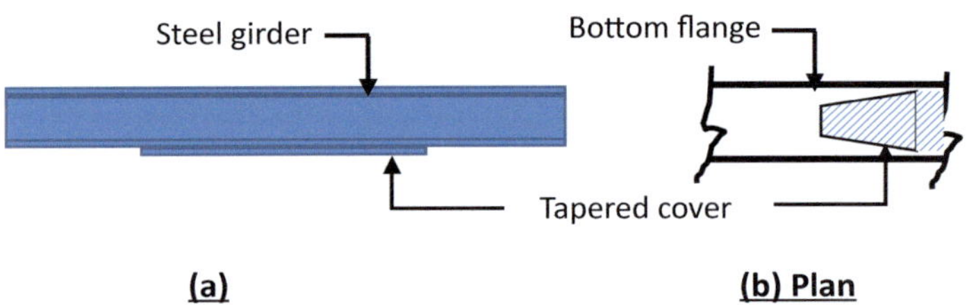

Fig. 19 Steel girder with a tapered cover plate

Fig. 20 Cross-section of a bolted splice connection

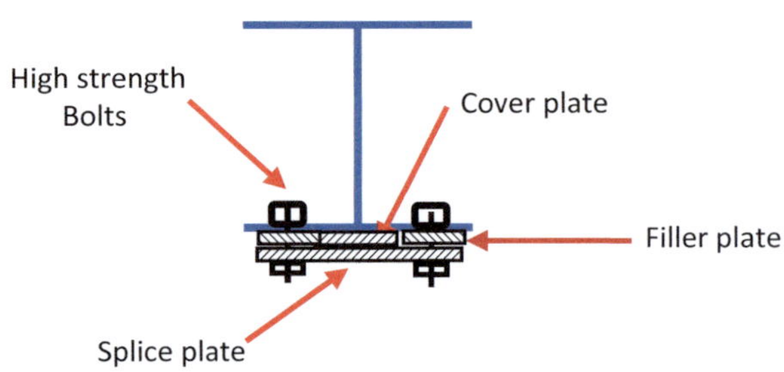

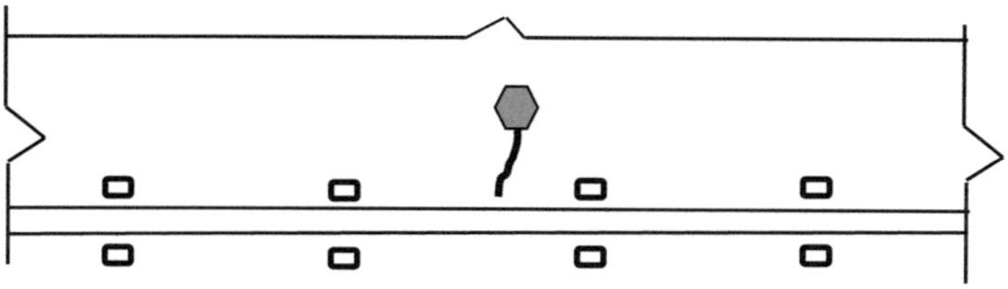

Fig. 21 Bolt installed at crack tip in a damaged girder

The same strategy can be adopted if the girder cracks before intervention as a consequence of fatigue stresses. In this case, the crack tip is drilled out, and a high-strength bolt installed (Fig. 21).

Typically, bridges have steel or concrete transverse diaphragms at the supports (abutments and piers). Diaphragms are added to improve the overall distribution of live and lateral loads and to contribute to the torsional strength of the bridge deck. If the transverse diaphragms are designed with capacity design principles to fail before other elements of the bridge, an important source of energy dissipation can be achieved. The lateral yield displacement of these elements must be reached before the yield displacement of the steel substructure. As a retrofit measure, two basic types of ductile diaphragms can be incorporated into bridge steel superstructures. One dissipates energy through the inelastic incursion of steel elements of the diaphragm, whereas the other incorporates special devices to dissipate energy. The use of energy devices has two advantages: (a) improves the system´s energy dissipation capacity and (b) concentrates damage in the device, which reduces or avoids damage in other elements. Figure 22 shows several ductile diaphragms for steel bridge superstructures.

The inclusion of ductile diaphragms is effective when the bridge superstructure is stiff; conversely, a flexible system may be inadequate. Careful attention must be provided during the design and building processes for the connection between the superstructure and the diaphragm, and lateral torsional buckling of the diaphragm elements must be avoided.

For slab-on-girder bridges, researchers [7, 8] recommend changing end-diaphragms of abutments and piers for ductile ones designed to yield before the strength of the substructure is reached. These ductile diaphragm systems are known as shear panels, eccentrically braced frames, triangular-plate added damping, and stiffness devices. For a more effective retrofit, the use of ductile end-diaphragms for larger bridges with a small number of girders is recommended.

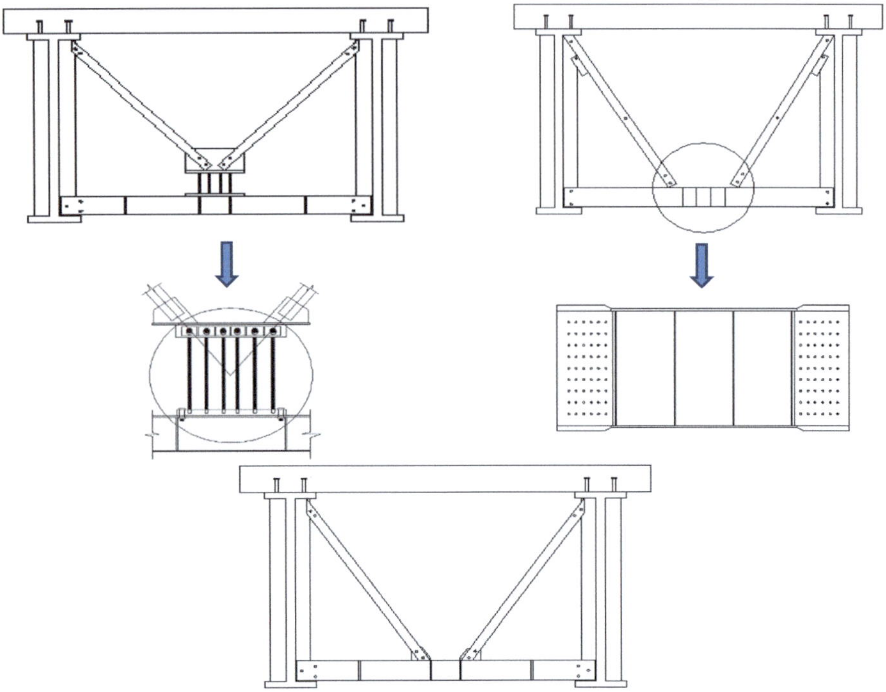

Fig. 22 Ductile diaphragms

2.2 Rehabilitation with Carbon Fibre-Reinforced Polymers

During the last two decades, researchers have dedicated increasingly more effort to better understand the structural and dynamic behaviour of metallic civil engineering structures reinforced with fibre-reinforced polymers (FRPs) used as plates, dry fibre sheets, or post-tensioning rods. FRP constituents are a polymer matrix reinforced with fibres; the polymer is generally an epoxy, vinylester, or polyester thermosetting plastic, whereas the fibres are made from materials such as carbon, glass, or an aramid. Because the mechanical properties of FRPs are similar to those of metallic structures, steel bridges have been reinforced with FRP material despite the lack of proper design regulations at the time of their implementation for the following reasons: the bridges exhibited aging deterioration, the bridges were important for historical reasons, and the infrastructure of the bridge needed to function better. Countries that initiated this retrofit practice and its research and implementation include the United Kingdom, United States, Canada, Italy, and Japan. Examples of this practice include the Hythe Bridge built in Oxford in 1874 and retrofitted with pre-stressed CFRP to improve its flexural capacity; the oldest cast iron highway in the world, the Tickford Bridge built in 1810 in Newport Pagnell, Buckinghamshire, UK, which was retrofitted with wet lay-up CFRP sheet system; and the King Street railway built in 1870 in Mold, Flintshire, UK using unidirectional carbon fibre and glass fibre reinforcement with an ultra-high modulus.

Several studies have reported the particular reinforcement details and additional examples, which are deeply discussed, in instances where FRP materials have been used as a retrofit system [9–16]. FRP materials offer important advantages over traditional techniques, such as adding steel plates or using external pre-stressed tendons, which are related to important reductions in the system's weight, time to complete rehabilitation, and a minimum or null traffic interruption during its installation process. Several important issues have been addressed experimentally and numerically, where one goal has been to determine the durability of the bond for bridges' girder retrofits; the durability is controlled by two important issues: environmental durability and fatigue resistance. Researchers understand several aspects of the problems involved, which allows for the development of design and practice guidelines for an FRP externally bonded system [9, 17–20], such as materials' characteristic and properties, installation techniques, conceptual designs, structural behaviour and analysis, bond models to calculate shear and peal stresses within the adhesive thickness, designs to reduce the stresses concentration at the end of the plates, quality control, and operation.

To understand the structural behaviour when implemented on steel elements, researchers have determined the most important properties required in design, namely the following: modulus of elasticity (normal and high), ultimate tensile strength, and ultimate tensile strain of the CRFP. Table 1 presents typical values reported in the literature [9, 21–25].

One of the advantages of FRPs is being corrosion resistant; however, in the case of carbon fibre-reinforced polymers (CFRP), a glass fibre reinforced-polymer (GFRP) must be used between the CFRP and steel element to prevent galvanic corrosion due to contact between two materials having dissimilar carbon contents.

The effectiveness of retrofitting steel structures with FRPs that are adhesively added relies in part on the adhesive bond. In industry, there are several types of adhesives, such as epoxies, polyurethanes, acrylics, and cyanoacrylates; however,

Table 1 Mechanical properties of FRP (Adapted from Zhao [1], Hollaway and Cadei [12])

FRP	Modulus of elasticity (GPa)	Ultimate tensile strength (MPa)	Ultimate tensile strain (%)	Coefficient of thermal expansion (10^{-6}/°C)
CFRP sheets	552	1175	0.20	−1.0 to −4.0
	230	2675	1.20	
CFRP plates	479	1607	0.36	
	338	1186	–	
	156	2691	1.72	
	171	2830	1.55	
GFRP	30	200–500	–	4.9
Aramid fibre	130	3000	2.3	−5.2

Table 2 Mechanical properties of adhesives (Adapted from Zhao [9])

Adhesive	Modulus of elasticity (MPa)	Ultimate tensile strength (MPa)	Ultimate tensile strain (%)
Araldite 2015	1750	14.7	1.51
Araldite 420	1828	21.5	2.89
FIFE-Tyfo	3975	40.7	1.11
Sikadur 30	11250	22.3	0.30
Sikadur 330	4820	31.3	0.75

the most commonly used adhesive is the polymer [12]; adhesives are conformed by a resin followed by polymerization. When selecting the adhesive, it is important to choose a resin compatible with the FRP's resin. Because of the adhesive chemistry, special attention must be given to the post-cure temperature because this can minimize bond strength reductions and creep effects. Zhao [9] reported the adhesive mechanical properties estimated by other researchers, which are reproduced in Table 2. However, it is important to remark that for design purposes, the materials' properties must be taken from the fabricant or by performing experiments on samples from the material to be used.

2.3 *Bond Joint Resistance*

The bond must resist several environmental conditions, such as deicing agents, chlorides from salt water, and extreme temperatures; the adhesive must overcome these challenges, where the most detrimental challenge is dealing with the presence of humidity or preventing liquid water [12]. To take into account these variables, the Italian design guidelines (CNR [19]) recommends the use of different design factors according to environment conditions. Several experimental studies have determined the failure modes from an adhesive bond between FRP and steel plates, e.g., adhesion failure between steel and the adhesive, cohesion failure on the adhesive layer, adhesion failure between the FRP and adhesive, FRP delamination, FRP rupture, and steel yielding [26–28]. Bond failure depends on the adhesive thickness and material properties; for example, CFRP with a normal elastic modulus can experience a cohesion failure with thin layers of the adhesive, whereas thick adhesives exhibit FRP delamination failure. Conversely, CFRP with a high modulus reaches FRP rupture when the bond length is long, whereas for a short bond length, the failure modes are cohesion failure and FRP delamination [9].

An important number of experimental and analytical studies on the bond strain have characterized the shape and parameters that govern the shear stresses and strains developed along the bond. These results have led to proposed simplified models [14, 28–30] that exhibit different behaviours for linear and non-linear adhesives, where a linear adhesive is characterized by bilinear behaviour and defined with a triangular shape of the stress-strain curve, whereas non-linear adhesives present a linear response that reaches a flat plateau followed by a decreased slope up to failure. The stress-strain relationship shows a symmetric distribution along the bond with three key parameters that define the bond strength: maximum shear stress, initial slip, and maximum slip. Based on these parameters, a bond slip model has been proposed to characterize the constitutive model of an adhesive; a summary of the details can be found in Zhao [9].

Another important issue to be defined for the retrofitted system is the effective length and bond strength. Different models and expressions used to estimate these parameters have been developed using the double-shear pull test, and approximate equations have been proposed [31]. These studies showed a stress concentration at the bond ends that leads to delamination of the FRP in these zones. To prevent this, the use of mechanical anchorage is recommended, where the FRP is wrapped at the laminated end, which provides a spew fillet of excess adhesive and tapers the edge of the bonded FRP laminate, or through the use of adhesives with large fracture energies [9].

Zhao [9] showed that temperatures below zero increased the brittleness of the adhesive. However, the elastic modulus, ultimate strength, and strain do not vary significantly; on the contrary, high temperatures significantly decreased the bond strength and the joint stiffness. Hart-Smith [32] proposed expressions that take into account the temperature effects to estimate the effective bond length and the ultimate load. Other works [33] showed that cyclic loading does not significantly reduce the bond capacity or the bond's global stiffness because it has been reported that the effects determined under impact loads more significant changed the overall mechanical properties. The Italian design guidelines [19] presents guidelines to evaluate the bond strength and strengthen fatigued elements.

2.4 *Strengthening of Bridges' Girders*

Steel bridges' girders are retrofitted with FRP with the aim to increase the load-carrying capacity, stiffness, and fatigue life. Using a high-modulus CFRP is recommended if it is desired to achieve important increases in the stiffness with the goal of either reducing deflections or increasing the buckling load capacity. The CFRP can also be pre-stressed to improve the effects of the retrofit system. During the design process, it is important to consider the possible failure modes that the retrofitted girder can exhibit. The inclusion of an FRP material moves the neutral axis towards the tension zone, which increases the stresses on the compression flange and web. These elements must be carefully evaluated to prevent local buckling in the plates. Other failures modes that must be reviewed are lateral buckling and end and intermediate debonding. Regulation codes, such as CNR-DT 202/2005, present expressions to assess failure modes, serviceability and ultimate limit states, and flexural capacity. In other countries, there are also design guidelines with similar criteria [17, 18, 34]. Because it is desirable to install FRPs without traffic interruptions, Moy and Bloodworth [33] studied the cycling loading effects under this condition. The authors reported that curing during cyclic loading affects the adhesive strength and can lead to complete failure of the bond if the deformation of the adhesive is too large. The adhesive flexibility reduces the effective properties of the retrofitted beam. The flexural and lap-shear tests showed that cyclic loading during the adhesive cure can reduce the strength and stiffness, which significantly affects the ends. As cautious actions, Moy and Bloodworth [33] recommend limiting the shear stress in the adhesive to a maximum of 1.0 N/mm^2.

Roach et al. [35] developed a bonded composite doubler repair method used in lieu of mechanically fastened metallic patches to reinforce or repair damaged structures. The basic parameters of the system are the ply lay-up, ply orientation, patch shape and taper, and the bond layer; the stress field and configuration of the structure determines the number of plies and the fibre orientation. Roach et al. [35] reported that bonded composite doubler repairs can extend the fatigue life by a factor of 100, which limits crack growth.

The recommendations for installing bonded FRPs are in general, extremely similar among the design guidelines and researchers' reports: (1) Prepare the surface with grinders or sandblasters to remove rust, paint, and primer from the steel element, (2) Pre-treat the bare steel with either Dow Corning Z-6040 silane adhesion promoter or ITW PC120 primer/conditioner according to the adhesive to be used, (3) Sand the bond side of the FRP to improve adhesion using medium-grit sand paper or a sandblaster and then clean with acetone, (4) Apply the adhesive on the pre-treated steel surface and FRP, (5) If using CFRP, place a glass fibre-fabric between the steel element and the CFRP to prevent corrosion, (6) Facilitate the installation using wood blocks that are temporarily glued to the FRP composites and clamp the FRP to the steel element at close intervals (25 mm is recommended) to provide a uniform pressure along the bonded surface for a uniform bond line, and (7) Remove clamps and wood blocks after the adhesive reaches full cure.

3 Substructure Rehabilitation Techniques

This section summarizes the state of practice of repairing and strengthening techniques for steel substructures; it describes the potential vulnerabilities and current practice for retrofitting piers damaged by corrosion, fatigue, buckling of slender elements, and inertial forces induced seismically.

Structural steel elements are common in bridges. Mostly, structural steel is used for girders, trusses, decks, and bearings; however, within the substructure, the use of steel piers is sparse, where the piles are the only structural steel elements of common use. Additionally, the repairing methods for substructures have focused on piling, and there are less research studies related to rehabilitating steel piers or columns.

Steel corrosion is the cause of the most significant deterioration exposed by steel substructures as well as buckling of slender structural elements under compression. Damage on steel bearings and brittle steel bracing failures have also been observed. Fatigue, overloading, vehicle impact, thermal strains, and stress concentrations are also causes of damage. In the case of piles, local buckling can result from pile-driving operations. Steel column failures caused by the occurrence of damaging earthquakes are infrequent because there are only a few bridges subjected to high seismic demands outside of Japan.

The causes of corrosion are well known, and in particular, moisture, industrial, ground, and seawater environments are critical aspects of corrosion. Corrosion reduces the transverse cross-section of the structural element, which leads to a loss of strength and stiffness of the structural member. The conventional strengthening and repairing techniques for a severely corroded pile include adding cover plates in the deteriorated areas of the element. The steel cover plates are welded or bolted to the pile; however, the labour can be intensive and costly. It is worth noting that welding causes residual stresses in the existing elements, which affects the buckling capacity and fatigue life of elements. If corrosion is localized, it is possible to repair the corroded material by removing the corrosion and maintaining the original material; if the deteriorated area reduces the strength of the element, the use of cover plates may help reduce or eliminate this effect.

In the case of piles, severe corrosion may be present in sections located close to the water line, which is caused by the continuous wetting and drying of the steel when it is in contact with the ground. Then, corrosion prevention in new and existing piles can be achieved by protecting the steel from water and ground exposure. Therefore, painting and watertight encasement are the first steps to prevent corrosion. The use of sacrificial anodes is another measure for protection against corrosion; however, their use has several disadvantages. Sacrificial anodes are located on the steel piles, where several of them are visible because they are located above the ground or the waterline, which exposes them to vandalism. Encasement of the anodes is a possible solution because encasement reduces the device exposure, which acts as additional protection against corrosion (Fig. 23). Another issue of concern is that anodes are welded onto the pile where access may be difficult in certain cases. Major corrosion protection can be achieved if sacrificial

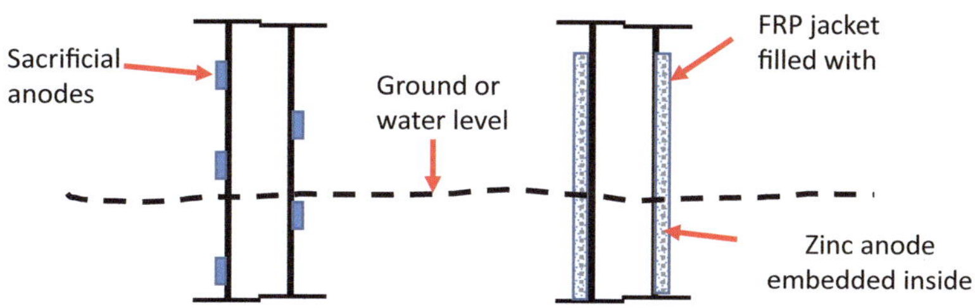

Fig. 23 Steel pile protected by sacrificial anodes

anodes are combined with fibreglass jackets. In this way, double protection is provided to the pile. Figure 23 shows how both elements protect an H-shaped pile.

Concrete or fibreglass jacketing of the steel pile is an appealing strategy (Fig. 24). The use of fibreglass is preferred because it is more impermeable than concrete. Fibreglass encasement is also useful to protect sacrificial anodes from vandalism. Concrete jackets improve the resistance and stiffness of the element and are less expensive than fibreglass jackets. Fibreglass is an expensive material used for jacketing piles and can be left permanently, which serves as additional protection for the concrete jacket (Fig. 25).

Fatigue is induced by traffic cyclic loading, which leads to cracks in the structural elements. Welding and/or adding cover plates are techniques commonly used for repairing cracks due to fatigue and also for repairing cracks due to vehicle collision. The addition of new structural elements (additional columns or walls) contributes to the redistribution of loads and reduces the effect of fatigue. Increasing the cross-section of the deteriorated area of the element is also a common measure to rehabilitate the damaged structure. Vehicle collision produces distortion, cracking, and tearing in the elements.

Retrofitting measures for improving the seismic resistance of steel piers deserves particular mention. Most of the damage and hysteretic energy dissipated during the occurrence of an earthquake is at the piers. Despite the high ductility capacity of steel elements, steel piers may be seismically vulnerable to inelastic buckling. If the compression member reaches its local buckling capacity, fracture occurs due to low cycle fatigue, which progressively leads the compressed member to collapse. This

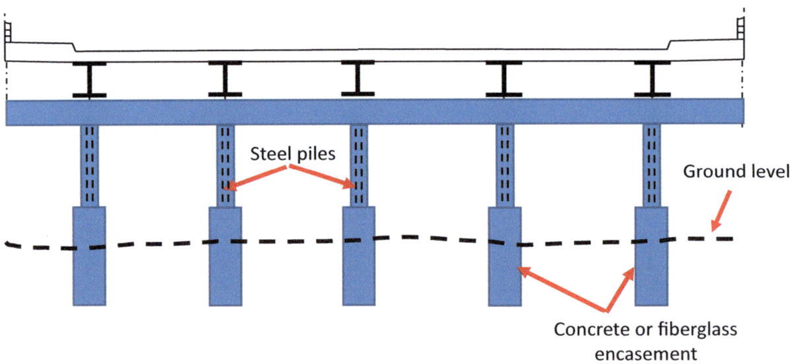

Fig. 24 Steel pile jacketing

Fig. 25 Cross-section of an H steel pile reinforced with a concrete jacket and permanent fibreglass form

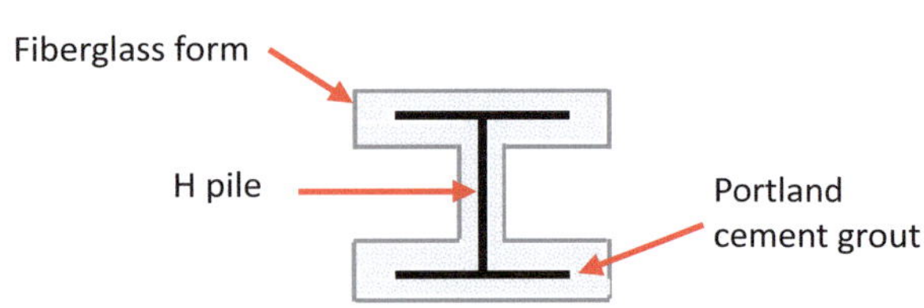

failure type is non-ductile, and the retrofitting focuses on preventing or delaying inelastic buckling.

Lateral torsional buckling typically is the governing action if the section is I-shaped [36], and the retrofit consists of enclosing the section with plates or channel sections (Fig. 26). Local buckling in circular piers also occurs at the region where the plate thickness changes. Member replacement, the addition of supplemental diagonal elements, cover plates, stiffeners, or the reduction of the unsupported length are all possible options for reinforcing bracing systems where the buckling capacity is lower than the yielding capacity.

Built-up members are commonly used for piers. If the built-up column is a non-compact section, it can undergo local buckling or lateral torsional buckling, which will reduce the pier ductility. Generally, retrofitting is oriented to conform to a close section (Fig. 27) to have a high torsional rigidity and flexural capacity, which reduces the susceptibility of premature local inelastic buckling.

When steel-braced frames are used as piers, the diagonal bracing is subjected to tension-compression forces derived from the lateral inertial forces, such that a brittle failure mode can occur if the diagonal members have large slenderness ratios, which leads to elastic buckling. The vertical elements are typically designed to carry the vertical loads, which is generally more resistant than the diagonal steel elements.

Deck-truss bridges are particularly vulnerable to seismic actions. The response of this type of bridge is improved by the use of different bracing system configurations incorporated to the end sway frames [37–43]. When bracing systems are incorporated to a sway frame, the stiffness is also increased, which increases the inertial forces and overloads the bearings and piers. As a result of the stiffness increase when bracing systems are adopted for reinforcing end sway frames, the capacity of the bearings, connections, and substructures must be assured; these

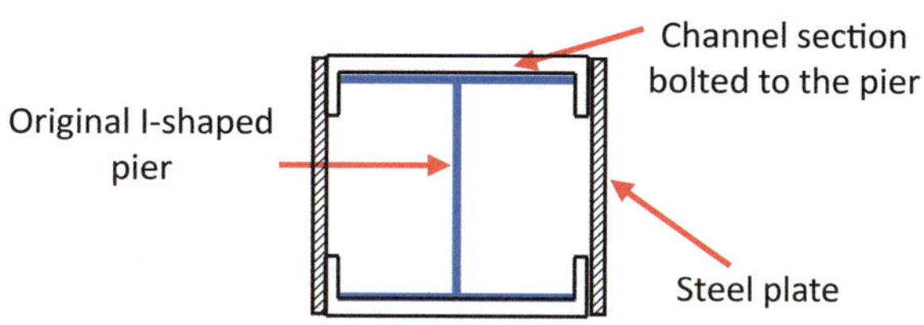

Fig. 26 I-shaped section retrofitted by channel sections and plates

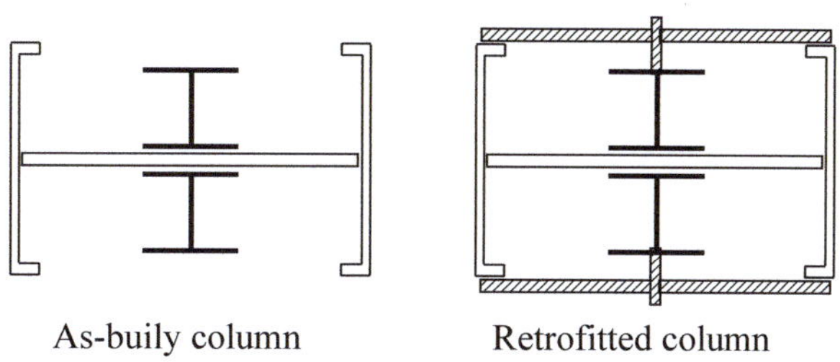

Fig. 27 Retrofit of as-built open section

elements must be strengthened. According to capacity design principles, all other members and their connections should behave elastically.

The most common type of failure observed after the occurrence of an earthquake is a loss of seat length. It is common to provide restrainers at the supports, at the abutments, and at the deck joints to prevent superstructures from collapsing during an earthquake. The use of these devices modify the inertial forces transferred to the substructure when subjected to seismic forces. Bridges retrofitted by restrainers alone, failed in past damaging earthquakes [44] demonstrating the need for retrofitting the bridge substructure as well. Then, a careful analysis should be conducted to estimate the new mechanical elements on piers when these restrainers are added to the superstructure. Again, capacity design principles should be kept in mind. In this framework, to avoid the need of retrofitting the foundation of multi-column bents, it is good practice to leave a clear space between the steel jacket and the foundation. Then, the moment transferred to the footing is lower than the moment capacity of the reinforced bent. The use of link beams is an appealing strategy for retrofitting multi-column bents. The link beams substantially increase the bent lateral stiffness in the transverse direction of the bridge and reduce the unsupported length of the columns.

The optimal strategy may be a combination of two retrofitting techniques; in the case of end sway frames, the bracing can be combined with supplemental damping systems to obtain a more efficient solution. The supplemental damping is a device that acts as a fuse by failing and dissipating energy, which protects the braces, connections, bearings, and piers from potential damage. The addition of shear links, viscous dampers, or another type of energy dissipation devices in braced steel towers is an attractive alternative that can be combined with traditional retrofitting measures to attain a more efficient solution.

Base isolation is an appealing strategy for reducing the seismic demand in piers and foundations. If base isolation is capable of reducing the seismic moment and shear force acting on the substructure to a magnitude less than the seismic capacity of the substructure, the columns would not require strengthening. Lead rubber bearings are the most common isolators that have been incorporated into existing bridges because of their low cost, reliability, simple modelling, and simple construction adaptability to the existing structure. The lead rubber bearings provide isolation and energy dissipation in only one device. Friction or sliding bearings, particularly the friction pendulum systems, are also commonly used for isolation and energy dissipation in bridge structures. High damping rubbers, which are obtained by the addition of fillers during the manufacturing process, increase the hysteresis and energy dissipation to high levels. The high damping rubbers are highly non-linear, and their applications are limited almost exclusively to Japanese bridges.

4 Selection of the Rehabilitation Technique

The selection of the best alternative to retrofit a steel bridge must consider the increase in capacity and the expected demands on the rehabilitated structure. If the bridge is located in a seismically active zone, one possible approach is by constructing fragility curves like the ones proposed by Jara et al. [45, 46] to evaluate the impact of the rehabilitation on the expected behaviour of the bridge. Fragility curves determine the probability of reaching or exceeding a limit state of behaviour for a specific intensity measure. Padgett [47, 48] and other authors [49–51] proposed a methodology based on fragility curves to evaluate different retrofit techniques on a family of bridge structures. One important conclusion in the analyses of multi-span simply supported steel girder bridges is that geometric uncertainties and ground motion uncertainties are more relevant than uncertainties in the modelling parameters. Among the retrofit techniques analysed, the replacement of steel bearings with isolation devices substantially reduces the bridge fragility.

In addition to fragility curves, loss models represent another tool that is used to better select the rehabilitation method and the bridges to be retrofitted [52]. Cost-benefit analyses can quantitatively assess the impact of the retrofit. Using a cost benefit ratio, Padgett et al. [53] demonstrated that among several retrofit possibilities on a multi-span continuous girder bridge, the use of loss models is an efficient methodology to select the best option.

5 Conclusions

This chapter discusses the most common techniques used to rehabilitate and retrofit steel bridges. The advantages and disadvantages of using traditional and innovative rehabilitation techniques are described. Fatigue cracks at the end of welded cover plates is a critical issue in bridges subjected to seismic loads. The addition of ductile transverse diaphragms in steel bridges is a plausible technique to improve the expected behaviour. Especial emphasis is placed on describing the use of carbon fibre-reinforced polymers. The expected performance of retrofitted steel elements with this technique highly depends on the bond joint resistance. The use of carbon fibres increases load-carrying capacity and improves the fatigue life of the structural elements making this procedure as one of the more appealing technique to retrofit steel elements. A critical issue in bridges subjected to seismic loads is the loss of seat length. The use of restrainers at the supports is a common practice to prevent this type of failure. However, a careful analysis must be conducted to consider the inertial forces transferred to the substructures when these devices are proposed. Finally, the best selection of the retrofit technique must be based on cost-benefit studies by using fragility curves and loss models methodologies.

References

1. ASSHTO: AASHTO LRFD Bridge design specifications, American Association of State Highway and Transportation Officials, Washington, D.C., 2007.
2. CALTRANS: Construction manual, California Department of Transportation, California, 2014.
3. Bruneau M. Performance of steel bridges during the 1995 Hyogoken-Nanbu (Kobe, Japan) earthquake-a North American perspective. Eng Struct. 1998;20(12):1063–78.
4. Hassan AF, Bowman MD. Fatigued crack repair of steel beams with tapered cover plates details. J Struct Eng. 1996;122:1337–46.
5. Sahli AH, Albrecht P, Vannoy DW. Fatigued strength of retrofitted cover plates. J Struct Eng. 1984;110:1374–88.
6. Dexter RJ, Ocel JM. Manual for repair and retrofit of fatigued cracks in steel bridges, U.S. Federal Highway Administration FHWA, Department of Transportation, Publication No. FHWA-IF-13-020, 2013.
7. Zahrai SM, Bruneau M. Ductile end-diaphragms for seismic retrofit of slab-on-girder steel bridges. J Struct Eng. 1999;125:71–80.
8. Celik OC, Bruneau M. Seismic behavior of bidirectional-resistant ductile end diaphragms with unbonded braces in straight or skewed steel bridges. Technical Report MCEER-07-0003, University at Buffalo, State University of New York, Buffalo, N.Y., 2007.
9. Zhao XL. FRP-Strengthened metallic structures. Cambridge: CRC Press; 2014.
10. Moy S. Case studies: FRP repair of steel and cast iron structures in Great Britain. FRP Int Off Newsl Int Inst FRP Constr. 2011;8(3):5–8.
11. Bastianini F, Ceriolo L, Di Tommaso A, Zaffaroni G. Mechanical and nondestructive testing to verify the effectiveness of composite strengthening on historical cast iron bridge in Venice, Italy. J Mater Civ Eng. 2004;16:407–13.
12. Hollaway LC, Cadei J. Progress in the technique of upgrading metallic structures with advanced polymer composites. Prog Struct Mater Eng. 2002;4:131–48.
13. Salama T, Abd-El-Meguid A. Strengthening steel bridge girders using CFRP, University Transportation Center for Alabama, UTCA Report Number 06217. Huntsville, Alabama, 2010.
14. Miller TC, Chajes MJ, Mertz DR, Hastings JN. Strengthening of a steel bridge girder using CFRP plates. J Bridge Eng. 2001;6:514–22.
15. Phares BM, Wipf TJ, Klaiber FW, Abu-Hawash A, Lee YS. Strengthening of steel girder bridges using FRP. In: Proceedings of the 2003 Mid-Continent Transportation Research Symposium Iowa State University, August 2003.
16. Mertz DR, Gillespie Jr JW. Rehabilitation of steel bridge girders through the application of advanced composite materials. IDEA Program Transportation Research Board National Research Council, University of Delawere, Contract NCHRP-93-ID011, Delawere, 1996.
17. Cadei JMC, Stratford TJ, Hollaway LC, Duckett WH. C595-Strengthening metallic structures using externally bonded fiber-reinforced composites. London: CIRIA; 2004.
18. Schnerch D, Dawood M, Rizkalla S, Sumner E. Proposed design guide lines for strengthening steel bridges with FRP materials. Constr Build Mater. 2007;21(5):1001–10.
19. CNR, National Research Council: Guidelines for the design and construction of externally bonded FRP systems for strengthening existing structures: Metallic structures, Preliminary study. Advisory Committee on Technical Recommendation for Construction, Rome, Italy, 2007.
20. JSCE: Advanced technology of repair and strengthening of steel structures using externally-bonded FRP composites (in Japanese). Hybrid Structures Reports 05. Tokyo: Japan Society of Civil Engineers, 2012.
21. Wu ZS, Yuan H, Niu HD. Stress transfer and fracture propagation in different kinds of adhesive joints. J Eng Mech. 2002;128(5):562–73.

22. Lanier B, Schnerch D, Rizkalla S. Behavior of steel monopoles strengthened with high modulus CFRP materials. Thin Walled Struct. 2009;47(10):1037–47.
23. Seracino R, Jones NM, Ali MSM, Page MW, Oehlers DJ. Bond strength of near-surface mounted FRP strip-to-concrete joints. J Compos Constr. 2007;11(4):401–9.
24. Fawzia S, Al-Mahaidi R, Zhao XL, Rizkalla S. Strengthening of circular hollow section steel tubular sections using high modulus CFRP sheets. Constr Build Mater. 2007;21(4):839–45.
25. Sena Cruz JM, Barros JAO, Gettu R, Azevedo AFM. Bond behavior of near-surface mounted CFRP laminate strips under monotonic and cyclic loading. J Comp Constr. 2006;10(4):295–303.
26. Colombi P, Paggi C. Strengthening of tensile steel members and bolted joints using adhesively bonded CFRP plates. Constr Build Mater. 2006;20:22–3.
27. Deng J, Lee MK, Moy SJ. Stress analysis of steel beams reinforced with a bonded CFRP plate. Compos Struct. 2004;65:205–15.
28. Yu T, Fernando D, Teng JG, Zhao XL. Experimental study on CFRP-to-steel bonded interfaces. Compos B. 2012;43(5):2279–89.
29. Xia SH, Teng JG. Behavior of FRP-to-steel bond joints. In: Proceedings of the international symposium on bond behavior of FRP in structures (BBFS, 29005), Hong Kong, December 2005.
30. Fawzia S, Zhao XL, Al-Mahaidi R. Bond–slip models for double strap joints strengthened by CFRP. Compos Struct. 2010;92(9):2137–45.
31. Yuan H, Teng JG, Saracino R, Wu ZS, Yao J. Full-range behavior of FRP-to-concrete bonded joints. Eng Struct. 2004;26(5):553–65.
32. Hart-Smith LJ. Adhesive-bonded double-lap joints. Minnesota Department of Transportation, Final Report, Department of Civil Engineering, University of Minnesota, 1973.
33. Moy SSJ, Bloodworth AG. Strengthening of a steel bridge girder using CFRP plates. Proc Inst Civ Eng Struct Build. 2007;I60(4898):81–93.
34. Moy S. ICE Design and practice guides FRP-composites life extension and strengthening of metallic structures. London: Thomas Telford Publishing; 2001.
35. Roach D, Rackow K, Delong W, Franks E. In-Situ repair of steel bridges using advanced composite materials. In: Proceedings of 10th international conference bridge and structures management, Transportation Research Board on the National Academies. Buffalo, New York, 2008.
36. Keady K, Alameddine F, Sardo T. Seismic retrofit technology. In: Chen WF, Duan L, editors. Bridge engineering handbook. New York: CRC Press; 2003. p. 11.1–11.30.
37. Kasai K, Popov EP. Cyclic web buckling control for shear link beams. J Struct Eng. 1986;112 (3):505–23.
38. Malley JO, Popov EP. Design considerations for shear links in eccentrically braced frames, UCB/EERC-83/24, University of California, Berkeley, California, CA, 1983.
39. Tsai KC, Chen HW, Hong CP, Su YF. Design of steel triangular plate energy absorbers for seismic-resistance construction. Earth Spect. 1993;9(3):505–28.
40. Nakashima M, Iwai S, Iwata M, Takeuchi T, Konomi S, Akazawa T, Saburi K. Energy dissipation behaviour of shear panels made of low yield steel. Earth Eng Struct Dyn. 1994;23 (12):1299–313.
41. Zahrai SM, Bruneau M. Impact of diaphragms on seismic response of straight slab-on-girder steel bridges. J Struct Eng. 1998;124(8):938–47.
42. Zahrai SM, Bruneau M. Ductile end-diaphragms for seismic retrofit of slab- on-girder steel bridges. J Struct Eng. 1999;125(1):71–80.
43. Alfawakhiri F, Bruneau M. Flexibility of superstructures and supports in the seismic analysis of simple bridges. J Earth Eng Struct Dyn. 2000;29(5):711–29.
44. Randall MJ, Saiidi MS, Maragakis EM, Isakovic T. Restrainer design procedures for multi-span simply-supported bridges, Technical Report MCEER-99-0011, Multidisciplinary Center Earth Eng Res. 1999.
45. Jara JM, Galván A, Jara M, Olmos BA. Procedure for determining the seismic vulnerability of an irregular isolated bridge. Struct Infrastruct Eng. 2013;7:516–28.

46. Jara JM, Madrigal E, Jara M, Olmos BA. Seismic source effects on the vulnerability of an irregular isolated bridge. Eng Struct. 2013;56:105–15.
47. Padgett JE, DesRoches R. Methodology for the development of analytical fragility curves for retrofitted bridges. Earth Eng Struct Dyn. 2008;37:1157–74.
48. Padgett JE, DesRoches R. Three-dimensional nonlinear seismic performance evaluation of retrofit measures for typical steel girders bridges. Eng Struct. 2008;30:1869–78.
49. Ramanathan K, DesRoches R, Padgett JE. Analytical fragility curves for multispan continuous steel girder bridges in moderate seismic zones. J Trans Res Board. 2010;2202:173–82.
50. Wright T, DesRoches R, Padgett JE. Bridge seismic retrofitting practices in the central and southeastern United States. J Bridge Eng. 2011;16:82–92.
51. Zakeri B, Padgett JE, Amiri GG. Fragility analysis of skewed single-frame concrete box-girder bridges. J Perform Constr Facil. 2014;28:571–82.
52. Frangopol DM. Life-cycle performance, management, and optimization of structural systems under uncertainty: accomplishments and challenges. Struct Infrastruct Eng. 2011;7(6):389–413.
53. Padgett JE, Dennemann K, Ghosh J. Risk-based seismic life-cycle cost–benefit (LCC-B) analysis for bridge retrofit assessment. Struct Saf. 2010;32:165–73.

Ground Reinforcement and Rehabilitation of Foundations Systems for Their Reuse

A. Viana da Fonseca and A. Pinto

1 Introduction

Foundation Reuse is an important activity in modern structural engineering. RuFUS Project in 2003 [1] provided a good overview of this reuse of foundations in contemporary construction. Among other themes, different methods of verifying pile integrity were discussed, as an example of the civil engineering actions involved in rehabilitation of the buildings in our highly populated towns and cities, where preservation of historical heritage is a priority. The objective of rehabilitation geotechnical engineering should be primarily the reuse of foundations even when an increase of ultimate bearing capacity is necessary. This is due when existing foundations can still be working safely, but also when old foundations are not capable to be reused but there is the necessity of he re-usage of the façade of a building whilst the interior parts of a building is rebuilt (see Fig. 1).

Reusing foundation for construction of new structures is not a new [2] describe how in Great Britain, ancient castles and cathedrals were constructed and then reconstructed upon old and existing foundations. The authors remind that during the Elizabethan times, new buildings were allowed if constructed on old foundations to avoid urban congestion. However, the incapacity of old foundations to stand for the new structures led the new buildings to be raised often on completely new

A.V. da Fonseca (✉)
CONSTRUCT—GEO, Department of Civil Engineering,
Faculty of Engineering, University of Porto, 4200-465 Porto, Portugal
e-mail: viana@fe.up.pt

A. Pinto
JETsj Geotecnia Lda, Rua Comandante Costeau, 20D, 1990-067 Lisbon, Portugal
e-mail: apinto@jetsj.com

A. Pinto
Instituto Superior Técnico, Civil Engineering Department, Technical University of Lisbon, Avenida Rovisco Pais, 1049-001 Lisbon, Portugal

A. Costa et al. (eds.), *Strengthening and Retrofitting of Existing Structures*,
Building Pathology and Rehabilitation 9, https://doi.org/10.1007/978-981-10-5858-5_12

Fig. 1 Underpinning for the re-usage of the façade of a building

foundations. Butcher et al. [1] state that most foundations have been recently re-engineered and successfully reused on projects such as railway bridges and major building projects.

The solution to be adopted for a reuse or rehabilitation of a foundation system demands for the elaboration of representative subsoil profiles based on competent soil investigation (in situ and laboratory testing). This will be interpreted based on the established and novel interpretation methodologies for the definition of parameters that can allow the design of the adapted or the new foundation under loading, and the transmission of stresses to the underlying mass of soil, taking in due consideration the particularities of the soil and rock masses involved. In view of these goals some reference books and manuals can be indicated [3–8] even being aware that other relevants can be omitted, which will help in the decision of which tests should be ideal tests to be selected, their interpretation and the design approaches based on them. Existing infrastructure in the ground and service tunnels can conditioned a lot the installation of new piles. To avoid this, ground investigations are decisive, namely when there is existing archaeology underneath the construction sites. Driven piles will be operationally limited operations, for what existing foundations may be utilized if less aggressive/intrusive techniques enable them to carry higher loads than the previously installed.

Soils generally present great variability of properties. Current design solution practice is based on calculations which depart from properties values representing the materials involved in the problem. A probabilistic treatment, in many cases, would bring meaningful advantages in the evaluation of safety [9]. This is implicitly considered in Eurocode 7 [10], for the geotechnical design rules, by adopting a concept for the characteristic value of the resistance concepts by deriving the characteristic of, for instance, an ultimate compressive resistance $R_{c,k}$ of a pile from measured values $R_{c,m}$, from load tests and from comparable experience, including one or more site specific pile load tests. An allowance shall be made for the variability of the ground and the variability of the effect of pile installation, by applying correlation factors that depend on the number of tests (details in Viana da Fonseca et al. [11]).

With such implicit consideration of safety approaches, investigation tests are compulsory but loadings from current structures can have a simplified treatment (residential and commercial buildings), adopting a deterministic approach regarding permanent and accidental loading [12], leaving more demanding probabilistic treatments to complex foundation systems (among others, tall buildings, bridges, industrial buildings, and, more important, sensitive monuments and heritage buildings). The correct evaluation of foundation loads and the way they interact with soils, as thought by primary design and the way they evolve during their history, as well as when load transference to the substrata is reequilibrate with the new structures for the novel layout, includes the identification of the failure mechanisms and deformation of the soil mass. Any doubts should be dealt by performing by numerical methods considering distinct scenarios of soil properties and load distributions.

An important element to include in these projects is the meticulous description of the execution processes of all parts of the systems, with emphasis to the connection between existing structural elements and the news reinforcement. A special emphasis should be addressed to problems associated to unclear mechanism developed in the distinct foundation elements and the way these can be overcome. These will depend a lot on the variability of depth of resistant subsoil layers, resulting in differential settlements and damage to the structure. The compatibility of settlements of different kinds of foundations have to be dealt with reinforcements that can demand the accessibility of equipment to certain areas of the building, such kinds of procedures being often very demanding.

Some options have to be attained such as the adoption of bearing capacity values for deep foundations based on empirical correlations considering penetration tests (SPT, CPT, PMT), without calibrating them with well documented tests for the more extensive type of foundations methods that can include regional practice. This has been one of the utmost advantages of French Method as described by Gambin and Frank [13], which calculation methods used for pile foundations have been validated against the results of static load tests and a model factor can introduced to reach a target safety. This is thoroughly described by the paper from Burlon et al. [14], focusing on the establishment of a model factor for the calculation of the ultimate limit state (ULS) bearing capacity of piles in compression from Ménard pressuremeter tests results, under the requirements of Eurocode 7. The model factors are meant to take into account the scatter of the calculation model, whereas the spatial variability of the ground properties is dealt with by means of statistical methods. Then an analysis of a database of 174 full-scale static pile load tests carried out by the Laboratoire Central des Ponts et Chaussées (LCPC, now called IFSTTAR), over the last 40 years, mainly in France, is performed and a new calculation model is established based on pressuremeter test results. Finally, the calibration procedure for deriving the model factor used in the recent French standard for pile design following Eurocode 7 is explained using two approaches: one compares the dispersions of the former and new calculation models and the other performs statistical analysis combining the scatter of the calculation model and the spatial variability of ground properties. This model factor was introduced into the recent French standard for pile design following Eurocode 7.

Some details have to be considered when the solutions are to be decided. For instance, Militinsky et al. [12] refer that knowing that for large bored piles there is usually the necessity of having large displacements (about 10% pile diameter) in order to fully mobilize the tip bearing capacity of the pile, results in the adoption of safe values regarding bearing capacity, but at a cost of settlements that are incompatible with the good performance of the structure. In cases of big differences in pile diameter under the same structure, such condition regarding different settlement to reach full strength mobilization may provoke important differential settlements and damage to the structure. Some attention has to be made to the determination of the loads acting in the foundations, specifically in constructions where there is no sufficient information of the original project (or the one that is available is insufficient), or there is some doubts on the options assumed in the design, which is very common in old buildings or/and in more recent special situations recognized as very sensitive: pre-cast structures, industry buildings, high-rise buildings, structures submitted to dynamic effects and shocks.

Another key point in the interventions for reusing/reinforcement of existing foundations to be strengthened to new loading conditions, is the intermediate conditions created during construction of the new solutions, as it will be illustrated in the case-histories that will be presented in this chapter. Adequate construction details have to be considered and explicitly drawn, such as the ones referred in Militinsky et al. [12]: (i) the link of the steel bars of the reinforced piles under tension and the reinforced concrete block topping the piles that support a structure loading point, resulting in no transference of structure loads to the foundations; (ii) the lack of details about insufficient concrete covering the steel bars for a given situation, specially relevant in cases of aggressive environment, or even absence of construction details, resulting in steel bars degradation and bad performance after some time period; (iii) adequate details of link of the steel piles under tension and the concrete block topping the pile. The development of new underpinning techniques has allowed the adoption of a wide number of solutions for reinforcement and rehabilitation of foundations systems for their reuse, progressively more adapted to the singularities and restraints of each scenario, especially when sensitive, old or historic, constructions founded on weak soils have to be underpinned. In this context, the solutions comprising micropiling and jet grouting techniques should be pointed out due to their versatility and advantages related to the vibrations and noise limitation, as well as the possibility to be adopted in small spaces with low head rooms and restricted access. These techniques also allow the soil improvement, minimising the soil disturbance due to the boreholes small diameter, drilled with suitable equipment (Bullivant and Bradbury [15]).

On the following chapters, after the presentation of the main aspects of these two techniques, some practical case histories over hodiern structures where micropiling and jet grouting underpinning techniques were used, are presented. In each case, the following topics are presented: scenario, main restraints, main conception and execution criteria, as well as main quality control and quality assurance procedures.

Important geotechnical interventions for the preservation of historic cities patrimonial monuments [16–18], where the demands of ancient and very sensitive masonry structures have involved different approaches, will not be addressed here. Some have not been published, a sit is the example of the work developed between 2001 and 2003 for the underpinning the "Grand Palais", in Paris, built in 1900 and protected as a National Heritage monument, by creating a new retaining wall for a future underground structure in the main hall. Other have been published, being an example of these the major renovation of Venice San Marco Bell Tower, complementing an original enlargement of the stone masonry in the early days of the 20th century, due to the signs of the development of some cracks in the '50's of that century. An inspection of the foundation block, indicated the detachment of the new masonry from the pre-existing one, which was solved by a recent intervention consisting of the insertion at two levels along the perimeter of the plinth prestressed titanium rebars, to increase the overall flexural stiffness of the foundation and stop further cracks opening. This intervention, described by the work of Macchi et al. [19] required excavations below the ground water level in soft/loose lagoon deposits within an extremely important monumental area and in presence of buried archeological remains, many of which were unknown. As described in the referred work, severe precautionary measures were taken to prevent even little settlements of the tower and surrounding monuments during excavation, dewatering and retrieval of the buried archeological remains. To achieve this goal Deep Cement Mixing columns (DCM), reinforced with steel pipes were selected to provide water tightness, lateral support as well as bottom stability to uplift of the seven pits ("*chambers*") that in turn will serve for the installation of the titanium bars [20, 21].

Other solutions involving monumental/heritage structures are not covered, such as underexcavations under prestigious buildings, as presented by Ovando and Santoyo [22], Jamiolkowski [16] and Burland et al. [23].

2 Ground Reinforcement and Rehabilitation of Foundations Techniques

2.1 Micropiling

Micropiling is a very old technique initially adopted with wood driven piles, which has been developed in the last years mainly due to the bearing capacity improvement (lateral friction at the bond length) related with the use of high pressure grout injection techniques (bigger than 4 MPa), and steel hollow tubes with high resistance and high versatility. The drilling machines are small and versatile, allowing the use of the solution on low head room and tight spaces, excellent features for reinforcement and rehabilitation of existent foundations (see Fig. 2). These improvements have allowed to design micropiles with steel hollow tubes (external diameter lesser than 130 mm) to carry axial service loads greater than 800 kN (Bustamante et al. [24]), and (Bustamante and Doix [25]).

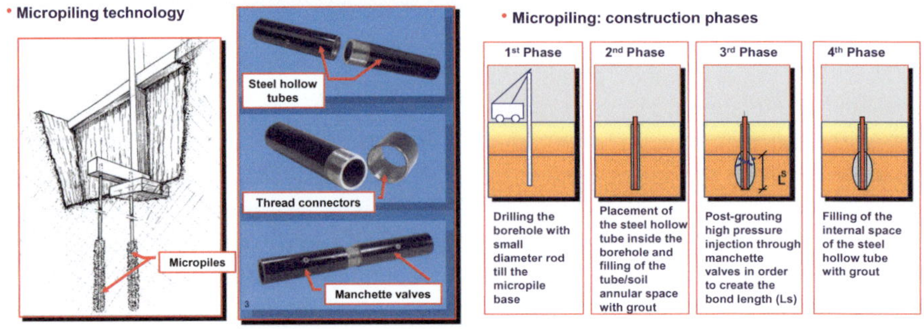

Fig. 2 Micropiling technology and most usual construction phases

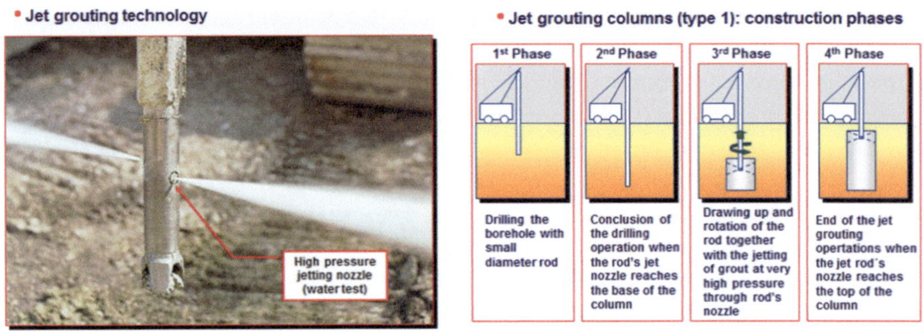

Fig. 3 Jet grouting technology and main construction phases

2.2 *Jet Grouting*

Jet grouting technology has initially been developed in Japan, the UK and Italy. For about 40 years it has been applied worldwide. In Portugal the technology has been applied in the last 20 years, initially on Lisbon Metro extension works to improve alluvial soils. Recently, vertical jet grouting solutions have also become competitive and advisable in several and more usual scenarios, like foundations, earth retaining and reinforcement and rehabilitation of existent foundations systems (Falcão et al. [26]; Greenwood [27]).

According to the definitions of the European Standard on Jet Grouting (CEN/TC 288), jet grouted structures consist of interlocking jet grouted elements. An element is the volume of soil treated through a single borehole, which may be a cylindrical jet grouted column or a planar jet grouted panel (Kutzner [28]).

Jet grouting has nothing to do with common grouting, as according to the jet grouting technique the soil is disintegrated by a jet of air, water or grout at very high pressure (bigger than 30 MPa), obtained through the transformation of the high pressure flow (potential energy) into the high speed jet directed to the soil (kinetic energy) due to the very small diameter nozzles effect, and is subsequently mixed with the grouting material (see Fig. 3).

A part of the mixed material, named spoil, returns to the surface along annular space around the drill rods or along neighboring boreholes, serving for necessary pressure relief. As with micropiles, the drilling machines are small and versatile, allowing the use of the solution on low head room and tight spaces, excellent features for reinforcement and rehabilitation of existent foundations (see Fig. 3).

3 Examples

3.1 Case A—Rehabilitation of the Rivoli Theater Building in Porto

Rivoli Theater opened in 1913 contributing to Porto urban centre modernization. In 1923 the new Rivoli Theatre appeared, remodelled, adapted to cinema, opera, dance, theater and music. The Architect and Engineer Júlio Brito designed it in Art Deco style. In the end of the 20th century the building was rehabilitated. This intervention had the purpose strengthening the stricture and its foundations due to the necessity of an expansion of the theatre, which took place in the late 90s (Fig. 4).

Originally the building structure was predominantly made of stone masonry with some structural elements in concrete. The expansion implied the maintenance of the main building, fully guarantying the integrity of the architecture both externally as in the interior, specially the auditorium, rooms, halls and accessing paths, while executing a basement below the original ground floor and the construction of two high floors on the existing structure (Fig. 5).

A more detailed investigation of the structure conducted during the construction phase allowed an assessment of real state of the construction materials and the degree of conservation. It was concluded that the structure was deteriorated and the

Fig. 4 Rivoli theatre building: external view when the rehabilitation started

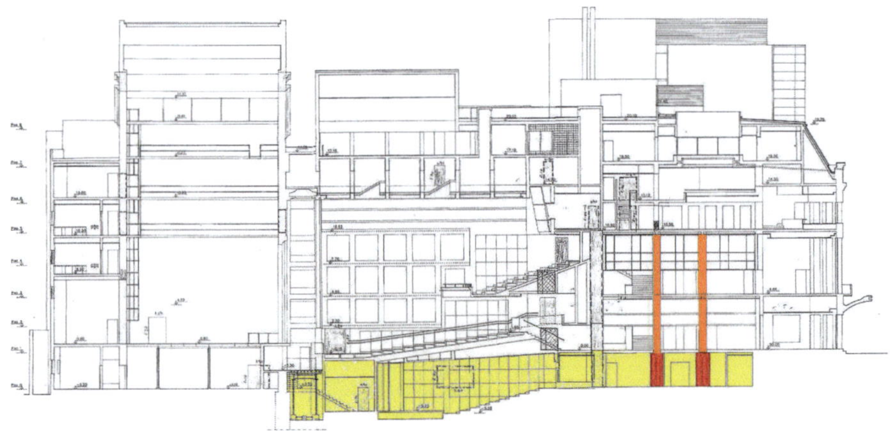

Fig. 5 Rivoli theatre building: a cut of the intervention late 20th century

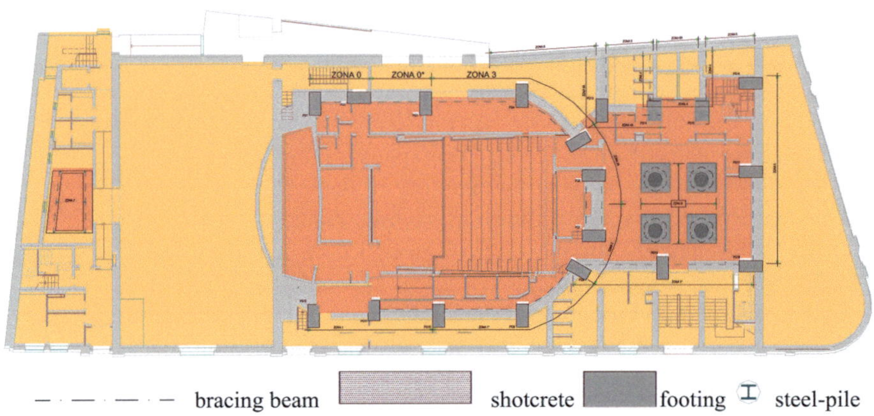

Fig. 6 Rivoli theatre building: foundations reinforcement—plant

resistance of the materials severely altered. The information obtained allowed the estimation of allowable deflection limits, a key factor in this type of interventions, particularly when the resilient structure includes materials such as masonry and stonework, with high susceptibility to bending/traction efforts. This conditioned a lot the adopted solution for bracing the new foundations supporting the sensitive structures (Fig. 6).

The basic criteria that presided over the design aimed at minimizing deformations, specially the distortions, between the initial phase and the final phase of construction. For this purpose it was necessary to reduce the number of successive load transfers in order to ensure that the cumulative displacements were effectively minimized (Fig. 7).

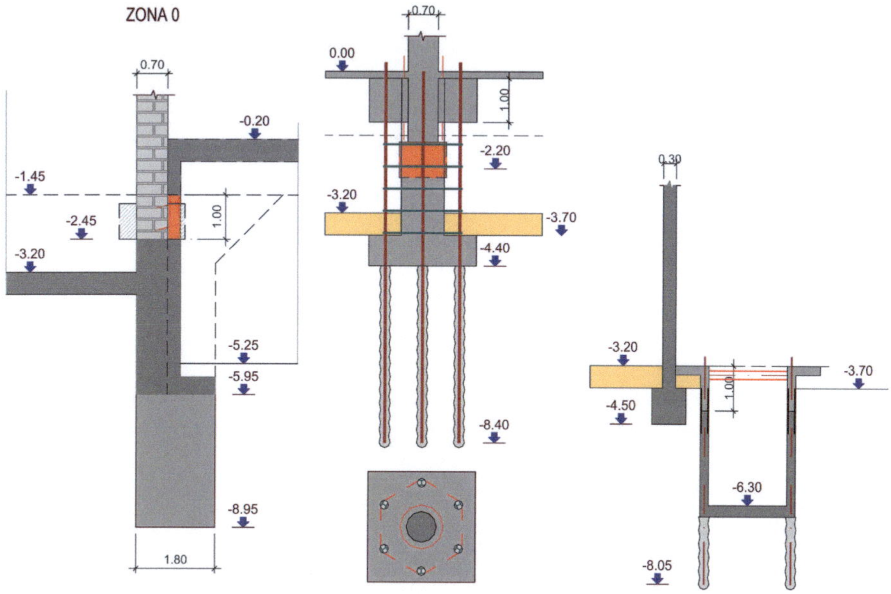

Fig. 7 Rivoli theatre building: details of load transfers to minimize settlements

Fig. 8 Rivoli theatre building: Rivoli theatre building: execution of H steel piles (drilled borehole of 0.20 m diameter)

The inspection revealed that the four support pillars of the foyer (Fig. 8), which supports the building, were made of concrete and were directly founded overlapping stone slabs, forming a sort of pyramid.

In this case, there was the benefit of the favourable ground conditions, since this area is dominated the rock masses of Porto granite. Although decomposed, weathering degree is moderate, being compact at small depths, about 6 meters. Thus, it was possible to use passive solutions for provisional foundation of the pillars (Fig. 8), while in less favourable foundation conditions would be necessary to use an active transfer system, more delicate and time-consuming, involving operations with hydraulic jacks for levelling the supporting beams.

A structural solution of H steel piles was adopted to transfer temporarily the loads to deeper levels aimed at constitute a high rigidity systems in order to achieve the stated objective, which assumed maximum displacement below 5 mm. The following pictures illustrate the process of underpinning for temporary opening space for the basement.

The solution adopted for provisional foundation of the pillars consisted on the execution of a metal frame supported by 6 mini piles arranged around each pillar, with sufficient clearance to prevent the crossing of the slabs that formed its original foundation. The minipiles, consisting of metallic profiles HEB120, were sealed with cement grout in predrilled boreholes with 0.20 m diameter, obtained by rotation from the existing floor, and conducted in depth to go beyond the future basement level. This technology was selected due to the circumstances of this delicate work, demanding a reduction to a minimum the levels of vibration induced in the structure.

However, it is not feasible in the crossing of the existing pillars foundation slabs—which by its thickness and strength require more powerful drilling equipment such as roto-percussion that would imply much higher vibration levels, or with rotational coring—imposing the clearance above described. Taking advantage of the soil (decomposed granite) resistance, the steel minipiles could be direct sealed in that ground. Its lower unit resistance (compared with a multiple and repeated injection, IRS, type) was compensated through the hole of larger diameter, to ensure equivalent resistance capabilities; however, the bearing capacity of the steel minipile is not determinant in this case, since the design is conditioned by the deformation.

The minipiles were headed by a reinforced concrete cap, responsible for transmitting the pillar efforts to the piles with very small deformations. After the concrete curing, it was possible to dig and demobilize safely the original footing, completing the first load transfer step (Figs. 9 and 10).

As the excavation progresses, H steel minipiles integrally stand for the loads of the building. The increase of the compression loads gradually imposes the bracing of the H steel minipiles to reduce its free length, thus reducing risks of instability to-bending-torsion (see Fig. 10).

The design of the H steel profiles had already considered this phenomenon, since HEB120 with 34 cm^2 cross section has an axial stiffness much higher than the tubular sections commonly used in micropiles—which, moreover, can have their resistance lowered to 75%, due to the reduction of the "net" section in the zone of the threaded amendments, thus conditioning its use when subject to bending or traction. The HEB120 profiles were easily integrated into the pillar suspension

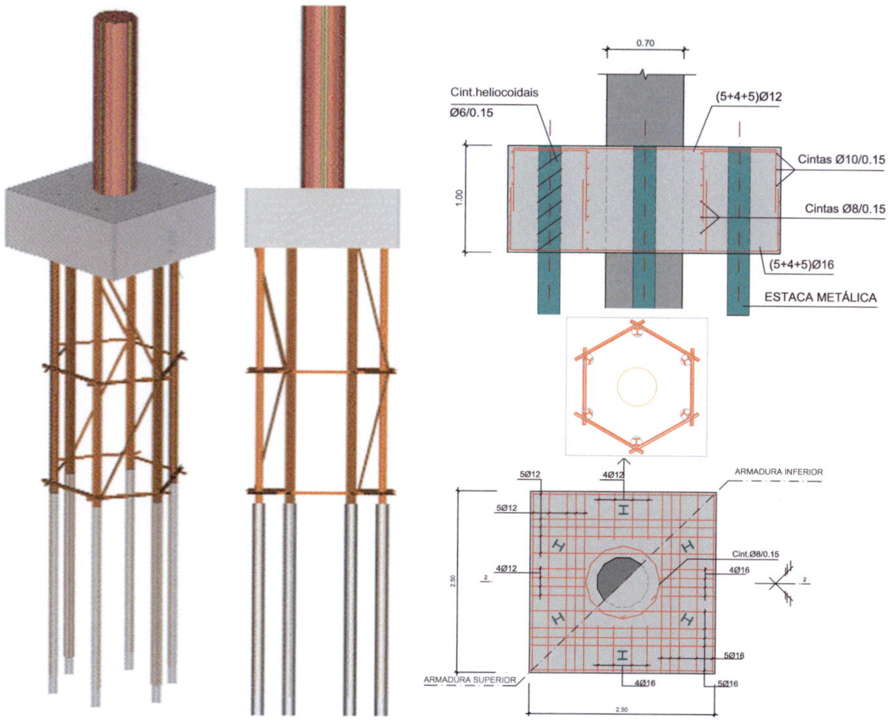

Fig. 9 Rivoli theatre building: suspension of the pillars of the foyer

Fig. 10 Rivoli theatre building: preparation of the pillar to connect the reinforced concrete minipiles' cap and the execution of the structure of the pillar suspension, concreting of the definite pillar with the bracing reinforcement

structure and demobilized after finishing the bottom of the pillar and the definite footing at the end of the load transference. The mini piles were integrated into the final structure, also contributing to the stiffness of the foundation, forming a kind of

Fig. 11 Rivoli theatre building: execution of wells for underpining

Fig. 12 Rivoli theatre building: view of the interior work already built

Pile-Raft structure. In the next figures, some pictures illustrate the final steps of the described process (Figs. 11 and 12).

3.2 Case B—Rehabilitation of the Sotto Mayor Palace in Lisbon

3.2.1 Scenario

Built at the beginning of the 20th century, with French classic style, brick and masonry structure, the Palace has 3 floors, an area of 30 × 30 m^2 and is located in the centre of Lisbon, being surrounded by important streets, like F. P. Melo Avenue (see Fig. 13). In order to face the new project, which demanded a 24 m average

Fig. 13 Palace localization and view from Palmeiras square at the beginning of the 20th century

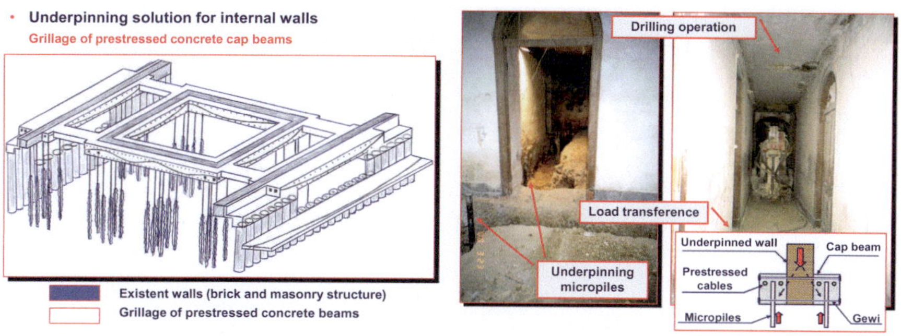

Fig. 14 Grillage of cap beams and internal underpinning works

depth excavation on Lisbon Miocenic soils, around the Palace (see Figs. 13 and 14), as well as the construction of one gallery bellow its original structure, the Palace was underpinned internally with micropiles and externally with contiguous bored piles (see Fig. 14). According to the new project, the Palace would become an hotel and the underground areas would be used mainly for parking and shopping purposes.

3.2.2 Main Restraints

As main restraints, it should be pointed out: the geological conditions, the surrounding conditions, the Palace structure and geometry and the construction schedule (see Figs. 15 and 16).

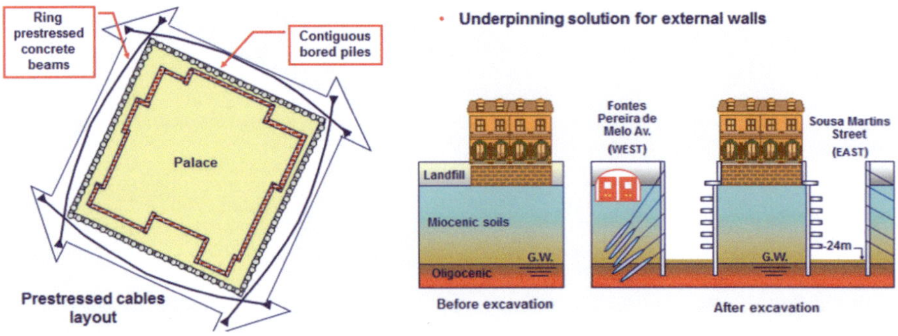

Fig. 15 Underpinning solution for external walls

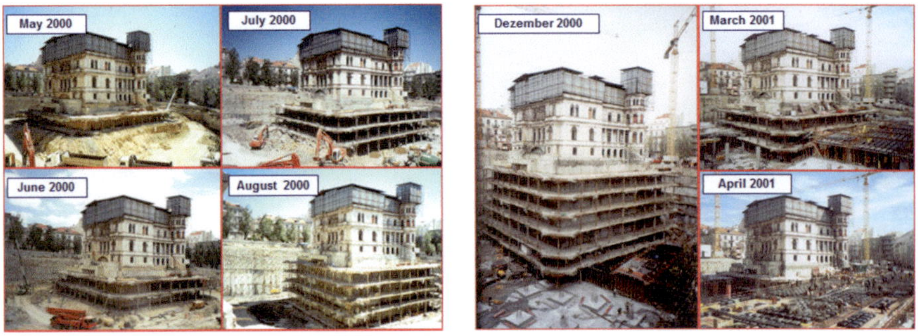

Fig. 16 Sequences of excavation and construction of the underground slabs around the Palace

3.2.3 Main Conception and Execution Criteria

Due to the Palace structure and geometry (low head room), the internal walls original foundations were underpinned with micropiles capped by a grillage of prestressed concrete beams, which were connected to the masonry walls with pairs of prestressed "Gewi" bars (see Fig. 15). The underpinning of external walls was done with a contiguous bored piles (∅0.80 m spaced 1.0 m) wall, connected to the masonry walls trough the piles cap beam. The piles were lined with sprayed concrete. Due to Palace geometry and the existence of the internal micropiles, the piles were braced at 6 levels by external concrete ring beams. These beams were cast against the ground and their levels were defined in order to agree with the final underground slabs. The beams were supported by vertical steel profiles in order to restrain their vertical deformation (see Fig. 16).

3.2.4 Quality Assurance

The Palace performance was analysed through a wide Monitoring and Survey Plan, comprising: topographic marks (reflective targets) and inclinometers (located inside the bored piles). During the excavation and basement works, data was collected once a week.

Figure 17 presents the results of the reflective targets installed at the Palace façades.

Figure 18 gives an overview of the final phases of excavation and construction of underground slabs around the Palace.

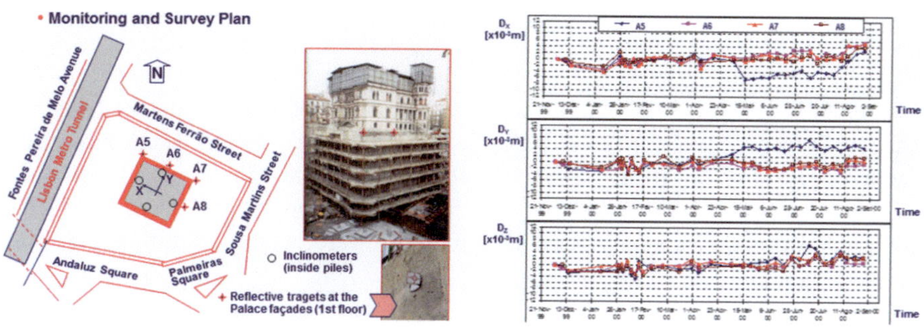

Fig. 17 Palace monitoring and survey plan

Fig. 18 Final phases of excavation and construction of underground slabs around the palace

3.3 Case C—Rehabilitation of the Aveiro Port Authority Building

3.3.1 Scenario

Built at the beginning of the 20th century, the Aveiro Port Authority Building is an important symbol of the "New Art" architecture, due to its important location in the Aveiro main down town navigation channel, as well as for its architectural quality. The building has a brick and masonry structure, with 2 floors, an area of $30 \times 15\ m^2$ and is located in the centre of Aveiro, being surrounded by important streets and buildings. During its exploration life the building had several differential settlements, due to foundations problems and affecting seriously the building performance. Those problems, lead to several reinforcement and rehabilitation of works over the original foundations: masonry arches resting over alluvial soils, but without a complete attainment.

The building rehabilitation allowed the construction of a new reinforced concrete structure, preserving the original main façade, facing the Aveiro main navigation channel, profiting from the existent raft slab capping micropiles, previously building during the demolishing works in order to support the retaining façade temporary structure. Those temporary structures were preserved during the foundations rehabilitation works (see Fig. 19).

3.3.2 Main Restraints

As main restraints, it should be pointed out: the geological conditions, the surrounding conditions, the façade temporary retaining structure, demanding the execution of the foundation rehabilitation works partially indoor and compatible with the existent mat slab and micropiles, as well as with the construction overall schedule. The geological conditions were a key issue, since the building original construction: fills, over alluvial soils, with about 24 m of overall thickness,

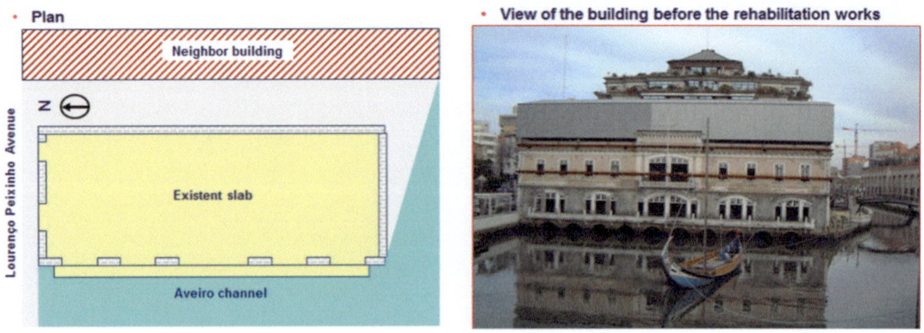

Fig. 19 Plan and view of the building before the rehabilitation works

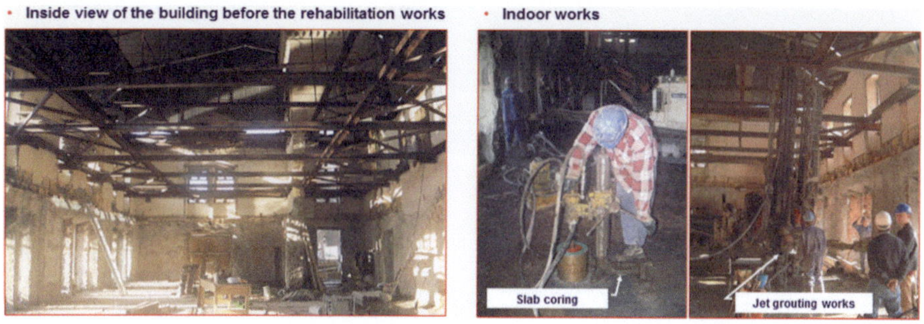

Fig. 20 Inside view of the building before the rehabilitation works and indoor sla

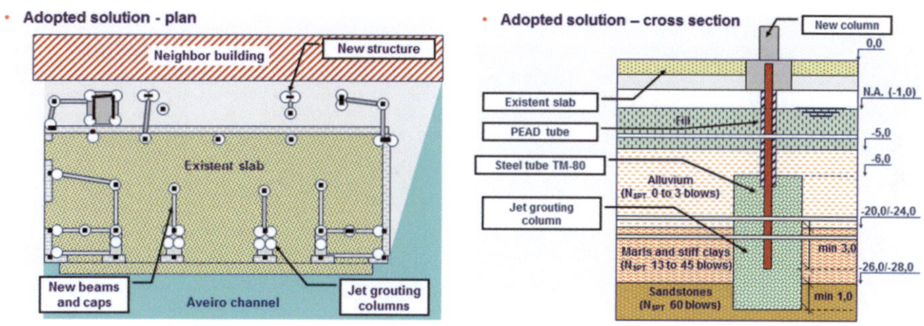

Fig. 21 Adopted foundations rehabilitation solution: plan and cross section

resting over the Aveiro Cretaceous materials: stiff and very stiff clays and sandstones (see Figs. 20 and 21).

3.3.3 Main Conception and Execution Criteria

Due to the new structure loads, the existent foundations as well as the main façade retention structure, lead to partially indoor and low head room works. In this scenario the foundations rehabilitation solutions comprised the execution of jet grouting (type 1) ∅1000 mm columns, reinforced with a steel TM-80 tube, resting at the Cretaceous sandstones. The reinforced columns were designed to resist to axial loads and were capped by the existent raft slab, as well as by new beams and caps. The capping elements were designed to resist to bending and shear loads (see Fig. 21).

The use of jet grouting solutions on a very sensitive urban location demanded the need to manage the indoor and the outdoor spoil (see Fig. 22).

The jet grouting columns diameter was limited to 1000 mm in order to better control the columns execution, as well as the behaviour of the surrounding structures and infrastructures. In order to increase the columns resistance, durability and

Fig. 22 Cleaning of the indoor and outdoor jet grouting spoil

Fig. 23 Topographic control and view of the building at the end of the refurbishment works

ductility, as well as to increase also the effectiveness of the connections to the capping reinforced concrete elements, the columns were reinforced with TM-80 steel tubes. Those tubes were protected against corrosion by a PEAD external tube, filled with cement slurry, at the unconfined top, specially when the fills and alluvium materials and, therefore, the columns head were not in contact with the existent mat slab bottom face (see Fig. 21).

As main advantages of the jet grouting foundations rehabilitation solution, comparing with the micropiles solution, should be pointed out: bigger stiffness and bigger cover of the steel tube, as well as better buckling resistance of the same steel tube. The main difficulty was the spoil management, mainly due to site location.

3.3.4 Quality Control and Quality Assurance

The new building as well as the surrounding buildings performance was analysed through a Monitoring and Survey Plan, comprising mainly topographic marks (reflective targets). During the jet grouting works, data was collected once a week (see Fig. 23) and also all the execution parameters were logged and analysed.

Before the jet grouting works started some trail columns were executed, allowing to take out cores for laboratorial UCS tests and to validate the columns geometry.

3.4 Case D—Requalification of "Bom Sucesso" Market in Porto

The building of "Bom Sucesso" Market was opened in 1952 with an impressive reinforced concrete structure (Fig. 24) designed by the Civil Engineer Joaquim Sarmento (Professor of Structures in the Civil Engineering Department of FEUP in the University of Porto), being in full service until 2010, fulfilling its original commercial function. The building was classified as Monument Public interest by the Ministry of Culture in 2011.

In view of the deterioration in his state in recent years, it imposed a significant intervention to repair a good number of anomalies installed, which extended even to the ground floor, where large settlement were observed (Fig. 24).

Thus, in 2011, the Bom Sucesso Market was the subject of an intervention that was not only limited to ensure the resolution of the identified damages but dared his rehabilitation, adding inside two buildings, a hotel and a set of offices (Fig. 25).

Fig. 24 The building of "Bom Sucesso" market in Porto and settlement of the ground level

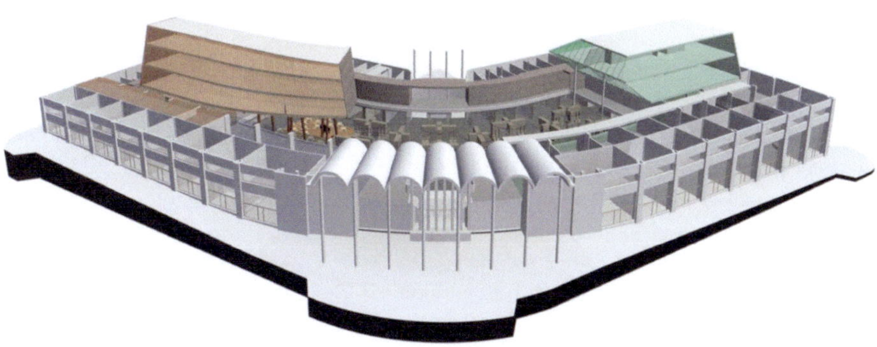

Fig. 25 Three-dimensional model of the building, after its rehabilitation

Fig. 26 Perspective of the hotel and the offices' building

While the hotel was designed independently of the existing structure, the set of offices involved the construction of a building which structure intercepted the original one (Fig. 26).

Although structural anomalies identified before the intervention did not indicted causes associated with the foundations of the building, the interconnection that was promoted between the office's building structure and the original one imposed the need to strengthen some structural elements of the original market, including some foundations, which were now subject to a different loading system when compared to the 1949 loading system.

The soil-structure interaction imposed by the new foundations, implied a good characterization of the involved ground, which geology is dominated by the Porto Granite, a highly heterogeneous profile [29], imposing an initial campaign of geological and geotechnical survey in 2008, supplemented by another in 2011, prior to redevelopment work phase. Simultaneously, documentary information was collected in order to recover the type, size and other characteristics of the foundations of the existing building. The results of the site investigation for ground characterization have identified the presence of two geotechnical units, with low strength and high deformability consisting of loose to very loose sandy soils, which extend to depths of 9 m order.

The general solution presented in the original project foundations plant, consisted on shallow foundations, apparently inconsistent with the findings the site investigation conducted for the rehabilitation program. This imposed an additional effort for the document collection phase, resulting in the discovery of another plant foundations, dated of 1951, where general workaround pile foundation (Fig. 27).

The doubts installed about the real adopted solutions conducted to the implementation of a series of polls to existing foundations. These studies have uncovered shoes of reinforced concrete resting on the ground. Unusual stone masonry consisting of successive layers oriented alternately in perpendicular directions, evolving into a horizon of granitic residual soil (Fig. 28). These surveys allowed to identify the presence of piles underlying reinforced concrete caps or, sometimes, in alternative, associated to stone masonry caps, similar to the ones referred (Fig. 28).

The structural design, made by the SOPSEC SA defined reinforcement of the existing foundations, involving the increase of the area in plan of the shallow

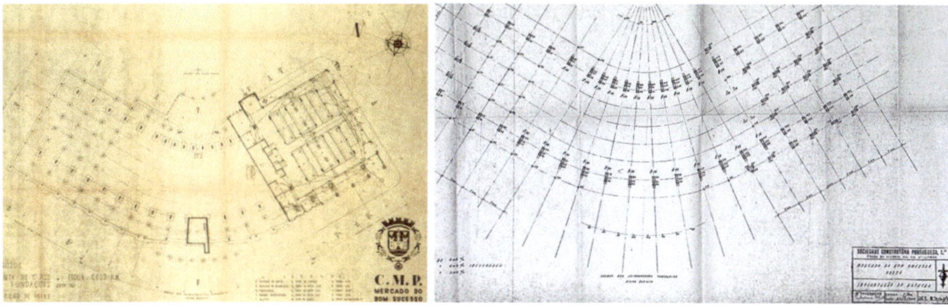

Fig. 27 Plant of the shallow and the pile foundations

Fig. 28 Footing laying over a stone masonry, capping the original piles

foundation and ensuring the link between the new concrete and the existing concrete and stone masonry elements through the installation of steel rods, sealed with epoxy resin. The pile foundation were reinforced by micropiles, headed by a massive reinforced concrete, connected to existing concrete using steel rods (Fig. 29).

For the foundations of new structures, although it had initially been conceived a generalized solution of pile foundation, due to a limited meagre right foot available micropiles were executed due to the high versatility of this technique for underpinning (Fig. 30).

In Fig. 31, existing foundations (in orange) are distinguished from the reinforced foundations (in green) and the new foundations (in blue).

The building, already requalified (Fig. 32), opened in April 2013 and was, in 2015, one of four European winners of the Global Awards for Excellence, sponsored by the Urban Land Institute, New York.

Fig. 29 Reinforcement of shallow foundations

Fig. 30 Foundation by micropiling

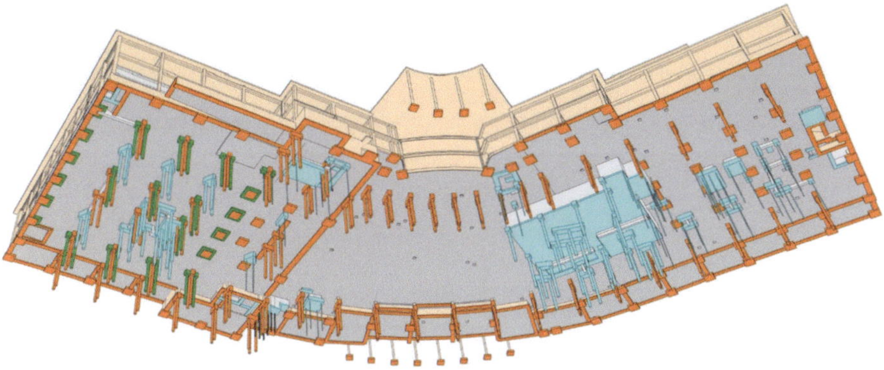

Fig. 31 Three-dimensional model of the building foundation with different solutions

Fig. 32 View of the interior of the building after requalification

4 Final Remarks

The demand for reinforcement and rehabilitation of foundations systems has increased steadily in the last years as renewals and refurbishment works have gained popularity. As example, the presented cases proved how the versatility of some underpinning techniques can fit the uniqueness and restraints of complex scenarios, involving old and historic sensitive buildings, sometimes founded on weak soils. On the figures below two comparative analyses between micropiling and jet grouting technologies (see Fig. 33) and between these techniques and the conventional ones are presented (see Fig. 34).

In this context, it is also important to point out that underpinning works requires expertise at the design and execution levels, along with safe working practices, especially when the underpinned building has an old structure and a special

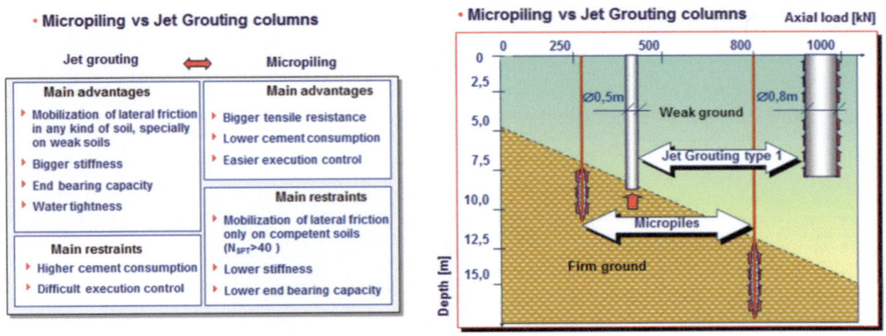

Fig. 33 Comparative analysis between micropiling and jet grouting underpinning techniques

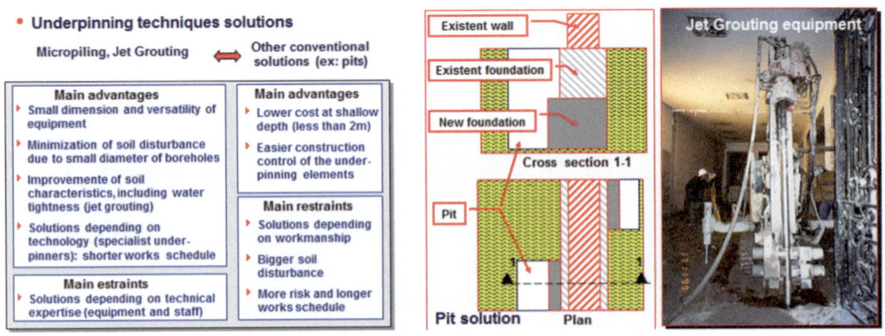

Fig. 34 Comparative analysis between micropiling/jet grouting and conventional techniques

architectural/historical interest and therefore is protected from full demolition or alteration. In these situations, considerable care is required on previous tasks, as for example: monitoring and survey, geological and geotechnical site investigation, stiffen, grout, shore strut, in order to prepare these old buildings and their original foundations for the underpinning works. As example, some of the presented cases proved how important is the role of the Monitoring and Survey Plan in this kind of works, mainly as a risk management tool, allowing to survey and predict the performance of the underpinned structures and, if necessary, to adjust in time the initial solution.

4.1 Quality Control and Quality Assurance

In order to assure a good survey of the consequences of the excavation, new foundations, retention walls and earthworks, a plan of instrumentation and monitoring has to be implemented. This plan will include a clear identification of the equipment to be installed and the periodicity of the registers and the subsequent interpretation, which has to be transmitted to the designers or their representing personnel in order to react accordingly. This instrumentation will include topographic targets and piezometers. The criteria for definition of "alert" and/or "alarm" situations have to be included in the execution projects similar to those defined below (adopted in the requalification of the "Bom Sucesso" Market, in Porto) and the corrective measures have to be well stated.

Values of displacement of 10 or 15 mm, are typically considered as reference for "alert" or "alarm" signs, respectively, taking into account the depth of the excavation, the foundations levels, their dimensions, and the distance to the adjacent structures. These values will be adjusted according to the survey of anomalies, depending on their type and severity.

The measures taken in the case of signs in the instrumentation being beyond the appointed limits, and therefore menacing the functionality of adjacent infrastructures and superstructures, will impose an increase of: (i) the frequency of instrumentation readings; (ii) the number of monitoring instruments; and, (iii) the installation of temporary stabilization elements. The adopted monitoring sequence of the topographic instruments can be the following: (i) on the installation, 2 initial readings; (ii) up to 30 days before starting of excavation work, 1 reading per week; (iii) during excavation, on the instruments located within a zone distanced by 30 m, the readings should have a periodicity of 4 readings/week; (iv) after the end of the excavation and until the completion of the buried structures, 2 readings per week; while, (v) until the completion of the work, 2 readings per month should be guaranteed. In case of "alarm" limit, the frequency of readings may be required to be daily. Changing of this frequency readings should be proposed to the building inspector and approved by the project team.

Acknowledgements Part of this work reports to research financially supported by Project POCI-01-0145-FEDER-007457—CONSTRUCT—Institute of R&D in Structures and Construction, funded by FEDER funds through COMPETE2020—Programa Operacional Competitividade e Internacionalização (POCI) and by national funds through Fundação para a Ciência e a Tecnologia (FCT). The authors wish to thank the contributions from the responsibles for the rehabilitionof the "Rivoli" Theatre (for the project and execution, Baldomiro Xavier and José Figueira, Teixeira Duarte, S.A., and Luís Neto, for the owner, Porto 2001 European Capital of Culture) and the rehabilitation of "Bom Sucesso" Market (for the project and execution, Hipólito Sousa and and José Filinto Trigo, Sopsec, S.A., and the owner Mercado Urbano—Gestão Imobiliária, S.A.).

References

1. Butcher AP, Powel JJ, Skinner HD. Reuse of foundations for urban sites. A best practice handbook. Bracknell: IHS BRE Press; 2006. ISBN: 978-1-86081-938-4.
2. Chapman T, Anderson S, Windle J. Reuse of foundations. London: CIRIA C653; 2007. ISBN: 978-086017-653-4.

3. Lunne T, Robertson P, Powell J. Cone penetration testing in geotechnical practice. London: Blackie Academic. EF Spon/Routledge Publishers; 1997.
4. Mayne PW, Christopher, DeJong J. Manual on subsurface investigations. Publication No. FHWA NHI-01-031. Washington, DC: National Highway Institute, Federal Highway Administration; 2001.
5. Schnaid F. In-situ testing in geomechanics: the main test. London: CRC Press, Taylor & Francis Group; 2008. ISBN: 9780415433860.
6. Foti S., Lai CG, Rix GJ, Strobbia C. Surface wave methods for near-surface site characterization. London: CRC Press, Taylor & Francis Group; 2014. ISBN: 9780415678766.
7. Monnet J. In situ tests in geotechnical engineering. Wiley-ISTE; 2015. ISBN: 978-1-84821-849-9.
8. Robertson PK, Cabal KL. Cone penetration testing for geotechnical engineering. Gregg Drilling & Testing, Inc.; 2015. http://www.cpt-robertson.com/doc/view?docid=xnhqTpmrnRdPTvYHHRSr6hcNdKJLWy.
9. Lacasse S, Nadim F, Rahi A, Guttormsen TR. Statistical description of characteristic soil properties. In: OTC-19117, Houston. 2007.
10. EN1997-1. Eurocode 7: geotechnical design—general rules. 2004.
11. Viana da Fonseca A, Buttling S, Coutinho RQ. Foundations: shallow and deep foundations, unsaturated conditions, heave and collapse, monitoring and proof testing. In: BBK Huat, DG Toll, A Prasad, editors. A handbook of tropical residual soil engineering. A.A. Balkema Book, CRC Press, Taylor & Francis Group; 2012. p. 283–412. ISBN: 978-0-415-45731-6.
12. Militinksy J, Schnaid F, Consoli N. Pathologies of foundations. São Paulo: Oficina de Textos; 2009. ISBN: 139788579751820 **(in Portuguese)**.
13. Gambin M, Frank R. Direct design rules for piles using Ménard pressuremeter test. Contemporary topics in in situ testing, analysis, and reliability of foundations. In: Iskander M, DF Laefer, MH Hussein, editors. Proceedings of the international foundation congress and equipment Expo'09 (IFCEE'09), 15–19 March, Orlando, Florida, vol 186. Reston: ASCE Geotechnical Special Publication; 2009. p. 111–8.
14. Burlon S, Frank R, Baguelin M, Habert J, Legrand S. Model factor for the bearing capacity of piles from pressuremeter test results—Eurocode 7 approach. Géotechnique. 2014;64(7): 513–25. doi:10.1680/geot.13.P.061.
15. Bullivant RA, Bradbury HW. Underpinning—a practical guide. Oxford, Blackwell Science; 1996. ISBN: 0632040041/9780632404049.
16. Jamiolkowski MB. The leaning tower of Pisa: end of an Odyssey, Terzaghi Oration. In: Proceedings of the XVICSMGE Istanbul, vol. 4. 2001. p. 2979–96.
17. Marchi M, Butterfield R, Gottardi G, Lancellotta R. Stability and strength analysis of leaning towers. Géotechnique. 2011;61(12):1069–79. doi:10.1680/geot.9.P.054.
18. Burghignoli A, Jamiolkowski M, Viggianni C. Geotechnics for the preservation of historic cities and monuments: components of a multidisciplinary approach. In: Proceedings of the 2nd international symposium geotechnical engineering for the preservation of monuments and historic sites, Naples. CRC, Taylor & Francis; 2013. p. 751–60.
19. Macchi G, Macchi S, Jamiolkowski M, Pastore V, Vanni D. Strengthening of the San Marco bell tower foundation in Venice. Geotechn Herit. 2013. doi:10.1201/b14965-12.
20. Vanni D, Macchi S, Pastore V. Deep soil mixing to help the restoration of Venice San Marco Bell Tower. In: 4th International conference on grouting and deep mixing. New Orleans: Deep Foundation Institute; 2012.
21. Vanni D, Siepi M, Macchi S, Pastore V. Deep soil mixing to help the restoration of Venice San Marco Bell Tower. Grouting Deep Mix. 2012. doi:10.1061/9780784412350.0060.
22. Ovando-Shelley E, Santoyo E. Underexcavation for levelling buildings in Mexico City: case of the Metropolitan Cathedral and the Sagrario Church. J Archit Eng. 2001;7:61–70.
23. Burland JB, Jamiolkowski M, Viggiani C. The stabilisation of the leaning tower of Pisa. Soils Found. 2003;43(5):63–80.

24. Bustamante M, Gianeselli L, Leoloux JL. Underpinning of the Bordeaux Pont-de-Pierre: a First Empire masonry. In: Viggiani C, editors. Geotechnical engineering for the preservation of monuments and historic sites. Rotterdam: Balkema; 1997. p. 551–9.
25. Bustamante M, Doix B. Une méthode pour le calcul des tirants et des micropieux injectés. Bull Liasion Lab Ponts Chass. 1985;140:75–92 (in French).
26. Falcão J, Pinto A, Pinto F. Case histories and work performance of vertical jet grouting solutions. In: Proceedings of the 4th international conference on ground improvement, geosystems, Helsinki, June 7–9. Building Information Ltd.; 2001. p. 165–71.
27. Greenwood D. Underpinning by grouting. Ground Eng. 1987;4:21–30.
28. Kutzner C. Grouting of rock and soil. Rotterdam: Balkema; 1996.
29. Viana da Fonseca A. Characterising and deriving engineering properties of a saprolitic soil from granite, from Porto. In: Tan TS, Phoon KK, Hight D, Leroueil S, editors. Characterisation and engineering properties of natural soils, vol. 2. Lisse: Swets & Zeitlinger; 2003. p. 1341–78. ISBN: 90 5809 537 1.

Code-Based Procedures for Seismic Safety Assessment and Retrofit

X. Romão and A. Penna

1 Introduction

Earthquake engineering experts, public authorities and general public agree on the idea that the seismic safety and performance of the built environment is a matter of high priority. Moreover, the widespread interest in methodologies which address the assessment and the retrofit of existing constructions reflects the global perception that such constructions are exposed to disproportionate levels of seismic risk. Rational and cost effective interventions on the built environment are therefore needed in order to mitigate such risk and reduce the expected level of losses in future earthquakes. Since there are significant differences between the design of a new structure and the assessment of the same structure after many years in service, these interventions must be based on adequate normative documents addressing the specific issues of seismic performance assessment of existing structures. Given that most structural standards and codes have been developed for the design of new structures, their procedures are often found to be inadequate for the assessment of existing constructions. Hence, over the past few years, several standards and

X. Romão (✉)
CONSTRUCT-LESE, Civil Engineering Department,
Faculty of Engineering, University of Porto,
4200-465 Porto, Portugal
e-mail: xnr@fe.up.pt

A. Penna
Department of Civil Engineering and Architecture,
University of Pavia, 27100 Pavia, Italy
e-mail: andrea.penna@unipv.it

A. Penna
European Centre for Training and Research in Earthquake Engineering,
27100 Pavia, Italy

A. Costa et al. (eds.), *Strengthening and Retrofitting of Existing Structures*,
Building Pathology and Rehabilitation 9, https://doi.org/10.1007/978-981-10-5858-5_13

guidelines addressing the problem of structural assessment and upgrading have been emerging for the specific case of earthquake loading.

In this context, the current chapter presents a review of existing international structural standards and codes that provide specific methodologies for the seismic safety assessment and the strengthening of existing constructions. The codes considered in this analysis are the Eurocode 8 Part 3 (EC8-3) [1], the United States International Building Code IBC 2012 [2], the International Existing Building Code IEBC 2012 [3], ASCE/SEI 41-13 [4] and relevant modifications in the upcoming 2015 editions of the IBC and IEBC, the Italian building code NTC08 [5], the Greek code of structural interventions KAN.EPE. [6], the Swiss standard SIA 269 [7], the Romanian code for seismic evaluation P100-3 [8], the Turkish standard for buildings in seismic zones [9], and the New Zealand NZSEE Recommendations [10]. The chapter presents a general description of the seismic safety assessment methods proposed by these standards, as well as an overview of the techniques they propose for seismic retrofit and strengthening. Following this comparison, the future evolution of code-based procedures for seismic safety assessment and retrofit is briefly discussed based on recently developed guidelines proposing more advanced approaches.

2 Integration of Seismic Assessment into Regulatory Frameworks

According to most standards, the seismic safety evaluation procedures are integrated into broader regulatory frameworks addressing the general safety and performance of existing buildings. The conditions under which the seismic assessment of a building is carried out (which may then trigger the need to retrofit or strengthen the building) usually falls into one of the following general categories:

- Seismic risk mitigation programmes under which building owners are required to carry out the seismic assessment of buildings and to retrofit/strengthen them, if needed.
- Interventions related to the use of the building, such as damage repair operations, alterations or additions to the building structure, or a change in the building use.
- Owner-requested seismic evaluation, and potential strengthening, of the building in order to meet the provisions of current standards or to upgrade its earthquake performance.

Most of the codes and standards considered herein either enforce or consider seismic evaluation scenarios which fall into the second category. On the other hand, explicit reference to the scenarios of the first or third categories is less frequent. Of the selected standards, only EC8-3 [1] and the NZSEE Recommendations [10] foresee that seismic evaluation may be enforced by a scenario of the first category.

Although not directly acknowledged by the Italian standard NTC08 [5], a scenario of the first category may be enforced in Italy in exceptional situations, as occurred after the 2002 M5.7 Molise earthquake. In this case, the scenario was set by the Premier Ordinance OPCM 3274 [11]. With respect to the third category, P100-3 [8], NTC08 [5], KAN.EPE. [6], IBC 2012 [2], IEBC 2012 [3] and ASCE/SEI 41-13 [4] explicitly acknowledge the possibility of this scenario.

3 General Seismic Assessment and Retrofit Principles Followed by the Selected Standards

A brief overview of the main principles governing the seismic safety assessment and retrofit procedures that are proposed by each of the selected international standards and codes is presented in the following.

3.1 Eurocode 8 Part 3 (2005)

The Eurocode project was developed in order to establish a set of rules covering the various aspects of structural design. These rules would then be common to the several countries which are part of the European Committee for Standardization (CEN) in order to provide more uniform safety levels. The Eurocodes include ten standards for structural design [12] and each Eurocode is divided in a number of parts covering particular technical aspects. Since different countries may have different requirements for constructions and their safety, the Eurocodes also allow each country to define the values of several safety-related parameters, as well as certain country-specific data. Such data is collectively known as the Nationally Determined Parameters and it is published in National Annexes complementing each Eurocode part.

For the case of earthquake loading, Eurocode 8 (EC8) establishes the criteria and rules for the seismic design of new structures and the assessment/retrofit of existing ones. For the latter, Part 3 of Eurocode 8 [1], EC8-3, provides criteria to evaluate the seismic performance of existing buildings and to design the necessary corrective measures. The scope of the EC8-3 procedures addresses buildings with RC, steel and composite, and masonry structures. In 2010, EC8-3 was updated to accommodate recent research results [13].

The methodology proposed by EC8-3 for the seismic assessment of existing buildings enables the analyst to consider a fully performance-based approach to determine the level of safety of a structure. Most of the procedures available in EC8-3 enforce a displacement- or deformation-based assessment approach while restricting the use of classical strength-based methods to a few situations only. In global terms, the EC8-3 assessment methodology involves a sequence of four stages of decision and analysis. The first stage corresponds to the selection of the limit states that will be considered for the structural performance assessment. The second stage addresses the level of knowledge available for the structure under assessment. The third stage involves selecting the structural analysis method that will be used to perform the assessment. The fourth and final stage of analysis corresponds to the safety verification stage where the conformity of each structural mechanism is checked involving procedures which depend on the nature of the mechanisms (i.e. ductile or brittle). A literal interpretation of the EC8-3 procedures indicates that a building is considered to be conforming to a given limit state only when the seismic demand of all the individual structural members does not exceed the corresponding capacity.

After defining the seismic safety assessment procedures, EC8-3 provides guidance and rules for the design of seismic strengthening and retrofit solutions. For the case of RC structures, EC8-3 covers member-level strengthening techniques to increase the flexural and/or shear strength, to increase the deformation capacity and ductility, and to improve the strength of deficient lap-splices. EC8-3 defines rules to achieve these enhancements involving concrete or steel jacketing and fibre-reinforced polymer (FRP) plating and wrapping, the latter being addressed more extensively. For the case of steel and composite structures, EC8-3 refers the possibility of using system-level retrofitting strategies to increase the capacity of the lateral force resisting systems and horizontal diaphragms and/or to decrease the seismic demand. However, rules to design strengthening solutions are only provided for member-level approaches which involve replacing damaged elements or techniques such as steel plating or RC encasing to enhance the stiffness, strength and ductility of the structural elements. Retrofitting beam-to-column connections is also addressed and rules are provided for weld replacement, weakening strategies and strengthening approaches. For the case of masonry structures, EC8-3 provides some guidance regarding the repair of cracks, the repair and strengthening of wall intersections, the strengthening and stiffening of horizontal diaphragms and the strengthening of walls. The suggested techniques depend on the retrofit purpose and can involve the addition of RC elements, steel elements or steel ties, steel plating, and RC, steel or FRP jacketing, among others.

Finally, it is noted that a 6-year work programme has recently been launched to develop the next generation of Eurocodes [12]. The new standards are expected to be published in 2020 and will address new technologies and market needs by extending the scope of the existing Eurocodes.

3.2 *United States Standards IBC 2012 (2012), IEBC 2012 (2012) and ASCE/SEI 41-13 (2014)*

The International Code Council (ICC) is a United States non-governmental organization dedicated to developing comprehensive and coordinated national construction standards. Among the several standards developed by ICC, reference is made to the 2012 editions of the International Building Code (IBC) and of the International Existing Building Code (IEBC), which are presently relevant for the analysis of existing buildings. Since the 2015 versions of these two standards are currently being finalised, the most relevant changes they will incorporate regarding earthquake safety are also referred herein.

The scope of the IBC 2012 [2] provisions is predominantly related to new buildings. Nonetheless, Chapter 34 contains general design requirements for existing structures, namely regarding aspects related to the alteration, repair, addition, or change of occupancy of existing structures, including historic buildings. The procedures provided in Chapter 34 reflect the existence of two different approaches: a prescriptive method and a performance-based method. However, for the particular case of seismic evaluation and rehabilitation, Chapter 34 can be seen to provide little practical guidance. In the new IBC 2015, Chapter 34 has been deleted and a reference to the upcoming IEBC 2015 replaces the chapter. All provisions previously available in Chapter 34 may be found in the IEBC 2015.

Given that existing older buildings, especially historic buildings, may have unique conditions that make it difficult to fully comply with standards devised for new buildings, ICC also developed the IEBC. The fundamental goal of IEBC is to provide alternative approaches to the remodelling, repair or alteration of existing buildings. Hence, IEBC 2012 [3] establishes three possible approaches for the rehabilitation of existing buildings: a prescriptive compliance method, a work area compliance method, and a performance compliance method. In the prescriptive compliance and the performance compliance methods, the procedures provided by IEBC 2012 are essentially those contained in Chapter 34 of IBC 2012 (which are also part of the IEBC 2015). In the work area compliance method, the IEBC 2012 provisions are defined according to the type of intervention. The standard establishes different compliance conditions for repairs, alterations (considering three possible degrees of alteration), change of occupancy, additions, building relocation and historic buildings. With respect to aspects related to the seismic evaluation and rehabilitation, IEBC 2012 establishes analysis scenarios which can be based on the procedures defined by IBC 2012, ASCE/SEI 31-03 [14] or ASCE/SEI 41-06 [15]. Since the ASCE/SEI 31-03 and ASCE/SEI 41-06 standards have been recently merged into the ASCE/SEI 41-13 standard [4], the 2015 edition of the IEBC has been harmonized with ASCE/SEI 41-13. Given these changes, only scenarios related to the seismic safety evaluation and retrofit based on the ASCE/SEI 41-13 procedures are addressed hereafter.

The ASCE/SEI 41-13 standard, entitled Seismic Evaluation and Retrofit of Existing Buildings, merges the procedures previously available in ASCE/SEI 31-03 and ASCE/SEI 41-06 and provides additional consistency between the evaluation and retrofitting procedures. In order to help with the comprehension of the detailed procedures it proposes, the standard also includes commentaries explaining some of the procedures in more detail. ASCE/SEI 41-13 proposes an evaluation methodology that is divided into three tiers of progressively more complex calculation and investigation. A building can be assessed according to only one level of evaluation or according to a combination of levels. Tier 1 is a fast screening level involving sets of checklists that evaluate structural, non-structural, and foundation/geologic hazard elements of the building and site conditions. To perform a Tier 1 analysis, the level of seismic hazard and the building performance level must first be selected. The Tier 1 screening process also involves several quick check analyses to obtain the stiffness and strength of certain building components and to determine if the building complies with certain evaluation criteria. Elements found to be nonconforming can then be evaluated according to the Tier 2 procedure or a seismic upgrade of those components can be defined instead.

The Tier 2 procedure is also a deficiency-based evaluation process but requires additional analysis and more complex evaluation procedures to analyse the potential deficiencies identified in Tier 1. Although Tier 2 is still a deficiency-based evaluation, a full-building analysis may be required in some cases. The evaluation procedure of Tier 2 also requires selecting the level of seismic hazard and the building performance level (usually the same levels selected for Tier 1) prior to carrying out the assessment. If a full-building analysis is required, only linear analysis methods are allowed for structural analysis and safety verifications are only required to be carried out for the potential deficiencies identified in Tier 1. Similar to Tier 1, members found to be nonconforming after Tier 2 can be evaluated according to the Tier 3 procedure or a seismic upgrade of those components can be defined instead.

Tier 3 is a calculation-intensive evaluation procedure which involves a full-building analysis. The general principles underlying a seismic evaluation that follows the Tier 3 procedures are similar to those of a detailed assessment methodology such as that of EC8-3. However, in addition to the direct effects of earthquakes, the standard also addresses the effects of local site geological hazards (e.g. liquefaction). The Tier 3 evaluation methodology involves a sequence of four stages. The first stage corresponds to the definition of the seismic hazard and the selection of the building performance level that will be considered for the assessment. The second stage addresses the information available to characterize the structure under assessment. The standard provides instructions to obtain as-built information, along with adequate default values if specific information is not available, and also includes detailed material testing requirements. The third stage involves selecting the structural analysis method that will be used to perform the assessment. The fourth stage corresponds to the safety verification and includes three types of acceptance criteria: (1) criteria to check the conformity of each structural member, using different procedures for deformation-controlled and force-controlled members; (2) criteria to check the safety of acceleration-sensitive and deformation-sensitive non-structural elements; (3) criteria to check the global performance of the building according to acceptable limits of interstorey drift demand.

With respect to retrofitting, ASCE/SEI 41-13 also provides some guidance for the design of seismic strengthening solutions for steel, concrete, masonry and timber constructions, including their foundations. Generally, ASCE/SEI 41-13 refers the possibility of designing system-level and member-level solutions for seismic retrofit. In terms of system-level approaches, ASCE/SEI 41-13 refers the possibility of increasing stiffness and strength by adding new elements (e.g. walls or bracing systems), increasing damping using supplemental damping devices, using seismic isolation devices, or decreasing mass. With respect to member-level approaches, ASCE/SEI 41-13 refers the local modification of components (beams, columns and joints) by weakening them or strengthening them using, for example, RC, steel or FRP jacketing. For the case of foundations, local solutions may involve modifying the existing element or adding additional elements (e.g. piles, tension tie-downs, grade beams). Design details and rules for these solutions are not included and ASCE/SEI 41-13 refers other documents and standards for that purpose.

3.3 *Italian Building Code NTC08 (2008)*

The impacts of the 2002 M5.7 Molise earthquake, namely the collapse of a school building that caused the death of 27 children, led to the development of an "emergency" code and a revised seismic hazard map covering the entire national territory [11]. This new standard was a simplified version of EC8 that was supposed to be applied together with the applicable standard at that time (that had been last revised in 1996) during a transition period of 18 months. However, this transition period was extended a number of times due to several reasons. During this time, while a group of researchers and professionals was working on the revision of OPCM 3274 and preparing the amendment OPCM 3431 [16], the Ministry of Infrastructures was preparing a unique and more consistent building standard that was issued in 2008. This new building code, the NTC08 [5] and its Commentary [17], incorporated the amended OPCM 3274 along with further corrections (e.g. mechanical property ranges for existing masonry typologies).

NTC08 [5] and its Commentary [17] include a chapter (Chapter 8) focusing on the assessment and the possible retrofitting/strengthening of existing structures, mostly referring to buildings and bridges. According to the standard, a structural safety assessment is mandatory when:

- The capacity of the structure has been reduced due to external environmental loads, to material and mechanical degradation effects, to accidental events (e.g. a fire or an explosion), or due to damages resulting from foundation settlements;
- A design error is found;
- There is a change in the loading conditions larger than a certain percentage;
- An intervention on non-structural elements that modifies the overall stiffness or mass distribution of the construction is carried out;
- An intervention on the structural elements that modifies the original behaviour of the construction is carried out.

Chapter "Structural Repair and Strengthening of RC Elements with Concrete Jacketing" only provides a general and conceptual overview of the assessment and strengthening of existing structures. More detailed descriptions about the methods and procedures for the assessment and strengthening can be found in the Commentary [17].

Section 8.7 of Chapter "Structural Repair and Strengthening of RC Elements with Concrete Jacketing" deals with seismic loading and contains requirements for different types of constructions, namely RC, steel and composite, and masonry structures. The NTC08 approach to assess the seismic safety of existing structures is similar to that of EC8-3 in terms of decision and analysis stages. However, NTC08 does not consider the first stage of the EC8-3 procedure since the assessment is only required to be carried out for the ultimate limit state. Still, the analyst and the owner can decide to assess the safety also for other limit states. Therefore, the first stage addresses the level of knowledge available for the structure under assessment. The second stage involves selecting the structural analysis method that will be used to perform the assessment. The third and final stage of analysis corresponds to the safety verification stage where the conformity of each structural mechanism is checked involving procedures which depend on the nature of the mechanisms. As for EC8-3, a construction is considered to be conforming when the seismic demand of all the individual structural members does not exceed the corresponding capacity. It is noted, however, that the conformity of an existing structure for seismic loading is not mandatory unless the construction undergoes a change in use or its structure underwent significant changes over time.

With respect to interventions, NTC08 [5] and its Commentary [17] refer that it is possible to define three different types of interventions: a retrofit to provide the existing structure with a capacity equal to that of a new one; a strengthening to increase the overall capacity of the existing structure up to a desired level; a local intervention which increases the capacity of certain elements of the existing structure up to a desired level. For the case of RC and masonry buildings, the Commentary [17] provides details on several system-level and member-level intervention approaches. For the case of masonry structures, these approaches include measures to increase the connection between the various elements of the construction (e.g. wall-to-floor, wall-to-wall connections or between the several leaves of a wall), to increase the stiffness of floors, to correct the in-plan distribution of vertical elements and to increase the strength of walls. Furthermore, specific measures for arches, vaults, pillars and columns, roof structures and foundations are also addressed. For RC structures, procedures are provided for the steel or RC jacketing of columns and walls to enhance stiffness, strength and ductility. The possibility of using FRP plating and wrapping is also suggested for beams, columns and walls to increase strength and ductility but reference is made to the design procedures in the technical guidelines Istruzioni CNR-DT 200/2004 [18].

Finally, it should be referred that, at the time of the preparation of this chapter, a new release of the Italian code is being prepared.

3.4 Greek Code of Interventions KAN.EPE. (2013)

Greece has developed a code for the seismic safety assessment of existing buildings and the design of required interventions KAN.EPE. [6]. The standard establishes the general principles and criteria for the seismic evaluation and for the development of interventions, and provides specific rules of application for RC buildings. An additional part of the standard containing application rules for unreinforced masonry buildings is also under development [19]. The standard has been developed in order to be harmonized with the relevant Eurocodes, in particular Eurocode 8, but contains procedures which are more detailed than those proposed by EC8-3. In order to help with the comprehension of the procedures, the standard also includes commentaries and remarks referring to issues of special interest or to specific applicability aspects of some of the procedures.

The seismic assessment methodology proposed by the Greek standard involves four stages of analysis that share the same general principles of those proposed by EC8-3. The first stage corresponds to the selection of the performance levels that will be considered for the structural assessment, as well as the corresponding seismic hazard levels. The second stage characterizes the reliability of the technical information available for the structure under assessment. The third stage involves selecting the structural analysis method that will be used to perform the assessment. The fourth stage of analysis corresponds to the safety verification stage where the conformity of each member is checked involving procedures for brittle and ductile failure mechanisms. Unlike EC8-3, the Greek standard clearly states that a building only conforms to a given performance level if the seismic demand in all the individual structural members does not exceed the corresponding capacity.

With respect to the design of strengthening interventions, the Greek standard covers techniques to increase the member-level flexural and/or shear strength, the stiffness, the deformation capacity and ductility, and to improve the strength of deficient lap-splices. Although similar to those provided by EC8-3, the strengthening design rules presented by the Greek standard are more detailed. For example, the standard includes rules for the strengthening of shear walls and addresses the safety of the force transfer mechanisms of interfaces between the original structural elements and the strengthening components. Furthermore, the standard also addresses system-level strengthening interventions such as adding interior unreinforced/reinforced concrete or masonry walls, converting frames to shear walls, strengthening existing masonry infills, adding bracing systems, or constructing additional shear walls.

3.5 Swiss Standard SIA 269 (2014)

The Swiss Society of Engineers and Architects (SIA) published the SIA 269 standard which addresses the safety assessment and retrofit of existing structures

[20]. The standard is composed by nine parts: SIA 269/0 which presents the principles, the terminology and the appropriate methodology to deal with existing structures, and eight additional parts dealing specifically with actions on existing structures (SIA 269/1), with RC, steel, composite, timber and masonry structures (SIA 269/2 to 269/6), with geotechnical aspects specific of existing structures (SIA 269/7), and with seismic safety evaluation (SIA 269/8). Even though the final version of SIA 269/8, covering the seismic safety evaluation, has not been formally published yet, its main aspects are addressed herein based on the 2014 draft version that was made available and on the information reported in [21]. Based on these documents, it appears that SIA 269/8 will not provide actual guidance regarding the selection or design of strengthening interventions.

According to SIA 269 [7], the safety of an existing structure needs to be analysed when there is a change of use/occupancy, a change in the requirements of the structure or when there is an alteration of the structure. For the particular case of the seismic safety assessment, the procedure defined by SIA 269/8 involves four main stages. In the first stage, the relevant building design data and the mechanical properties of the building materials must be established according to the available information and based on the general guidance provided by SIA 269/8. The second stage involves selecting between a force-based or a displacement-based approach to analyse and assess the structural behaviour of the building. According to SIA 269/8, the seismic intensity level for which the building performance must be analysed corresponds to the design level considered for a new building of the same typology as defined by the SIA 261/1 standard [22]. The third stage corresponds to the safety assessment stage where a compliance factor α_{eff} representing the ratio between a capacity parameter and a demand parameter is determined for each structural and non-structural member. The verifications that are required depend on the type of method of analysis that was selected in the second stage. In the fourth and final stage, the results of the third stage are examined in order to determine if the building requires an intervention. According to SIA 269/8, the global level of safety of the building is measured by the lowest α_{eff} value obtained in the third stage. In order to decide if a building needs to be strengthened or if its current condition can be accepted, the standard provides criteria that reflect the efficiency of potential strengthening measures based on cost-benefit considerations combined with minimum requirements for individual and collective risks to persons. These criteria depend on the building class, on its average occupancy, on its remaining useful life and are defined for ranges of the α_{eff} factor. These ranges define situations where:

- The structural safety of the building is not met but the safety level can be accepted ($\alpha_{adm} < \alpha_{eff} < 1$, where α_{adm} is a sufficient threshold of α_{eff}). In this scenario, it is expected that strengthening the building would not be *proportional* (i.e. the strengthening cost would be too high with respect to the level of seismic risk reduction that would be achieved);
- The structural safety of the building is not met and the building should be strengthened if the costs are *proportional* ($\alpha_{\min} < \alpha_{eff} < \alpha_{adm} < 1$, where $\alpha_{\min}$ is the lowest acceptable value of α_{eff});

– The structural safety of the building is unacceptable and the building must be strengthened ($\alpha_{eff} < \alpha_{\min}$).

To determine if a certain required strengthening intervention is *proportional* or not, simplified risk and cost-benefit analyses need to be carried out. The cost-benefit analysis includes components related to the loss of lives, to structural and non-structural damage and to business interruption. The procedure involves determining the expected level of risk reduction after the intervention as a function of the average occupancy of the building and the change in risk to the people as a result of the intervention. The annual cost of increasing the safety of the building (i.e. the cost of the intervention) is then determined considering the remaining useful life of the building and a pre-set discount rate. This cost might include other components other than the actual construction costs. The efficiency of the considered intervention is then determined by the ratio between the cost of the intervention and the risk reduction. This efficiency is measured in monetary units per live saved. According to SIA 269/8 [7], a certain intervention is considered *proportional* if this ratio does not exceed 10 million CHF per live saved. If a certain strengthening intervention is not *proportional*, the current level of safety of the building can still be accepted as long as $\alpha_{eff} \geq \alpha_{\min}$. If $\alpha_{eff} < \alpha_{\min}$, the building must be strengthened irrespective of the cost. Despite this statement, SIA 269/8 refers that the level of safety of the building can still be accepted in this case if the individual risk and other risks can be reduced by organizational measures (e.g. by limiting the occupancy of the building).

3.6 *Romanian Code for Seismic Evaluation P100-3 (2008)*

Over the past years, Romania has been updating its design standards to be harmonized with the Eurocodes. For the particular case of earthquake loading, the update of the Romanian standard for seismic design and assessment, termed P100, was carried out to include some of the more recent earthquake engineering concepts found in Eurocode 8 [23]. The P100 standard is divided into eight parts, in which P100-3 [8] provides rules and criteria for the seismic evaluation and repair of existing buildings. The scope of the P100-3 procedures addresses the seismic safety assessment and retrofit of buildings with RC, steel, masonry and timber structures. In addition, procedures to check the safety of non-structural elements are also provided. The standard also refers that the safety assessment procedures may also be applied (up to some extent) to monuments and historic buildings. However, a final part of the standard, the P100-8, that is expected soon [24], will specifically address the seismic evaluation and repair of historic monuments and buildings of architectural value.

The P100-3 assessment methodology involves a sequence of several stages of decision and analysis. After the first two stages, which are similar to those of EC8-3, a seismic safety evaluation is then carried out. This stage includes both qualitative and quantitative evaluation procedures. The qualitative procedure analyses several building parameters and defines the R_1 and R_2 indices. Index R_1 is a percentage score representing the degree of fulfilment of several criteria related to the structural configuration of the building, the interaction between structural and non-structural elements, and to the structural detailing. Index R_2 is a percentage score that reflects the current level of damage of the building. The quantitative procedure involves the numerical analysis of the structure, as well as the explicit comparison of the building capacity with the corresponding seismic demand to define the R_3 index. Depending on the type of structure under assessment, this quantitative procedure can be carried out by one of three different methodologies which involve different methods of structural analysis and capacity/demand verifications. Based on the values obtained for R_1, R_2 and R_3, a seismic risk class is assigned to the building which defines the need for retrofit or strengthening operations. This seismic risk class is defined by the lowest value of the selected index obtained for all the members or for a group of members that are essential for the stability of the structure.

With respect to the design of repair or retrofit interventions for existing buildings, Annex F of P100-3 provides an extensive and detailed presentation of procedures for RC, steel, masonry and timber structures, non-structural elements, and includes guidelines for the design of energy dissipation systems and base isolation systems. The presentation of the procedures is preceded by a conceptual discussion addressing several aspects that must be accounted for when defining an intervention in existing buildings. This initial part discusses the possible objectives of interventions, e.g. to reduce the seismic demand in terms of strength or deformation, to increase the construction capacity in terms of strength or ductility, or to enhance structural behaviour by reducing vertical and/or in-plan irregularities. In addition, other important aspects are also highlighted such as the need to account for the available economic, logistic and technological resources, and for the possible need to vacate the construction when the intervention is being carried out. The standard also addresses the possibility of enhancing the seismic behaviour of a building by reducing the mass in its upper storeys. The standard suggests that a reduction of the mass can be achieved by replacing the existing roof or other structural and non-structural elements with lighter components, by removing heavy elements or machinery from the top floors or by demolishing upper storeys of the building.

The standard addresses member-level and system-level strengthening interventions. In terms of member-level strengthening techniques for RC members, the standard refers to the use of concrete or steel jacketing and FRP plating and wrapping. With respect to system-level approaches, the standard addresses the possibility of converting frames into shear walls, strengthening existing masonry infills, adding bracing systems, or constructing additional shear walls. Furthermore, aspects related to the strengthening and enhancement of foundations and of the foundation soil are also reported. For the case of steel structures, several member-level repair techniques are also addressed (e.g. to repair screwed connections, to repair cracks by welding or to straighten deformed members). Strengthening approaches involve replacing damaged elements, techniques such as steel plating to enhance the stiffness, strength and ductility of the structural elements or constructing additional bracing systems. Aspects related to improving the behaviour of members to buckling are also discussed. For masonry structures, the standard provides guidance regarding the repair of cracks, the repair and strengthening of wall intersections, the strengthening and stiffening of horizontal diaphragms and the strengthening of walls. The strengthening techniques that are addressed depend on the retrofit objective and involve the addition of RC elements, steel elements or steel ties, steel plating, RC, steel or FRP jacketing, reinforced mortar jacketing, among others. Finally, for the case of timber structures, the standard also provides some guidance regarding the repair of degraded floor members, the strengthening of floor-to-wall connections, the strengthening and stiffening of horizontal diaphragms, and the strengthening of walls, columns and roofs.

3.7 Turkish Standard for Buildings in Seismic Zones (2007)

After the 1999 earthquake sequence that occurred in Turkey, the Turkish seismic design code was revised. In 2007, the new Turkish earthquake engineering standard [9] included a chapter addressing the seismic safety assessment and retrofit of existing buildings. The scope of the assessment and retrofit procedures proposed by the Turkish code addresses buildings with RC, steel, masonry structures, timber and adobe structures.

The seismic assessment methodology proposed by the Turkish standard involves four stages of analysis that are similar to those of EC8-3. The first stage establishes the global target levels of performance that must be considered for the assessment as a function of the type of building. The second stage addresses the level of knowledge available for the structure under assessment. The third stage involves selecting the structural analysis method that will be used to perform the assessment. With respect to this stage, it is noted that this standard is the only one that explicitly foresees the use of multi-mode nonlinear static (pushover) analysis. The fourth stage of analysis corresponds to the seismic performance evaluation which involves member-level evaluation procedures as well as building-level performance acceptance criteria. The member-level evaluation procedures are established as a function of the type of analysis selected in the third stage. On the other hand, the building-level performance acceptance criteria are established as a function of the target levels of performance defined in the first stage. For each global level of performance, these criteria are defined by limits on the acceptable interstorey drift demand and by admissible levels of member damage. This latter criterion reflects the fact that a given global level of performance can be accepted even if the seismic demand imposed in part of the individual structural members exceeds their corresponding capacity. For each performance level, the standard provides specific conditions to determine the type and number of admissible nonconforming members.

With respect to the design of strengthening interventions, the Turkish standard includes guidance and details for the design of conventional member-level approaches such as concrete or steel jacketing, and FRP plating and wrapping. Furthermore, the Turkish standard also provides guidance regarding system-level interventions such as converting frames into shear walls, strengthening existing masonry infills or constructing additional RC frames or shear walls. Additionally, the standard also refers the possibility of reducing the mass in upper storeys of RC buildings as a procedure that will reduce the seismic demand, thus increasing the performance and safety of the building. As the Romanian P100-3 standard, the Turkish standard suggests that a reduction of the mass can be achieved by replacing the existing roof or other structural and non-structural elements by lighter components, by removing heavy elements from the top floors (e.g. elevated water tanks) or by demolishing upper storeys of the building.

3.8 New Zealand NZSEE Recommendations (2006)

In 2004, a new Building Act [25] was introduced by the New Zealand government that established a new set of provisions dealing with earthquake-prone buildings. Prior to this legislation, only unreinforced masonry buildings were considered earthquake-prone constructions in New Zealand. Currently, buildings of any material can be determined to be earthquake-prone. After the implementation of this new legislation, the New Zealand Society for Earthquake Engineering (NZSEE), supported by the Department of Building and Housing, produced a set of guidelines termed *Assessment and Improvement of the Structural Performance of Buildings in Earthquakes* (NZSEE, 2006) outlining recommended procedures to assess the seismic performance of existing RC, steel, masonry and timber buildings, including heritage buildings. The guidelines have been updated four times since their first publication [26–29] to accommodate recent research results and technical changes [30]. A fully revised version of the guidelines that will include procedures reflecting lessons learnt from the Canterbury earthquake sequence, the Canterbury Earthquakes Royal Commission recommendations, as well as additional research results, is also expected soon [30].

To analyse the seismic performance of existing buildings, the guidelines provide an initial assessment procedure and a detailed evaluation procedure. The former is a fast preliminary evaluation procedure based on an exterior visual inspection of the building with the fundamental objectives of indicating the likely seismic performance of the building and of identifying critical structural weaknesses. The expected seismic performance of a building is analysed by taking into account its type and age of construction, local seismicity and ground conditions. Based on this procedure, the building is rated according to the severity of these weaknesses and the assessment result is expressed as a compliance percentage with respect to the expected performance of a new building that is termed %NBS (i.e. a percentage of the performance required according to the applicable New Building Standard).

The detailed evaluation procedure can be carried out according to either a force-based or a displacement-based approach. In both cases, the assessment methodology involves three stages. The first stage corresponds to the selection of the seismic performance level that will be considered for the evaluation. The second stage involves selecting the structural analysis method that will be used for the assessment. The final stage corresponds to the safety verification stage where the conformity of the structure is checked. The procedure that needs to be carried out to check the building conformity for the selected performance level depends on the type of evaluation procedure that was initially selected, i.e. a force-based, a displacement-based or a nonlinear pushover analysis-based approach. In the force-based approach, deformation and strength member-level mechanisms are first checked. If any member is found to be unsafe regarding the member-level mechanisms, the building needs to be retrofitted. If conformity is found for all members, a system-level verification that checks the interstorey drift demand is also enforced. If the building does not conform to this additional verification, the building needs to be retrofitted. Unlike this approach, the displacement-based and the nonlinear pushover analysis-based procedures start by establishing the global lateral deformation capacity of the structure by combining member-level deformation and strength capacities with maximum interstorey drift limits set by the standard. The building conformity is then checked by comparing this capacity with the expected deformation demand of the structure. If the building does not conform to this verification, the building needs to be retrofitted.

With respect to the design of strengthening interventions, the NZSEE guidelines include guidance for the selection of conventional system-level and member-level approaches for RC, streel and masonry structures. The presentation of the procedures is preceded by a conceptual discussion addressing available strategies to improve the structural performance of structures under earthquake loading. Among other aspects, this conceptual discussion covers the local modification of components, removing or reducing existing structural irregularities and discontinuities, global structural strengthening and stiffening, the use of seismic isolation, the use of supplementary energy dissipation elements, reducing the mass, widening seismic joints and linking buildings together across seismic joints.

For the strengthening of RC and steel structures, the recommended system-level approaches are, among others, constructing additional RC, steel frames or shear walls or adding steel V-braced, cross braced or yielding braced frames. Regarding the member-level approaches, the proposed strengthening solutions appear to be mostly for RC elements and include concrete or steel jacketing, FRP plating and wrapping, and external post-tension, among others. Additional guidance is also provided regarding the strengthening of foundations and floor and roof diaphragms. For the case of masonry structures, the suggested procedures cover the strengthening of walls for in-plane, out-of-plane and combined in-plane and out-of-plane loading, the strengthening of diaphragms and the behaviour enhancement of chimneys, towers and appendages. For in-plane loading, the guidelines suggest the strengthening of walls by adding concrete or FRP overlays and RC or V-braced frames, for example. For the strengthening of masonry piers, the guidelines suggest adding flexural rods or axial post strengthening. For out-of-plane loading, the guidelines suggest adding floor, roof and ceiling level ties, FRP flexural strips or elements providing additional buttressing or propping, among others. For combined in-plane and out-of-plane loading, solutions involve the use of vertical and/or horizontal post-tensioning, deep drilling and reinforcing of walls, grouting rubble filled walls or concrete overlays in walls. For the strengthening of diaphragms, the guidelines suggest adding boundary connections, diaphragm chords, drag ties, steel flat overlays, and concrete topping overlays, among others.

4 Closing Remarks

According to the general overview of the standards presented in the previous section, it can be seen that, in general, the selected standards establish similar principles for the seismic safety assessment of existing buildings. Even though a more comprehensive analysis of these standards has not been presented herein, it is nonetheless referred that the implementation details of the evaluation procedures proposed by each standard exhibit significant differences. Furthermore, by analysing existing application and validation studies carried out over the years to analyse the reliability of the procedures proposed by some of these standards, several features are seen to require further developments (e.g. see [31–43]). Among the several issues being referred, two stand out in particular.

The first one deals with the fact that procedures established by these standards are unable to adequately account for the many different sources of uncertainty that characterize the seismic safety assessment problem and may lead to highly variable assessment results for a given structure. The second one is related to the way structural performance is quantitatively measured with respect to the qualitative descriptions of limit states or performance levels that are enforced. This issue is seen to involve aspects related to how system-level performance can be expressed from member-level performance and to how engineering demand parameters can be related to practical decision variables such as economic losses. With respect to this second issue, it is noted that the Swiss standard SIA 269/8 addresses part of the problem by establishing that retrofit decisions are made using criteria based on admissible risk and repair costs.

These two issues are not entirely new and have been the focus of research over the past years. Several approaches have been proposed to address these issues, but their quantitative and practical implementation (in order to be suitable for a standard-based procedure) still present some problems. In this context, reference is made herein to two recently developed frameworks that explicitly address these issues in a consistent seismic safety assessment framework. The first document is the recent Italian Provisions for Probabilistic Seismic Assessment [44]. According to [45], after the extended use of current Italian seismic safety assessment standards showed their inability to provide consistent assessment results, the Italian National Research Council started to develop a new framework for seismic safety assessment. This new framework is based on the general concept of performance-based earthquake engineering (PBEE) and is characterized by a fully probabilistic approach that is able to account for all types of uncertainties and provides measures of performance in terms of mean rates of exceedance for the selected limit states. The purpose of this framework is to provide higher-level methods for special seismic safety assessment applications by overcoming some of the limitations of current Italian standards. However, it is believed that some of its provisions might be included in future revisions of those standards [45].

The second document is the FEMA P-58 Seismic Performance Assessment of Buildings [46] framework which resulted from a 10-year cooperation between the United States Federal Emergency Management Agency and the Applied Technology Council to develop advanced PBEE methodologies. The FEMA P-58 framework is a comprehensive implementation of PBEE involving procedures to assess the likely performance of buildings based on their site, structural, non-structural, and occupancy characteristics. The performance measures are characterized on a probabilistic basis and include potential casualties, direct economic losses and potential loss of use. Among other aspects, the FEMA P-58 procedures enable the evaluation of the risks of collapse and casualties, direct economic losses to repair damage or replacement of collapsed or demolished buildings and repair time (which is indexed to repair costs) [47]. Aside from the seismic safety assessment or upgrade of existing buildings, the FEMA P-58 methodology and procedures are also applicable to the performance-based design of new buildings. In addition to the procedures, the FEMA P58 framework provides a library of damage and consequence functions to evaluate losses in common building systems. Furthermore, a software called PACT (Performance Assessment Toolkit) is also available to apply the procedures and facilitate their practical use by design professionals.

Even though every aspect of the seismic safety assessment problem is not entirely solved by these two documents, they involve significant advancements and improvements for the practical application of modern PBEE concepts. Although their application is expected to unveil the need for further refinements and improvements of the assessment methodologies, they are believed to provide a preliminary outlook to what the future of reliable standard-based seismic safety assessment procedures might be like.

Acknowledgements Part of this work reports to research financially supported by Project POCI-01-0145-FEDER-007457—CONSTRUCT—Institute of R&D in Structures and Construction, funded by FEDER funds through COMPETE2020—Programa Operacional Competitividade e Internacionalização (POCI) and by national funds through Fundação para a Ciência e a Tecnologia (FCT).

References

1. CEN. EN 1998-3. Eurocode 8: design of structures for earthquake resistance—part 3: assessment and retrofitting of buildings. Brussels: European Committee for Standardization; 2005.
2. ICC. International building code. Washington, DC: International Code Council; 2012.
3. ICC. International existing building code. Washington, DC: International Code Council; 2012.
4. ASCE. Seismic evaluation and retrofit of existing buildings (ASCE/SEI 41-13). Reston: American Society of Civil Engineers; 2014.
5. NTC08. Norme Tecniche per le Costruzioni. Decreto del Ministero delle infrastrutture, Supplemento Ordinario n.30 alla Gazzetta Ufficiale della Repubblica italiana n.29 del 4/02/2008. 2008 (in Italian).
6. EPPO. Code for structural interventions. Earthquake Planning and Protection Organization of Greece, Ministry of Environment Planning and Public Works. 2013.
7. SIA 269/8. Swiss Standard SIA 269/8: existing structures—earthquakes. Enquiry version. Zurich: Swiss Society of Engineers and Architects; 2014.
8. MTCT P100-3: Seismic Design Code—Part 3: Provisions for seismic assessment of existing buildings. Ministry of Transport, Construction and Tourism, Bucharest, Romania (in Romanian), 2008.
9. MPWH. Standard for buildings in seismic zones. Ministry of Public Works and Housing. Official Gazette; 2007 (in Turkish, also available in English).
10. NZSEE: Assessment and improvement of the structural performance of buildings in earthquake. Recommendations of a NZSEE Study Group on Earthquake Risk Buildings. New Zealand Society for Earthquake Engineering, New Zealand, 2006.
11. OPCM 3274. Primi elementi in material di criteri generali per la classificazione sismica del territorio nazionale e di normative tecniche per le costruzioni in zona sismica. Ordinanza del Presidente del Consiglio dei Ministri n° 3274/2003, Supplemento Ordinario n.72 alla Gazzetta Ufficiale della Repubblica italiana n.105 del 8/5/2003, e successive modifiche ed integrazioni. 2003 (in Italian).
12. JRC. EN Eurocode Parts and Packages. Joint Research Centre, European Commission. 2015. http://eurocodes.jrc.ec.europa.eu/showpage.php?id=13. Accessed Dec 2015.
13. CEN. EN 1998-3:205/AC. Eurocode 8: Design of structures for earthquake resistance—Part 3: Assessment and retrofitting of buildings. Corrigendum. European Committee for Standardization, Brussels, Belgium, 2010.
14. ASCE. Seismic evaluation of existing buildings (ASCE/SEI 31-03). Reston: American Society of Civil Engineers, Virginia, USA, 2003.
15. ASCE: Seismic rehabilitation of existing buildings (ASCE/SEI 41-06). American Society of Civil Engineers, Reston, Virginia, USA, 2007.
16. OPCM 3431. Ulteriori modifiche ed integrazioni all'Ordinanza n.3274 del 20/3/2003, recante 'Primi elementi in materia di criteri generali per la classificazione sismica del territorio nazionale e di normative tecniche per le costruzioni in zona sismica'. Ordinanza del Presidente del Consiglio dei Ministri n° 3431/2005, Supplemento Ordinario n.85 alla Gazzetta Ufficiale della Repubblica italiana n.107 del 10/5/2005. 2005 (in Italian).
17. MIT. Circolare del Ministero delle Infrastrutture e dei Trasporti, n.617/2009, Supplemento Ordinario n.27 alla Gazzetta Ufficiale della Repubblica italiana n.47 del 26/02/2009. 2009 (in Italian).
18. CNR. Istruzioni per la progettazione, l'esecuzione ed il controllo statico mediante l'utilizzo di compositi fibrorinforzati. CNR-DT 200/2004. Consiglio Nazionale delle Ricerche. 2004 (in Italian).
19. Chronopoulos P, Zigouris N, Asteris P. Investigation/documentation and aspects of seismic assessment and redesign of traditional masonry buildings in Greece. In: EACS 2012—5th European conference on structural control genoa. 2012.

20. Brühwiler E, Vogel T, Lang T, Lüchinger P. Swiss standards for existing structures. Struct Eng Int. 2012;22(2):275–80.
21. Wenk T. Seismic assessment based on cost-benefit considerations. In: Second European conference on earthquake engineering and seismology, Istanbul. 2014.
22. SIA 261: Swiss Standard SIA 261/1: Actions on Structures, Swiss Society of Engineers and Architects, Zurich, Switzerland, 2014.
23. Lungu D, Craifaleanu I. Seismic assessment and retrofitting of existing structures in Romania: background, programs, regulatory basis. In: Soares CG, editor. Safety and reliability of industrial products, systems and structures. London: Taylor & Francis Group; 2010.
24. Ligia A. International and Romanian regulations in regard to interventions on historic fabric—a cross-reading study. Ann Oradea Univ Fascicle Manag Technol Eng. 2012;XI(XXI) nr2:5.1–6.
25. NZPCO: The Building Act. New Zealand Parliamentary Counsel Office, 2004.
26. NZSEE: Assessment and improvement of the structural performance of buildings in earthquake. Recommendations of a NZSEE Study Group on Earthquake Risk Buildings. Corrigenda No. 1. New Zealand Society for Earthquake Engineering, New Zealand, 2008.
27. NZSEE: Assessment and improvement of the structural performance of buildings in earthquake. Recommendations of a NZSEE Study Group on Earthquake Risk Buildings. Corrigenda No. 2. New Zealand Society for Earthquake Engineering, New Zealand, 2012.
28. NZSEE: Assessment and improvement of the structural performance of buildings in earthquake. Recommendations of a NZSEE Study Group on Earthquake Risk Buildings. Corrigenda 3 (Section 3, Initial Seismic Assessment). New Zealand Society for Earthquake Engineering, New Zealand, 2014.
29. NZSEE: Assessment and improvement of the structural performance of buildings in earthquake. Recommendations of a NZSEE Study Group on Earthquake Risk Buildings. Corrigenda 4 (Section 10, Revision Seismic Assessment of Unreinforced Masonry Buildings & Section 14, New Section—Geotechnical Considerations). New Zealand Society for Earthquake Engineering, New Zealand, 2015a.
30. NZSEE: New Zealand Society for Earthquake Engineering, New Zealand. 2015b. http://www.nzsee.org.nz/publications/assessment-and-improvement-of-the-structural-performance-of-buildings-in-earthquake/. Accessed on Dec 2015.
31. Franchin P, Pinto P, Rajeev P. Confidence factor? J Earthq Eng. 2010;14(7):989–1007.
32. Romão X, Guedes J, Costa A, Delgado R. A comparative application of different EC8-3 procedures for the seismic safety assessment of existing structures. Bull Earthq Eng. 2010;8 (1):91–118.
33. Caprili S, Nardini L, Salvatore W. Evaluation of seismic vulnerability of a complex RC existing building by linear and nonlinear modeling approaches. Bull Earthq Eng. 2012;10 (3):913–54.
34. Tondelli M, Rota M, Penna A, Magenes G. Evaluation of uncertainties in the seismic assessment of existing masonry buildings. J Earthq Eng. 2012;16(sup1):36–64.
35. Kim I, Hagen G. Seismic assessment and retrofit of existing RC buildings: case studies from Degenkolb Engineers. ACI Spec Publ. 2014;297–9:9.1–9.12.
36. Romão X, Delgado R, Costa A. Probabilistic performance analysis of existing buildings under earthquake loading. J Earthq Eng. 2014;18(8):1241–65.
37. Sucuoglu H, Lin W. Pseudo-dynamic testing, performance assessment, and modeling of deficient reinforced concrete frames. ACI Struct J. 2014;111(5):1203–12.
38. Bech D, Tremayne B, Houston J. Proposed changes to steel column evaluation criteria for existing buildings. In: Leon R, editor. Improving the seismic performance of existing buildings and other structures 2015. Reston: American Society of Civil Engineers; 2015.
39. Binici B, Yakut A, Ozcebe G, Erenler A. Provisions for the seismic risk evaluation of existing reinforced concrete buildings in turkey under the urban renewal law. Earthq Spectra. 2015;31 (3):1353–70.
40. Chen Y. Evaluation of simplified seismic vulnerability assessment procedures for reinforced concrete buildings. M.Sc. Dissertation. University of Canterbury; 2015.

41. Kam W, Jury R. Performance-based seismic assessment: myths and fallacies. In: 2015 New Zealand society for earthquake engineering annual technical conference. 2015.
42. La Brusco A, Mariani V, Tanganelli M, Viti S, De Stefano M. Seismic assessment of a real RC asymmetric hospital building according to NTC 2008 analysis methods. Bull Earthq Eng. 2015;13(10):2973–94.
43. Ojala D, Osteraas J. Performance-based engineering and the evaluation of existing buildings—potential legal ramifications. In: Leon R, editor. Improving the seismic performance of existing buildings and other structures 2015. Reston: American Society of Civil Engineers; 2015.
44. CNR. Istruzioni per la valutazione affidabilistica della sicurezza sismica di edifici esistenti. CNR-DT 212/2013. Consiglio Nazionale delle Ricerche, Italy. 2014 (in Italian).
45. Pinto P, Franchin P. Chapter 3—existing buildings: the New Italian provisions for probabilistic seismic assessment. In: Ansal A, editor. Perspectives on European earthquake engineering and seismology, vol. 1. Berlin: Springer; 2014.
46. FEMA. Next-generation methodology for seismic performance assessment of buildings. Report No. FEMA P-58, prepared by the Applied Technology Council for the Federal Emergency Management Agency, Washington, DC. 2012.
47. Krawinkler H, Deierlein G. Challenges towards achieving earthquake resilience through performance-based earthquake engineering. In: Fischinger M, editor. Performance-based seismic engineering: vision for an earthquake resilient society. Berlin: Springer; 2014.

Evaluation of Strengthening Techniques Using Enhanced Data Envelopment Analysis Models

Isabel M. Horta and Celeste Varum

1 Introduction

Recent studies [1, 2] point to noticeable discrepancies between the technical and functional quality of the existing building stock and the future demands of the population. Substantial efforts are required in order to assure its adjustment to the prospective needs in terms of architecture, functionality, safety and so forth. Not surprisingly, most developed economies are providing incentives for rehabilitation and retrofitting of the existing building stock, with the rehabilitation sector registering a boom in recent years.

Rehabilitation of the building stock comprises a wide range of interventions, not only at architectural, functional and physical level, but also at structural level [1–3]. Structural interventions include the rehabilitation of existing structures to make them more resistant to seismic activity and ground motion. The need of seismic retrofitting of the existing building stock is nowadays well acknowledged. This is mainly due to the increasing knowledge about the seismic impact on structures and the recent experiences with large earthquakes that occurred in urban centres [2, 4].

Project teams and managers need to select the most suitable retrofitting strategy among several solutions, which constitutes a difficult task. Beyond the criteria related with the technical characteristics, the selection of the strengthening method for a particular case should also consider the associated costs [5–9]. The research presented herein extends the previous attempts as it develops a new decision

I.M. Horta (✉)
Department of Industrial Engineering and Management,
Faculdade de Engenharia, Universidade do Porto, 4200-465 Porto, Portugal
e-mail: imhorta@fe.up.pt

C. Varum
Department of Economics, Management and Industrial Engineering,
GOVCOPP, University of Aveiro, 3810-193 Aveiro, Portugal
e-mail: camorim@ua.pt

A. Costa et al. (eds.), *Strengthening and Retrofitting of Existing Structures*,
Building Pathology and Rehabilitation 9, https://doi.org/10.1007/978-981-10-5858-5_14

making model to evaluate strengthening solutions based on Data Envelopment Analysis (DEA) specified with a directional distance function.

The method proposed is illustrated with data from a study in which five strengthening solutions were used (for further details see Diz et al. [5]). The input of the model is the intervention cost and the outputs correspond to the most common strengthening characteristics: total drift, inter storey drift, stress concentration and energy dissipation. The original data includes 30 observations which correspond to the behaviour of five strengthening solutions, all five tested in each of the six houses.

The chapter is organized as follows. Section 2 discusses the advantages of using DEA in comparison with other methods. Section 3 presents the new DEA based model developed. Section 4 shows the empirical application, including the description of the data used and the discussion of the results. The last section concludes and raises issues for future research.

2 Cost-Benefit Analysis: The Advantages of the DEA

Recent seismic events around the world enabled to underline the importance of seismic retrofitting of existing structures. Seismic retrofitting of a building may be necessary if a seismic risk assessment reveals deficiencies in how an existing structure responds to an earthquake. This may depend on building characteristics, site location, and seismic characteristics of the surrounding area.

Existing experimental and numerical studies provide solid results about the technical performance of different strengthening techniques. Similarly, the literature in the field registered great advances in what regards the methods for technical assessment of different strengthening techniques (see, for example, Costa [10]). From a technical point of view, the best solutions are those that improve the global performance of the structures. For the case of seismic retrofitting, it might be, for example, to reduce structural fragility, structural damage, stress distribution or to increase energy dissipation. There is usually more than one indicator that matters, and none of them dominates over the others.

A final decision on which technique should be used must involve an analysis of the technical characteristics and the costs. In presence of budget constraints (which is often the case), the aim should be to maximize the technical performance with lower costs [7, 11]. Few studies in the literature assess the effectiveness in the mitigation of the construction's seismic vulnerability accounting for the cost of different retrofitting techniques. In particular, the studies developed by Williams et al. [7], Costa et al. [8], Varum [6], Vicente et al. [9] and Diz et al. [5] are notable contributions on this regard. They suggest a cost benefit analysis to support the decision on which strengthening technique should be used for seismic retrofit of structures.

Williams et al. [7] consider that a seismic retrofit of a building is a viable option if some economic benefit can be gained as a result of the retrofit. Hence, in their study a framework is outlined to compute the expected economic benefit resulting from a retrofit. A parametric analysis is performed to determine the effects that achieved loss reduction, investment return period and retrofit cost have on the economic feasibility of seismic retrofitting. Wang and Lai [11] analyse the cost-effectiveness of bridge seismic retrofit using lead-rubber bearings. An Economic Index is proposed to identify the most cost-effective solution, i.e., the solution with the greatest benefit at the least cost. The Economic Index developed accounts for two major parameters: seismic load reduction as a result of the rehabilitation and retrofitting costs. Vicente et al. [9] evaluate different strengthening techniques to improve the global structural behaviour of four traditional masonry buildings. To compare the efficiencies of the retrofitting solutions studied, the authors analyse graphically the reduction of the horizontal displacement at the top level of the walls and the ratio between the cost of the retrofitting solution and the patrimonial value of the solution. Costa et al. [8] report results from on an in situ experimental test campaign. The authors developed an equation that accounts for the displacement, strength, energy dissipated and intervention cost that enables ranking each strengthening solution. Diz et al. [5] analyse the case of real structures with different geometrical and physical characteristics, by establishing a comparison between the seismic performance of reinforced and non-reinforced structures. This work analysed the reduction of displacement and the maximum stress of each scenario (benefits) relative to the original structure in comparison with the cost of the intervention of the selected retrofit schemes. The examination of the performance obtained in each case, in relation to the cost of implementing the reinforcement technique, allowed to draw conclusions about the best options for different situations.

Previous studies have, however, several limitations. First, cost-benefit analysis (also called cost analysis economic evaluation, cost allocation, cost-benefit analysis, or cost-effectiveness analysis by different authors) is currently a controversial set of methods for performance evaluation. One reason for the controversy is that these terms cover a wide range of methods, and are often used interchangeably. Typically, these methods use two different approaches for measuring the performance of a particular solution. One approach is to select one-dimensional measure of outcome, leading to the comparison of different solutions on the basis of a cost-benefit ratio. However, this result in a limited assessment as most real world cases often require the use of a set of variables to adequately evaluate the whole spectrum of costs and benefits. The other approach is to use an index resulting from the aggregation of different variables. On this line, Costa et al. [8] and Wang and Lai [11] created an index consisting of two major parameters seismic load reduction as a result of the rehabilitation and retrofit cost. Nonetheless, the aggregation of multiple variables is based on a subjective system of weights that could vary according to the decision maker preferences. This is a common problem even in other methods used for performance evaluation such as the multi criteria decision making methods. For further details see the study developed by Caterino et al., [12] that investigate the applicability of different multi criteria decision making methods (e.g., ELECTRE, VIKOR) for selecting the best option to seismically upgrade an existing building.

A method of circumventing these problems is proposed in this chapter. It consists of measuring relative performance of a given intervention in a way which is well-known from the efficiency analysis literature, the DEA method, first introduced by Charnes et al. in 1978 [13].

The use of DEA method is not new to the construction industry. Recently, the literature on performance measurement describes successful applications of DEA to the construction industry. For instance, DEA was used to support the evaluation of general contractors [14], the selection of bids [15], the purchase of construction materials [16], or the selection of project location [17]. There are a relatively large number of studies applying the DEA method to evaluate the efficiency of firms in the construction industry. On this regard, Lee [18] provides a survey of DEA applications in measuring the efficiency of construction organizations.

To the best of our knowledge, this is the first study that uses the DEA method for the evaluation of strengthening techniques. Nonetheless, DEA presents many advantages to support the selection of the best strengthening solution. In particular, DEA enables the estimation of an overall performance score for each strengthening solution based on multiple inputs (e.g., costs) and multiple outputs (e.g., strengthening characteristics). In addition, DEA derives the weights for the different inputs and outputs directly from the data, eliminating the subjectivity involved in the selection. The weights are estimated recurring to optimization which attributes to each strengthening solution the best possible score. Other major advantage of DEA is that it specifies improvement targets for the inefficient solutions to behave efficiently. This information is derived based on a comparison with the other solutions in the sample. In the next section it is illustrated how this method is likely to be applied to compare alternative strengthening interventions.

3 The DEA Model Developed

DEA uses linear programming to evaluate the relative efficiency of a sample of organizational units, such as firms, cities, or buildings in their use of multiple inputs to produce multiple outputs.

Traditional DEA assessments assume that the outputs are measured in a scale where higher values correspond to better performance. Thus, traditional DEA does not allow modelling directly the production of undesirable outputs. In the literature, there are two main approaches for handling undesirable outputs in DEA models.

One approach transforms the measurement scale of the undesirable outputs to treat them as standard outputs. For instance, Scheel [19] proposed to incorporate the undesirable outputs as normal outputs in the form of their additive inverses ($-y_{und}$). Golany and Roll [20] proposed to treat undesirable outputs in the form of their multiplicative inverses ($1/y_{und}$). Seiford and Zhu [21] proposed to add to the additive inverses of the undesirable outputs a sufficiently large positive number ($-y_{und} + M$). The limitations associated with these approaches were highlighted by several authors.

The other approach treats the outputs in their original form, but it requires the use of modified DEA models. Herein, we follow this latter approach as it has the advantage of preserving data interpretability. In particular, we use a DEA model specified with a directional distance function. The evaluation of efficiency based on a directional distance function was first proposed by Chambers et al. [22]. The model developed by the authors allows to simultaneously expand outputs and contract inputs according to a directional vector. Chung et al. [23] extended this approach to allow including undesirable outputs in the efficiency evaluation.

The model proposed to measure the efficiency of strengthening interventions follows the formulation of Fare et al. [24] and is presented in (1). It involves a similar treatment of all indicators that should be minimized, irrespectively of their intrinsic nature being an input (e.g., cost) or an undesirable output (e.g., stress concentration).

$$
\begin{aligned}
& Max\ \beta \\
& \text{s.t.} \sum_{j=1}^{n} y_{rj}\lambda_j \geq y_{rjo} + \beta g_y \quad r = 1,\ldots,s \\
& \sum_{j=1}^{n} x_{ij}\lambda_j \leq x_{ijo} - \beta g_x \quad \text{i} = 1,\ldots,m \\
& \sum_{j=1}^{n} b_{sj}\lambda_j \leq b_{sjo} - \beta g_b \quad \text{s} = 1,\ldots,l \\
& \lambda_j \geq 0 \quad \text{j} = 1,\ldots,n
\end{aligned}
$$

In model (1), x_{ij} ($i = 1,\ldots m$) are the inputs used by unit j ($j = 1,\ldots,n$) to produce y_{rj} ($r = 1,\ldots,s$) desirable outputs and b_{sj} ($s = 1,\ldots,l$) undesirable outputs. The components of vector $g = (g_y, -g_b, -g_x)$ indicate the direction of change for the desirable outputs, undesirable outputs and inputs, respectively. Positive values of the components are related to expansion of the desirable outputs and negative values to contraction of undesirable outputs and inputs.

The decision variables of model (1) are the λ_j ($j = 1,\ldots,n$) and β. Model (1) is an optimization model that allows constructing a frontier of best practices against which all units are evaluated. The units that define this frontier are considered the benchmarks of the sample. For the units deemed inefficient, located in the space below the frontier, the model identifies the peers that can be used as a reference to search best practices. The peers are assigned a value of λ_j greater than zero at the optimal solution. The targets to become efficient can be obtained by a linear combination of the peers identified. These correspond to the left-hand side of the constraints of model (1) for each indicator.

The factor β indicates the extent of units' inefficiency. An inefficiency score equal to zero corresponds to the best performance observed in the sample, whilst values greater than zero indicate the existence of performance levels less than those observed in the peers. In this study, we specified a directional vector that searches for reductions in cost, stress and displacement effects, and for increments in energy dissipated. Thus, the directional vector was specified as $g = (y_{rjo}, -b_{sjo}, -x_{ijo})$. The values observed in the unit under assessment (jo) were used in the components of the directional vector to allow a proportional interpretation of the changes yielded by the optimization model. In the remainder of this text, the value of β is called the performance score.

4 Empirical Study

4.1 Data and Empirical Context

In this study, we use data from an in situ experimental test campaign carried out in The Azores archipelago. Further details about the data used can be found in Diz et al. [5]. The archipelago is a seismic prone region with a vast cultural heritage, presenting a building stock mainly constructed in traditional stone masonry. It is known that this type of construction exhibits poor behaviour under seismic excitations, but it is extensively used in seismic areas.

The experimental campaign tested five strengthening solutions applied in the walls of six houses. This enables us to analyse a sample of 30 observations, representing the behaviour of the strengthening technique at a house level. Note that the five strengthening solutions are all tested in the same houses, which allows undertaking a fair comparison between strengthening solutions.

Concerning the characteristics of the six houses analysed, four houses (H1, H2, H3, H4) are detached houses located in a rural area, having similar structural characteristics (stone masonry in the exterior walls and wooden structures in the roof and floor). The other two houses (H5, H6) are located in an urban area and correspond to buildings with better construction and higher quality constructive materials using for instance concrete masonry blocks.

The strengthening solutions considered are as follows: (i) reinforced plaster system that consists on the application of a steel mesh made of galvanized or stainless steel in the walls (called hereafter as reinforced plaster), (ii) restraining connection between parallel walls and the floor/roof using steel plates and wood beams as connecting elements (called hereafter as strengthening of connections), (iii) reinforced concrete beams at the foundation level to anchor the steel mesh (called hereafter as foundation beam). Table 1 presents the five strengthening solutions evaluated in this study.

Table 1 Strengthening solutions compared in the assessment

A	Reinforced plaster
B	Reinforced plaster + Foundation beam
C	Strengthening of connections
D	Strengthening of connections + Reinforced plaster
E	Strengthening of connections + Reinforced plaster + Foundation beam

In order to illustrate how the performance of the strengthening solutions could be measured, we use the cost of the intervention (as the input) and four variables representing the strengthening characteristics as outputs. In particular, the undesirable outputs (higher values of these outputs correspond to worse performance) are the total drift (in meters), inter storey drift (in meters), and stress concentration (in MPa). Note that we use the maximum values obtained in the walls tested to represent these strengthening characteristics for each house. The desirable output (higher values of this output correspond to better performance) is the quantity of energy dissipated (in kNm) in each house. Table 2 reports the original values, the average and the standard deviation of the variables used in the assessment. It is possible to observe that the characteristics of the strengthening solutions are relatively heterogeneous concerning the variables analysed (see for instance the large values of the standard deviation). It is also interesting to observe that the strengthening solution C exhibits the lowest costs when compared with the other solutions.

4.2 *Results and Discussion*

In this section, we report the results related to the performance assessment of different strengthening solutions tested using six houses. Figure 1 presents the performance score of each strengthening solution in each house (β). Note that according to model (1) a performance score of zero indicates the best performing strengthening solution.

From Fig. 1, it can be observed that the strengthening solution C tested in house H6 and the strengthening solution D in house H6 were considered fully efficient. These solutions were both tested in house H6, which could be expected as house H6 exhibits higher construction quality. These solutions presented lower costs, implied lower displacements and stress concentration, and higher energy dissipation when compared to the other solutions analysed.

The solution C concerns the strengthening of the connections between parallel walls and the floor/roof, whereas the solution D includes a combination of the strengthening of the connections and the reinforced plaster system.

Table 2 Original values and descriptive statistics of the variables

House_Strengthening	Cost (10^3 €)	Total drift (m)	Inter storey drif (m)	Stress (MPa)	Energy (kNm)
H1_A	14.86	0.078	0.078	1.69	25.20
H1_B	46.63	0.077	0.077	1.84	33.42
H1_C	2.67	0.071	0.071	1.49	20.59
H1_D	17.53	0.077	0.077	2.05	32.14
H1_E	49.30	0.059	0.059	1.54	19.78
H2_A	24.44	0.059	0.029	1.87	12.43
H2_B	58.31	0.052	0.025	1.68	10.25
H2_C	4.39	0.050	0.027	1.54	17.57
H2_D	28.84	0.032	0.016	1.33	9.82
H2_E	62.70	0.033	0.017	1.38	9.43
H3_A	17.94	0.026	0.014	1.97	7.72
H3_B	42.54	0.024	0.012	1.88	6.93
H3_C	3.22	0.012	0.007	0.85	6.06
H3_D	21.16	0.012	0.008	0.77	4.34
H3_E	45.76	0.011	0.008	0.73	4.15
H4_A	30.81	0.038	0.021	1.75	12.89
H4_B	68.31	0.034	0.021	1.65	11.76
H4_C	5.54	0.027	0.022	1.36	12.51
H4_D	36.35	0.013	0.011	0.84	5.73
H4_E	73.84	0.014	0.011	0.81	5.57
H5_A	47.41	0.026	0.011	2.41	6.72
H5_B	82.39	0.026	0.011	2.38	6.66
H5_C	8.52	0.026	0.008	1.87	22.15
H5_D	55.93	0.028	0.008	2.04	20.29
H5_E	90.91	0.027	0.008	1.97	16.29
H6_A	64.75	0.120	0.045	5.68	161.98
H6_B	99.40	0.117	0.044	5.66	158.11
H6_C	11.63	0.082	0.031	3.50	91.48
H6_D	76.38	0.076	0.029	3.30	166.78
H6_E	111.03	0.075	0.028	3.24	159.65
Mean	43.45	0.047	0.028	2.04	35.95
St. Deviation	30.66	0.030	0.023	1.21	52.67

In addition, we can observe that solution C was the most efficient strengthening solution considering the six houses analysed. In turn, strengthening solution D only managed to be the most efficient solution (alongside solution C) in house H6. In particular, the average performance score of solution C is 0.228, whereas the average score of solution D is 0.476. Overall, this can indicate that solution C—the

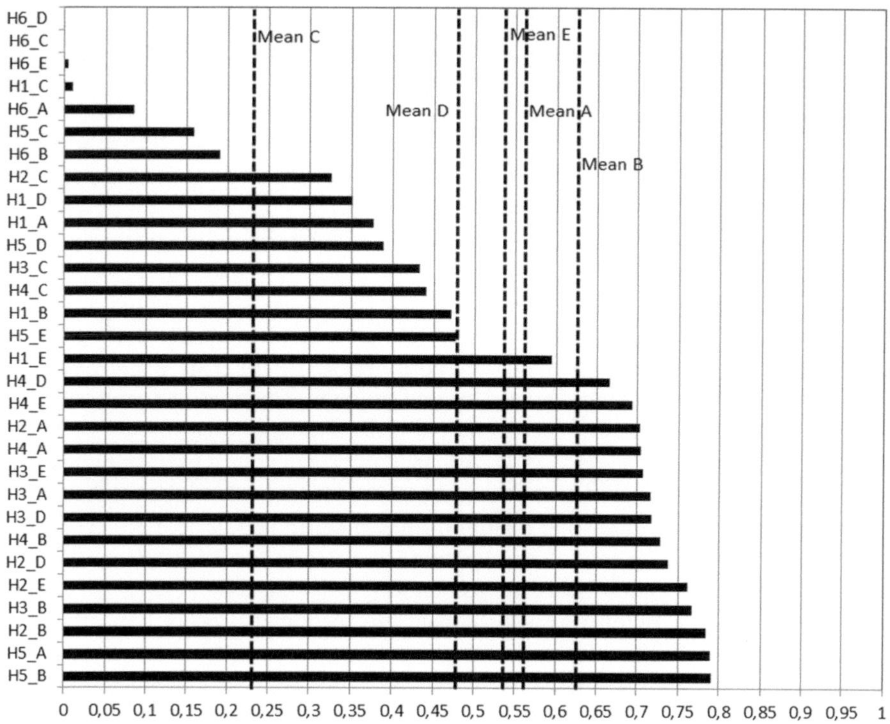

Fig. 1 Performance score of each strengthening solution and house

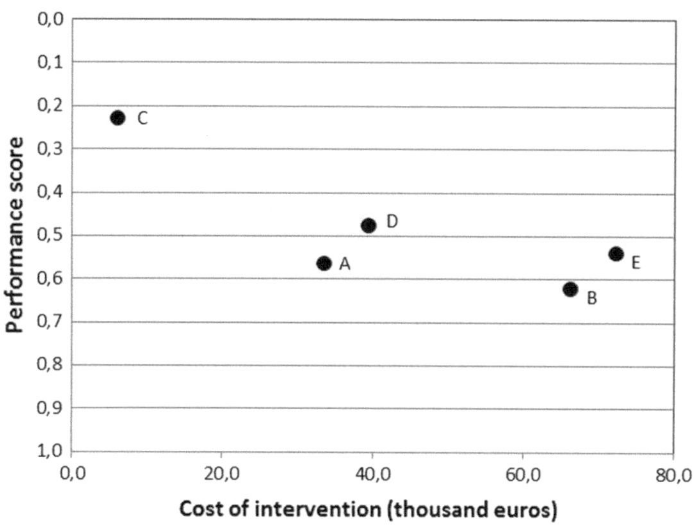

Fig. 2 Strengthening solutions performance score vs cost of intervention

strengthening of the connections between parallel walls—is the best strengthening solution to be adopted taking into account both the cost of intervention and the strengthening characteristics.

In order to depict how the strengthening solutions stand against each other in ms of cost and technical characteristics, Fig. 2 shows the position of each strengthening solution considering the average performance score obtained (vertical axis) and the average cost of the intervention (horizontal axis).

From the analysis of Fig. 2, it is possible to confirm that the best option corresponds to the strengthening solution C as it exhibits the lowest average costs and the highest average performance score. The intermediate strengthening solutions correspond to solutions D and A, as they present intermediate average values of performance score and cost of intervention. Solutions B and E are perceived to be the worst options. They tend to have similar average performance scores as solutions D and A, but exhibiting considerably higher average costs.

Next, we analyse the targets for performance improvement of each strengthening solution. The targets are important aspects to be analysed as they provide an indication on the dimensions that require further enhancement. Table 3 presents the percent reduction for the cost of intervention (input), and for the displacement and stress concentration (undesirable outputs), as well as the percent increase for energy dissipated (desirable output). These outputs are technical aspects typically considered in strengthening interventions. These changes in cost and strengthening characteristics would enable each strengthening solution to behave efficiently. The percent change is calculated as the target value minus the observed value divided by the observed value.

From Table 3, we can observe that, on average, for solutions A, B, C and D the largest scope for improvement concerns the inter drift displacement (see the highest percent reduction required in this dimension). In turn, the largest scope for improvement for solution E lays in cost of intervention.

Finally, we perform a sensitivity analysis on the models developed in order to check the importance of including the variable cost in the selection of the best strengthening solution. This would allow comparing the benefits of the model proposed with the traditional approaches that typically do not take into account the economic dimension. In particular, we run model (1) using the variable cost equal to a constant value for all observations (i.e., equal to one). From this analysis, it is possible to conclude that if the assessment only focuses on technical characteristics (displacement, stress and energy), the average performance scores of the five strengthening solutions would be quite similar. In particular, the average performance score of solutions A, B, C, D and E would be equal to 0.60, 0.59, 0.52, 0.51, and 0.54, respectively. This confirms that the cost of intervention is an important factor that enables to distinguish the best strengthening solution.

Table 3 Targets for performance improvement

House_Strengthening	Targets (% change)				
	Cost	Total drift	Inter drift	Stress	Energy
H6_D	0	0	0	0	0
H6_C	0	0	0	0	0
H6_E	−34	−2	1	−2	0
H1_C	−1	−74	−90	−47	1
H6_A	−9	−9	−9	−18	9
H5_C	−16	−32	−13	−59	16
H6_B	−19	−20	−18	−28	19
H2_C	−33	−59	−71	−42	33
H1_D	−35	−59	−85	−35	35
H1_A	−38	−68	−88	−38	38
H5_D	−77	−54	−40	−73	39
H3_C	−43	−45	−62	−66	43
H4_C	−44	−44	−75	−53	44
H1_B	−52	−71	−89	−47	47
H5_E	−88	−60	−49	−76	48
H1_E	−71	−75	−91	−59	59
H4_D	−88	−67	−85	−78	67
H4_E	−94	−68	−85	−77	69
H2_A	−70	−78	−83	−70	70
H4_A	−70	−71	−80	−72	70
H3_E	−93	−71	−85	−81	71
H3_A	−72	−72	−79	−84	72
H3_D	−84	−71	−84	−81	72
H4_B	−86	−73	−83	−76	73
H2_D	−74	−75	−80	−74	74
H2_E	−88	−77	−83	−76	76
H3_B	−87	−77	−82	−87	77
H2_B	−86	−84	−87	−78	78
H5_A	−88	−79	−81	−90	79
H5_B	−93	−79	−81	−90	79
Mean A	−58	−63	−70	−62	56
Mean B	−70	−67	−73	−68	62
Mean C	−23	−42	−52	−44	23
Mean D	−60	−54	−62	−57	48
Mean E	−78	−59	−65	−62	54

5 Concluding Remarks

A big challenge to successfully intervene in the building sector is to find effective strategies for retrofitting of existing buildings. This challenge also applies to the context of seismic retrofitting. It is recognized that our knowledge about the technicalities of the different strengthening solutions improved substantially, being possible to evaluate the different options against a number of technical criteria in each context. In this chapter we highlight that the selection of the strengthening method for a particular case should also consider the associated costs. Reducing budgets and increasing safety as much as possible, requires thorough knowledge of the cost effectiveness of different options.

In this chapter, we suggest an enhanced DEA model specified with a directional distance function to support strengthening selection accounting for both the strengthening technical characteristics and the costs, which are usually not considered together in most of the current research in this field. Its relevance is not only derived from academic purposes but also from the practitioner's urgent need to apply better decision support tools.

We used an example to demonstrate the applicability of the model proposed. In particular, we used data from the application of five strengthening solutions in the walls of six houses. The analysis conducted enabled to find the most suitable strengthening solution to be adopted when considering simultaneously the technical characteristics and the associated costs. In addition, it was possible to identify the priority areas in need of improvement and their extent for each strengthening solution to behave efficiently. Moreover, from the analysis conducted we concluded that the cost of the intervention has a major impact on the selection of the best solution. The influence of the cost on the final choice was clearly highlighted by showing that the best solution changes if costs are considered to be equal.

Finally, we do not attempt to say that the cost is a paramount criterion, but neglecting the cost can be considered unrealistic. The consideration (or not) of the cost in the decision making will depend also on the building itself and its book value. As highlighted by Diz et al. [5] 'It isn't cost/effective to spend as much to reinforce a rural building, as to retrofit a monument or a distinguished building, because their book values are very different'. For future research it would be interesting to include in the model developed other factors reflecting the intrusiveness and reversibility of the techniques that may also affect the final decision. We also suggest the possibility of extending the sample to test the impact of the house characteristics on the efficiency results. Another topic for future developments, it would be the possibility of incorporating information on decision-maker preferences about the relative importance of particular strengthening characteristics, or accounting for appropriate trade-offs between the strengthening characteristics. This could be done by extending the model applied using weight restrictions.

References

1. Andeweg MT, Brunoro S, Verhoef LGW, editors. COST C16 improving the quality of existing urban building envelopes: state of the art. Amsterdam: IOS Press BV; 2007.
2. Marini A, Passoni C, Riva P, Negro P, Romano E, Taucer F. Technology options for earthquake resistant, eco-efficient buildings in Europe: research needs. European Commission, Joint Research Centre Scientific and Policy Reports, 2014.
3. Alexander MG, Beushausen H-D, Dehn F, Moyo P. Concrete repair, rehabilitation and retrofitting III: 3rd International Proceedings of the 3rd International Conference on Concrete Repair, Rehabilitation and Retrofitting (ICCRRR) Cape Town, South Africa, September 3–5, 2012.
4. Fardis MN. Seismic Design, Assessment and Retrofitting of Concrete Buildings: based on EN-Eurocode 8. Springer, 2009.
5. Diz S, Costa A, Costa AA. Efficiency of strengthening techniques assessed for existing masonry buildings. Eng Struct. 2015;101:205–15.
6. Varum H, Vicente R, Rodrigues H, Silva JAR. Seismic evaluation of old masonry buildings: Performance and strengthening. In: Topping BHV, Montero G, Montenegro R (eds) Proceedings of 8th international conference on computational structures technology, CST2006. Las Palmas de Gran Canaria, Spain. Civil-Comp Press, Stirlingshire, UK, 2006, p. 12–15.
7. Williams RJ, Gardoni P, Bracci JM. Decision analysis for seismic retrofit of structures. Struct Saf Risk Accept Risk Commun. 2009;31(2):188–96.
8. Costa AA, Arêde A, Costa A, Oliveira CS. In situ cyclic tests on existing stone masonry walls and strengthening solutions. Earthq Eng Struct Dyn. 2011;40(4):449–71.
9. Vicente R, Rodrigues H, Varum H, Mendes da Silva J. Evaluation of strengthening techniques of traditional masonry buildings: case study of a four-building aggregate. J Perform Constr Facil. 2011;25(3):202–16.
10. Costa AA. Experimental testing of lateral capacity of masonry piers. An application to seismic assessment of AAC masonry buildings. Ph.D. Thesis. European School For Advanced Studies In Reduction Of Seismic Risk, Rose School, Universitá degli Studi di Pavia, 2007.
11. Wang EH, Lai SH. Cost-effectiveness of bridge seismic retrofit using lead-rubber bearings. In: Proceedings of the 3rd international conference on bridge maintenance, safety and management—bridge maintenance, safety, management, life-cycle performance and cost. Porto; Portugal; 16 July through 19 July, 2006, p. 273–5.
12. Caterino N, Iervolino I, Manfredi G, Cosenza E. Comparative analysis of multi-criteria decision-making methods for seismic structural retrofitting. Comput Aided Civ Infrastruct Eng. 2009;24(6):432–45.
13. Charnes A, Cooper WW, Rhodes E. Measuring efficiency of decision making units. Eur J Oper Res. 1978;2(6):429–44.
14. Horta IM, Camanho AS, Moreira da Costa J. Performance assessment of construction companies: a study of factors promoting financial soundness and innovation in the industry. Int J Prod Econ. 2012;137(1):84–93.
15. El-Mashaleh M. Decision to bid or not to bid: a data envelopment analysis approach. Can J Civ Eng. 2010;37:37–44.
16. Castro-Lacouture D, Medaglia A, Skibniewski M. Supply chain optimization tool for purchasing decisions in B2B construction marketplaces. Autom Constr. 2007;16(5):569–75.
17. Cheng E, Li U. Exploring quantitative methods for project location selection. Build Environ. 2004;39(12):1467–76.
18. Lee Y-S. A survey of DEA applications in measuring the efficiency performance of construction organizations. Korea J Constr Eng Manag. 2014;15(5):103–14.
19. Scheel H. Undesirable outputs in efficiency valuations. Eur J Oper Res. 2001;132(2):400–10.
20. Golany B, Roll Y. An application procedure for DEA. Omega Int J Manag Sci. 1989;17 (3):237–50.

21. Seiford LM, Zhu J. Modeling undesirable factors in efficiency evaluation. Eur J Oper Res. 2002;142(1):16–20.
22. Chambers RG, Chung Y, Fare R. Benefit and distance functions. J Econ Theory. 1996;70 (2):407–19.
23. Chung YH, Fare R, Grosskopf S. Productivity and undesirable outputs: a directional distance function approach. J Environ Manag. 1997;51(3):229–40.
24. Fare R, Grosskopf S, Pasurka CA. Potential gains from trading bad outputs: The case of U.S. electric power plants. Resour Energy Econ. 2014;36:99–112.

MIX
Papier aus verantwortungsvollen Quellen
Paper from responsible sources
FSC® C105338

If you have any concerns about our products,
you can contact us on
ProductSafety@springernature.com

In case Publisher is established outside the EU,
the EU authorized representative is:
Springer Nature Customer Service Center GmbH
Europaplatz 3, 69115 Heidelberg, Germany

Printed by Libri Plureos GmbH
in Hamburg, Germany